The History of
Air Intercept (AI) Radar
and the British
Night-Fighter 1935–1959

This book is dedicated to the memory of Professor Edward Bowen, CBE, FRS, the father of airborne radar in Britain and the United States

and

to Flying Officer Glyn 'Jumbo' Ashfield of the Fighter Interception Unit and his crew of Flying Officer Geoffrey Morris and Sergeant Reginald Leyland, who destroyed the first enemy aircraft by radar interception on the night of the 22/23 July 1940.

The History of
Air Intercept (AI) Radar
and the British
Night-Fighter 1935–1959

Ian White

Pen & Sword
AVIATION

First published in Great Britain in 2007 by
PEN & SWORD AVIATION
an imprint of
Pen & Sword Books Ltd
47 Church Street
Barnsley
South Yorkshire
S70 2AS

ISBN 978-1-84415-532-3

Typeset by Concept, Huddersfield, West Yorkshire
Printed and bound in Great Britain by Biddles Ltd, King's Lynn

Pen & Sword Books Ltd incorporates the Imprints of
Pen & Sword Aviation, Pen & Sword Maritime, Pen & Sword Military,
Wharncliffe Local History, Pen & Sword Select,
Pen & Sword Military Classics and Leo Cooper.

For a complete list of Pen & Sword titles please contact
PEN & SWORD BOOKS LIMITED
47 Church Street, Barnsley, South Yorkshire, S70 2AS, England.
E-mail: enquiries@pen-and-sword.co.uk
Website: www.pen-and-sword.co.uk

Contents

Acknowledgements

It was as the result of reading Jimmy Rawnsley and Bob Wright's book *Night Fighter* when I was in my penultimate year at school, that I began to take an interest in air intercept radar and the history of night-fighting. Some ten years later in 1973 as a newly appointed Assistant Executive Engineer I was posted to the Post Office Air Defence Group, that was located in offices just around the corner from the old Public Records Office in Holborn, Kingsway. It was during my lunch-breaks that I first began the researches that led to this book. The information I gleaned in the PRO was supplemented by contacts with a number of RAF night-fighter crews found amongst the membership of the 604 (County of Middlesex) Squadron Association, particularly its Secretary John Annals, Chairmen John Davies, the late Sqadron Leader Jeremy Howard Williams, DFC, and Flight Lieutenant George Evans, DFM and Wing Commander Jack Meadows, DFC, AFC, AE.

In addition to these I would wish to acknowledge the help and support of Ted Cooke-Yarborough and Professor John Griffiths who read my first very rough draft and provided constructive criticism, the late Professor Robert Hanbury Brown, FRS, to Dr Paul Hensel, my boss in BT, and Chris Stevens who got me writing the book on an ancient Olivetti computer and to my colleagues and friends in BT who I bored rigid during my researches. Also, to Douglas Fisher who gave me full access to his photographic archive, to his son Robert for scanning the photographs and his wife, Joanna, for her lunches that lasted most of the afternoon, to Judith Last who got me out of many a spot with *Word* and in conjunction with my son James, helped me to master the basic concepts of *Photoshop*. I am grateful to the late Bryan Philpott for his help in providing information on the Meteor night-fighters, to Flight Lieutenant Ken Wright late of Nos 29 and 151 Squadrons and Peter Verney of Nos 39 and 152 for writing their experiences and opinions of night-fighting in the post-war era and answering my questions, to Flight Lieutenant Richard 'Jimmy' James, DFM, for describing his time as Guy Gibson's R/O and as Navigator Leader with No.89 Squadron, to John Rosenfeld, CB, Lieutenant Commander George Davies, DSC, Lieutenant Commander Ken Holme and the late Maj Skeets Harris, OBE, DSC, RM, for their help with the naval aspects of UK air defence. Further, to Ned Cartwright, another of my former BT colleagues, for his input to Appendix 1, to Dr Don Watts, one of my lecturers at Anglia Polytechnic University (APU) for reading the Introduction, to R. Cargill Hall, Andy Slater, Colonel Will W. Gildner Jr, USAF and Mrs Marcie T. Green of the USAF Historical Research Agency for their help in trying to unravel the background to the APS-57 radar, to Emma Jones, one of my fellow students at APU, for her help in tracing a number of the publishers quoted in the bibliography, to Nigel Eastaway of the Russian Aviation Trust for his help in acquiring information on the development of Soviet strategic bombing, to Squadron Leader Chris Goss for access to his knowledge of the *Luftwaffe*'s aircraft losses and to Squadron Leader Mike Dean for guiding me through the post-war ground radar 'environment'.

In respect of copyright, I would wish to acknowledge the assistance of Ken Ellis the editor of *Flypast* for permission to quote from the late Jeremy Howard Williams' article *Head-Up, Fifty Up* and to Gill Howard Williams for permission to quote extracts from her husband's book, *Night Intruder*, to the family of the late George Evans for permission to quote from his book *Bless 'em All*, to Mike Allen the current Chairman of the 604 Squadron Association for extracts from *If You Want Peace, Prepare for War*, to Christine Woollett and Gill Jackson at the Royal Society for the Society's biography of Derek Jackson and the photographs of its fellows, to Peter Waller of Ian Allan Publishing for *Kampfgeschwader Edelweiss*, to Linda Nicol at the Cambridge University Press for *Chance & Design*, to Christine Smith of Hayes Publishing for *Meteor*, to Sue Slack of the Cambridgeshire Library and Peter Lofts for photographs of Sir Alan Hodgkin, to Alan W. Hall, Ray Sturtivant and Reg Auckland for their help in trying, unsuccessfully, to trace the family of the late Mike Keep whose drawings I used to illustrate Appendix 3[1], to Dr John Navis and Zoe Arey for © 1987 from *Radar Days* by E.G. Bowen, © 1991 from *Boffin* by R. Hanbury Brown and © 1991 from *Echoes of War* by Sir Bernard Lovell, which are reproduced by permission of Routledge/Taylor & Francis, Group, LLC, to Wampre De Veer of Sutton Publishing for *Venom* and *Blitz on Britain*, to Gill Richardson at Crecy Books for *Mosquito*, the Canadian National Research Council, to Andrew Renwick of the Photographic Section of the RAF Museum and to Yvonne Oliver of the Photographic Archive of the Imperial War Museum.

Finally, to my editor at *Pen & Sword*, Peter Coles, for his confidence in taking on what is my first commercial book, to my wife Carol, who read and corrected the first draft and to my editorial team of Roger Lindsay, Lew Paterson and Phil Judkins, who have provided constructive criticism, support and advice throughout the year it has taken to write this book. To them all, I express my sincere gratitude and thanks and express my regret for any mistakes I may have made, or people I have failed to acknowledge.

Ian White
Martlesham Heath
February 2007

Note

1. Should anyone know the whereabouts of the late Mike Keep's wife and/or family, I would appreciate being put in contact with them.

Introduction

Defining the Problem

From the time of the Spanish Armada, through the Napoleonic Wars, and on to the end of the nineteenth century, the principal threat to the security and independence of the British Isles lay in a sea invasion from the European Continent. Britain's geographical position as an island and the protection afforded to it by the men and ships of Their Majesties' Royal Navy, ensured Britain's survival for nigh on 500 years. This apparent state of grace lasted throughout the nineteenth century and on into the first decade of the twentieth, when the first signs of a change became apparent. In 1908 the development of the airship reached the point where it could begin its transformation into a useful weapon of war, whilst Louis Blériot's crossing of the English Channel in an aeroplane the following year, reduced that barrier to a little more than forty minutes.

To reinforce the 'island fortress' mentality, British foreign policy was geared towards the avoidance of treaties that would involve the Army in military expeditions on the European mainland. Consequently, Britain had no peacetime alliances, with the single exception of her guarantees to Belgium, choosing instead to rely on the Empire and the Royal Navy for her sustenance and protection. From the defence viewpoint, it was the responsibility of the Royal Navy's Grand Fleet to maintain supremacy in home waters, whilst relying on its cruisers to protect the country's supplies of food and raw materials from attacks by commerce raiders. Home air defence (hereafter, Home Defence) was, therefore, viewed as the primary responsibility of the Royal Navy, with the Army's territorial forces being allocated the lesser task of tackling any small scale incursions that might otherwise penetrate the naval screen. As a consequence of this policy, the Army took second place to the Royal Navy in terms of finance, manpower and equipment procurement.

When war broke out in August 1914, the threat to the country came not from the battleships of Germany's Imperial Navy, but from the rigid[1] airships (Zeppelins) of its Airship Division and the submarines of its U-boat arm. By the war's end in 1918 the strategic bombing threat to Great Britain had changed significantly. The possibility of an invasion by the Imperial Navy had disappeared by 1916, with Germany's failure to gain possession of the North Sea and the aeroplane bomber had emerged as one of the two principal threats to the nation's well being (the other being the U-Boat). The recognition of the threat posed by strategic bombing, and in particular its effect on the civil population, led to the creation of a properly managed and co-ordinated air defence system. Equally, the realisation that the enemy was also vulnerable to bombing, precipitated the creation a bomber force to strike at Germany's towns and cities and an independent Air Force to manage it. At the beginning of the war, the nation's first arm of defence was the Royal Navy. By the end of that war, the ownership had passed to the RAF, where it would remain for the next fifty years.

The system that evolved to protect Great Britain by 1918 was the best in the world and one that incorporated most of the elements which would be in operational use twenty years later to protect the country for a second time, namely: a command, control and reporting network with its headquarters located in London,[2] specialist fighters equipped with radio telephony (R/T) sets and flown by crews with night-fighting training, an embryo method of fighter control, an intelligence organisation, anti-aircraft (AA) gun batteries, searchlights and balloon curtains to provide blocking barrages, an inland observer reporting system and a rudimentary early warning capability, all of which were supported by a national telecommunications network. Notwithstanding these facilities, the defence frequently failed to intercept the enemy for two principal reasons. First, the 'reach' of the early warning network was restricted to the range of human observers based on picket vessels operating in the North Sea and at locations around the English coast and second, the ability of pilots to see their targets in the night sky in anything but the best weather conditions, was poor. Only on nights of reasonable moonlight would they stand any chance of intercepting the enemy, and even then, they needed to know the direction from which they were coming.

Overall, the inability of the defence to provide an early warning of the approach of enemy aircraft, from say 60 miles (95 km), and provide an airborne instrument that would enable aircrew to see further than 400 feet (120 metres) in the dark sky, were the principal failings of the 1918 system. As aeronautical technology advanced over the following two decades and investment in air defence declined, the situation steadily deteriorated before it was finally halted in the 1930s by a radio-based device that would ultimately fulfil both requirements.

Notes

1. A rigid airship is one that has a metal or wooden skeletal hull and contains gas ballonets for lift and a keel to which the engines, crew and passenger gondolas are attached.
2. This was located in Carlton Gardens, London, and commanded by Major General E.B. Ashmore, the Officer Commanding the London Air Defence Area.

CHAPTER ONE

The Electronic Solution

1930–1937

Having been advised by the Air Staff in 1925 that there was no known defence against air attack, the British Government placed an increasing reliance on universal disarmament and the League of Nations,[1] as instruments of international security. These factors conspired to induce a malaise within the country, which accepted that bombing was indefensible, and caused many people to repudiate war and the armed forces that waged it. To some extent the RAF was responsible for this situation. It was their proposition that a defence against the manned bomber was not feasible, and the only means by which Britain might be protected lay in the field of massive retaliation. This policy, sometimes referred to as the 'knock-out blow' or 'counter-bombing', was to feature prominently in Air Staff doctrine and Government thinking throughout the 1920s.

By the time of the opening of the League of Nations conference on World Disarmament, in February 1932, the fear of air bombardment, fuelled by politicians and writers of fiction,[2] had reached manic proportions. Five days before the opening of the Conference, fighting broke out in Manchuria between China and Japan, from where newsreel film and photographs showed the appalling destruction of Shanghai by Japanese bombers. Stanley Baldwin, the Lord President of the Council in Ramsey MacDonald's Government of National Unity,[3] was deeply shocked by the images and later described them as a 'nightmare that would not fade'. The failure on the part of the League of Nations to act decisively against Japanese aggression, induced Baldwin to make his infamous statement in the House of Commons on 10 November 1932:

> I think it is well also for the man in the street to realise that there is no power on earth that can prevent him from being bombed. Whatever people may tell him, the bomber will always get through. The only defence is offence, which means you will have to kill more women and children more quickly than the enemy if you want to save yourselves.[4]

By coincidence, or by design, Baldwin's assertion that 'the bomber will always get through', complied exactly with the Air Staff's policy on retaliation and reinforced the RAF as a bomber oriented service. From the public's viewpoint, the statement fuelled the civilian's fear of bombing and raised the prospect of 'appeasement' as the only means to prevent widespread devastation. Therefore, by the end of 1932, the Government, the Air Staff, and the alarmists, were pulling roughly in the same direction. In the meanwhile, the British Government continued to press for international disarmament at the Geneva Conference.

However, since neither the United States (US), nor the Soviet Union, belonged to the League of Nations, the proposals to outlaw aeroplane bombers came to nought. Accordingly, the Conference was abandoned in November 1934, having done little to reduce the arms race or prevent war.

1922–1933

Until the late 1920s, France was regarded as Britain's principal 'enemy', as far as the air defence of the United Kingdom (UK) was concerned, and the direction from which any attack was most likely to come. To this end, as during the First World War, the Government continued to sponsor the development of sound locators. In 1922, an experimental station run by the Army was moved from its location at Joss Gap, near Dover, to The Roughs, near Hythe, where a 20 feet (6 metre) diameter sound 'mirror', built as a solid concrete casting, was set against the cliff face. With the sound collected at the mirror's focus and piped by an elaborate stethoscope to a collocated observer, the system was declared operational during the early months of 1923 and, in September, detected an aircraft at a range of 12 miles (19 km).

In 1925, Dr W.S. Tucker was appointed as the Director of Acoustical Research and two years later proposed the installation of a chain of 20-feet sound mirrors along the south coast. Only two were completed, the first at Abbott's Cliff, near Dover, and a second at Denge on the Dungeness peninsula. Before these were completed in 1928, Dr Tucker finalised the design of a more sophisticated 30 feet (9 metre) mirror, that incorporated the lessons learned from the previous experiments. Angled slightly upwards and with more protection for the operating staff, two of the new mirrors were built at Hythe and Denge and completed by the spring of 1930. On trials the mirrors demonstrated a range capability comparable to the 20-feet version, but did show an improvement in accuracy, especially in the vertical plane.

Tucker's final design was intended as a long-range device and since the targets would be approaching at low angles of elevation, the height of the mirror was reduced to 26 feet (8 metres), but increased in length to 200 feet (61 metres) with a curvature of 150 feet (46 metres). The instrument's size precluded the use of stethoscopes, so twenty static microphones were employed to catch the sound. One example of these giant mirrors was built at Denge in 1930, and employed, along with the others, in the annual air defence exercises until 1935. In 1932, the 200 feet mirror detected aircraft at 30 miles (48 km), when the unaided ear could only manage 5½ miles (8 km). Although a Thames Estuary scheme was proposed by Tucker and approved, the order was cancelled and no work was undertaken in light of the developments in radar.[5]

From 1933 onwards, however, France, through the neglect of her armed forces and the transition towards a static defence, had fallen by the wayside in the bomber race and passed the mantle to its erstwhile enemy, Germany. Although banned from maintaining bombers and submarines under the terms of the Versailles Treaty, Germany nevertheless established the beginnings of a clandestine air force, the *Luftwaffe*, within the small Army (*Reichswehr*) permitted by the Treaty. With secret Government funding and the active co-operation of the Soviet State, and through the sponsorship of civil aviation and sports flying, Germany was able to recreate the basic structure of an air force by 1930.

With the appointment of Adolf Hitler as Chancellor in January 1933, Germany embarked on a programme of rearmament and industrial expansion and established a foreign policy based on the reclamation of land lost during the First World War. In all three areas the emerging *Luftwaffe* was destined to play a significant role. With Herman Goering at the head of a new Air Ministry (*Reichsluftfahrtministerium* – RLM), the *Luftwaffe* was established covertly during 1933, with the ex-general manager of the state airline, Lufthansa, Erhard Milch, as the State Secretary responsible for organising the new arm and *Generalleutnant*[6] Walter Wever, as its first Chief of Staff and the custodian of its doctrine and strategy.

In 1933, the response in Britain to the possibility of the re-emergence of a rearmed Germany, was, on the whole, greeted with scepticism by Parliament and public alike. Nevertheless, Baldwin's warning on the possibilities of strategic bombing was particularly relevant to Britain in two respects. First, the country's geography placed the major conurbations and much of its industry within 70 miles (113 km) of the coast and, second, the improvement in the speed of bombers was beginning to erode the fighter's performance advantage.[7] By 1933, the margin of safety had very nearly reached zero and had thus dramatically reduced the period during which the bomber could be intercepted and destroyed before reaching its target.

1934

In these circumstances, the need for some form of early warning mechanism to alert the defences to an impending attack, coupled with the availability of fast climbing fighters, was of paramount importance to the country's survival. These criteria were ably demonstrated during the 1934 air defence exercises, held during the late summer, which took the form of night attacks on London, and ironically, Coventry, as it was one of the country's most important industrial targets. The exercises amply demonstrated Baldwin's theory, when only two out of every five bombers were intercepted, even though they were required to fly with their navigation lights switched on, and on the last night when half the raiding force reached their targets unmolested.

In reality, however, the *Luftwaffe*'s strength was less than the Government supposed and its bombing capability was almost non-existent against the shorter-range European cities, let alone London. Whilst Germany possessed relatively large numbers of aircraft, few of them were formed into cohesive units and many were deficient in operational equipment. In 1934, therefore, the *Luftwaffe* might best be considered more a collection of pilots and aeroplanes, than an effective air force. Faced with the RAF, whose equipment, experience and organisation was more effective, Britain's position was exposed, but not desperate. Provided the country looked towards its defences, the situation was redeemable.

The results of the 1934 exercises prompted the Air Officer-in-Chief (AOC-in-C) of the Air Defence of Great Britain (ADGB), Air Marshal Sir Robert Brooke-Popham,[8] to agree to the formation of a special sub-committee of the Committee of Imperial Defence (CID), to examine the strength of London's defences. Earlier that year, a junior scientist, Dr A.P. 'Jimmy' Rowe, the Personal Scientific Assistant to the Air Ministry's Director of Scientific Research (DSR), Dr Harry Wimperis, although barred from becoming involved in radio or armaments research, undertook a thorough search of the Ministry's files to see what references were available on air defence. Rowe's trawl unearthed fifty-three files, which showed that whilst the RAF had expended some considerable effort on the design of fighter aircraft, they had neglected to apply scientific analysis to the problem of air warning. Rowe submitted his report to Wimperis, with a recommendation that an approach be made to the Secretary of State for Air, Lord Londonderry, advising him that unless some method of early warning be developed, the country stood a very good chance of losing the next war, if it began within the following ten years!

Interest in the air defence problem outside Government circles was broadly confined to Winston Churchill and his scientific advisor, Professor Frederick Lindemann,[9] who considered the adoption of a defeatist attitude in the face of any threat, without an examination of the scientific alternatives, to be short-sighted. To this end, he proposed the whole of the Government's not inconsiderable scientific resources be made available to resolve the issue.

Pressure from these groups was eventually brought to bear on the Secretary of State for Air to find a solution. On 12 November 1934, Wimperis drafted a document in which he reviewed current technology, including an outline of the transmission of radio energy 'along clearly defined paths', and proposed that a group of scientists be formed to assess and evaluate the possible alternatives. Since the Admiralty and the War Office had an interest in Anti-aircraft (AA) problems, Wimperis recommended that the group be established under the auspices of the CID, on which he would sit and represent the Air Council's interests. Wimperis' paper, forwarded to Lord Londonderry, the Chief of the Air Staff (CAS), Marshal of the RAF (MRAF) Sir Edward Ellington,[10] and Sir Christopher Bullock, the Permanent Secretary at the Air Ministry, may rightly be considered as the document that brought radar into being.

Londonderry approved Wimperis' paper in late November, and formally invited Professor Henry Tizard of Imperial College, London,[11] to chair the group that would ultimately comprise himself, Professor A.V. Hill[12] of London University, Professor P.M.S. Blackett[13] of Birkbeck College, London, Wimperis and Jimmy Rowe as secretary. The first meeting of what was entitled 'The Committee for the Scientific Survey of Air Defence', but better known today as 'The Tizard Committee', was scheduled for 28 January 1935.

In the meanwhile, Wimperis contacted the Superintendent of the Radio Research Laboratory at Slough, Mr Robert Watson Watt,[14] and invited him to the Air Ministry on 18 January to discuss the feasibility of 'death rays'. Following the meeting, Watson Watt returned to Slough and tasked his assistant, Arnold Wilkins,[15] to calculate exactly how much radiated power would be required to raise the temperature of a given quantity of water, to a stated level, at a stated distance. Prior to undertaking the calculation, Wilkins noted the initial temperature of the water was slightly above that of human blood and the stated temperature to be that of a man with a fever. He, therefore, guessed his calculations were related to a death ray! As expected, the calculation showed the radiated electrical power required to raise the normal body temperature to a dangerous level, was enormous and, therefore, totally impractical as a weapon.

Having proven that which they already believed, Wilkins recalled how aircraft were interfering with the radio waves from the General Post Office's (GPO) radio station at Daventry and causing them to fade.[16] Taking Wilkin's suggestion as the basis of an idea, Watson Watt suggested he calculate how much power would be required to produce detectable signals from an aircraft at a given range. In order to complete his calculation, Wilkins assumed the target was a bomber, with a typical wing span of 82 feet (25 metres) and a height of 11 feet 6 inches (3.5 metres) and its shape would simulate a conventional half-wave dipole (see Appendix 1 for a description of basic radar theory) and would radiate in a similar fashion. Using these criteria he was able to confirm that it would be possible to detect a bomber-sized aeroplane by radio means. In his subsequent report, Watson Watt was able to reassure Wimperis of the practicality of detecting aeroplanes by radio means. Arnold Wilkins' calculations might, therefore, be regarded as the first scientific proof in Great Britain of the feasibility of a radio-based aircraft detection system.

1935

Whilst it is not the intention of this book to provide a detailed history of the development of ground radar, a general appreciation of the events that led to the introduction of the Chain Home (CH) radio direction finding (RDF) system, is pertinent to the development of air-borne radar.

The Tizard Committee met, as scheduled, on 28 January 1935, in Room 724 of the Air Ministry, to consider Watson Watt's findings. In his later, and more detailed report, dated 12 February, and entitled 'The Detection and Location of Aircraft by Radio Methods', Watson Watt suggested it would be possible to 'illuminate' an aircraft with radio waves and detect the minute amounts of energy that would be reflected by it. In this respect, received signal levels of the order of one ten million, million, millionth of the power transmited, were predicted, or a factor of 1 in 10^{-19}.

On 14 February, Tizard, Sir Christopher Bullock and Wimperis met to discuss Watson Watt's paper, before briefing Air Vice Marshal Sir Hugh Dowding,[17] the Air Council's Member for Research & Development, the following day. Dowding was not at first convinced of Wilkins' calculations, but, nevertheless, agreed to a practical experiment, after which he would make the sum of £10,000 (£1,800,000)[18] available to begin research into radar.[19] On 26 February, a Handley Page Heyford bomber from the Wireless & Electrical Flight of the Royal Aircraft Establishment (RAE), Farnborough, flown by Squadron Leader R.S. Blucke, was directed to fly on a course that would intersect the beam of the Post Office Radio Station at Daventry, whilst Wilkins, Watson Watt and Rowe witnessed the event by the rise and fall of a trace on the screen of a cathode ray tube (CRT).

Watson Watt's demonstration of the radar principle, was far from being a practical system that would provide indications of the range, bearing and elevation of a target aircraft. The demonstration was, however, sufficiently convincing for a delighted Dowding to release the funding and begin preliminary research. Due in part to the disinterest of the Slough Radio Station's owners, the National Physical Laboratory (NPL), and Watson Watt's keen personal interest in the new science of air navigation, it was agreed that the Slough Group be transferred to the Air Ministry's payroll. To ensure the utmost secrecy, the Ministry moved the Group on 13 May, to a more secure and secluded location on the old bombing range at Orfordness, on the Suffolk coast.

Within a few days of their arrival at Orfordness, the Group, under Wilkins' leadership and comprising L.H. Bainbridge-Bell, Dr Edward 'Taffy' Bowen[20] and George Willis, Bainbridge-Bell's technical assistant, had established themselves in a number of First World War vintage huts, and with the help of two additional staff, Joe Airey[21] and Alec Muir, had erected the transmitter and receiver and connected them to mains power. The original Orford transmitter gave a very healthy 20–25 kiloWatts (kW) on a wavelength of 50 metres (a frequency of 6 MegaHertz [MHz]), which by the summer, had been pushed to 100 kW. On 17 June, the equipment recorded its first clear target, a Short Scapa flyingboat, which was detected at a range of 17 miles (27.5 km) and tracked to its base at Felixstowe. With the active co-operation of the Aeroplane & Armament Experimental Establishment (A&AEE), at nearby Martlesham Heath, who provided the target aircraft, the range was gradually increased in incremental steps to 80 miles (130 km) by December and to 100 miles (160 km) early in 1936. During August, the first attempts at height finding were undertaken, to be followed in October by the detection of an aircraft at 15 miles (24 km), and at an altitude of 7,000 feet (2,135 metres), with an error of $\pm 1,000$ feet (305 metres).

The original 50 metre wavelength had been chosen as it corresponded very conveniently with the wingspan of most of the large aircraft of the day and would, therefore, achieve good resonance and a more powerful return signal. However, Wilkins was forced to abandon this wavelength, due to its interfering with long-distance commercial radio communications. A change was, therefore, made to 26 metres (3.85 MHz) without any apparent reduction in performance. This was further reduced to between 10 and 15 metres (20–30 MHz), which became the standard early warning wavelength to the end of the Second World War and beyond.

The success of the Orfordness Group was such that by September 1935, the Treasury, had in principle, approved the necessary finances to fund the installation of a radar warning chain extending from Southampton to the Tyne. Money for the first phase, comprising five CH stations sited around the south-east coast between the North Foreland and Bawdsey, was released by the Treasury on 19 December. However, it was quickly realised that the task of building the chain was beyond the capability of the small team at Orfordness, whose principal task was research. To provide the necessary room for an expansion of the research capacity and a site that could also accommodate the 360 feet (110 metres) CH transmitter and 240 feet (73 metres) receiver towers and provide the requisite spacing between them, the Air Ministry began the search for new premises.

The initial investigation proved fruitless, until Wilkins discovered that a site owned by Sir Cuthbert Quilter at Bawdsey was ideally suited, since it had ground that rose 70–80 feet (21–24 metres) above sea level – a rare commodity in Suffolk. Coincidentally, the site, Bawdsey Manor and its surrounding 180 acres, was up for sale. The Ministry responded quickly and purchased the Manor and its grounds for £23,000 (£822,940).

1936

The move, to what was now retitled the 'Bawdsey Research Station', with Watson Watt as its first Superintendent, began in March 1936. Laboratories were set-up in the White Tower and the Stable Block, whilst the CH transmitter and receiver towers were erected on the higher ground behind the Manor. During the same month, Wilkins completed another part of the radar puzzle by devising a method that would establish a target's bearing. This employed some of the principles he had learned from his early work on the measurement of the downward angles of incoming transatlantic radio signals. Later, a more accurate method that compared the angle of arrival of the received signal in the azimuth plane using the horizontal pair of a crossed dipole and a goniometer was devised. The target height was established in a similar manner by comparing the angle of arrival in the vertical plane using a pair of vertical dipoles at different heights on the array and a goniometer. Therefore, by March 1936, the Orfordness Group, with the active support of the Tizard Committee, had met Watson Watt's requirements for a basic radar system; namely, the ability to calculate the range, bearing and height of a target aircraft.

At this point, the development of an early warning system was interrupted by politics and a change of Government. In December 1934, Lindemann, with Churchill's support, had been pressing for a full CID committee to investigate the political and financial aspects of air defence. Unaware of the existence of Tizard's CID Sub-Committee, Lindemann's request was granted in April 1935, when Prime Minister Ramsey McDonald appointed Sir Philip Cunliffe-Lister, later the Earl of Swinton, to chair the CID air defence committee, the Swinton Committee, on which Tizard had a seat. In June Churchill was invited to join the Committee, which took over the ownership of the Tizard Committee, and in turn appointed Lindemann as its sixth member.[22] Before his departure from government later in June and his replacement by Stanley Baldwin as Prime Minister, Ramsey McDonald appointed Swinton to the post of Secretary of State for Air.

Lindemann's appointment to the Tizard Committee disrupted the congeniality of its members and introduced a number of the Professor's more esoteric (cranky) solutions to the air defence problem. Not withstanding his ideas on aerial mines, wires suspended from balloons and infra-red (IR) detection, the two basic facts affecting air defence were clearly identifiable:

1. The need to locate and track enemy raiders in good time to effect an interception by day or by night (with greater accuracy being required in the case of the latter).
2. The ability to destroy the enemy once he had been found, in daylight or in darkness.

As far as daylight interception was concerned, radar offered by far the greatest potential for development, with night interception, at that time, being impossible. The night problem was, however, being pursued under the codename *Silhouette*, which was based on the illumination of the cloud-base (when one existed), against which a bomber would be 'silhouetted' in outline. This line of investigation was eventually terminated, because of the vast amounts of power required to drive the numerous cloud-lighting searchlights. This left only two alternatives; first, the development of a radar set light enough and compact enough, to be installed in an aeroplane, and second, the detection of an aircraft's exhaust gases by infra-red (IR) means.

Coincidentally, one of Lindemann's post-graduate student's, R.V. Jones,[23] was conducting research into the IR spectrum and building a detector, but it would be a few years before a reliable system was capable of being installed in an aircraft.[24] The matter had also been occupying Tizard, who, in a letter to Dowding dated the 27th, placed his ideas before the Air Member for Research & Development. In this paper Tizard argued that whilst it was possible with the CH system to detect aircraft during the hours of daylight and to give indications as to their range and bearing with reasonable accuracy, the same was not true of interceptions at night, where the pilot's visibility was restricted to a few hundred yards/metres. To 'see' further required the installation of a radar set in the intercepting fighter. He was also convinced that the early warning chain then being built, would cause the *Luftwaffe* such losses during the hours of daylight, that they, like their forebears in the Imperial Army and Navy, would be forced into bombing by night. These prophetic words, written in 1936, would be turned into reality in 1940.

Tizard copied his letter to Watson Watt at Bawdsey. With much of the Station's effort and resources being directed towards the construction of the CH chain, Watson Watt chose to divide his staff by passing the responsibility for CH installations to Wilkins and the night-interception problem to Bowen. Bowen began his task by seeking the advice of two resident engineers at Martlesham Heath, Fred Roward and N.E. Rowe, who had a reputation for sound, practical engineering and were also privy to the radar secret. He also made a number of visits to the Air Ministry and HQ Fighter Command at Bentley Priory, to seek the advice of any officer having knowledge of night interception techniques. From these discussions, Bowen drew-up a set of guidelines for the design of an airborne radar set:[25]

- A total weight not to exceed 200 lb (91 kg).
- The maximum space not to exceed 8 cubic feet (0.22 cubic metres).
- The maximum power consumption not to exceed 500 watts.
- An aerial system no greater than 1 foot (305 mm) in length.
- The system to be capable of operation by the pilot, or an observer.

The need to constrain the power consumption to 500 watts was dictated by the total power capable of being generated by aircraft of the day. This was usually not far short of 500 watts at 12 volts, the majority of which was absorbed by the aircraft's electrical systems, leaving very little for a radar set. The objective of aircraft designers constantly to reduce aerodynamic drag, necessitated the use of 1 foot dipole aerials, which corresponded to a wavelength of approximately 0.6 metres (500 MHz).

Bowen had an immediate stroke of luck. By means unknown to this day, but most probably through the 'back door' of the EMI Company, he was presented with a tuned radio receiver, designed for the BBC's forthcoming television service. This 'EMI' receiver, comprising seven or eight valves mounted on a chassis measuring 3 inches (76 mm) in width, by 15 or 18 inches (380–460 mm) in length, operated on a frequency of 45 MHz (6.7 metres), with a bandwidth of 1 MHz. Although its sensitivity has not been recorded, it was said by Bowen to be 'far and away better than anything which had been achieved in Britain up to that time'.[26] When combined with a simple CRT indicator, the EMI amplifier was to form the basis of airborne radar development for the next two years. The overall weight of the receiver and indicator unit was some 20 lb (9 kg) and the power consumption was regarded as being minimal. However, its wavelength at 6.7 metres meant, that for the time being, the equipment could not be carried in an aircraft.

Bowen was allocated laboratory accommodation in the roof spaces of the Red and White Towers at Bawdsey and three more staff, Scientific Officer Dr Gerald Touch[27] from the Clarendon Laboratory and two experienced engineers from the radio industry, Sidney Jefferson[28] and Percy Hibberd. These four were to be joined later in the year by Keith Wood,[29] the first of many very talented Technical Assistants. Touch became Bowen's second-in-command of what was now entitled the 'Airborne Group' of the Bawdsey Research Station, and later assumed the responsibility for the design of air-to-surface vessel (ASV) radar. Hibberd, after designing the first airborne transmitter, transferred to the Ministry of Aircraft Production (MAP), where he played a vital role in supervising the manufacture of radar sets.

Having, for the time being, resolved the receiving element of the system, the Group set about the task of building a pulsed transmitter along the lines of the original Orfordness set. This equipment operated on a wavelength of 6.7 metres, with a pulse width of 3–4 microseconds (μsecs) and an output of 30–40 kW. Appreciating that it would not be possible to install the transmitter and receiver in an aircraft at this stage, Bowen elected to gain some experience by placing the receiver and indicating unit in the aircraft and leaving the transmitter on the ground. The first ground trials of the system, with the transmitter driving a dipole aerial located on the roof of the Red Tower and working to a similar aerial and receiver on the White Tower, demonstrated a maximum range of 40–50 miles (64–80 km) against the target aircraft.

With the co-operation of Group Captain A.C. Maund,[30] the Officer Commanding A&AEE, at Martlesham Heath, Bowen was able to procure the use of a Handley Page Heyford bomber for the airborne trials of the receiver and display. The Heyford was an ideal choice.

It was large enough to carry the receiving dipole aerials mounted between the undercarriage legs and, being powered by Rolls-Royce (R-R) Kestrel engines, it possessed a well designed ignition harness that provided a moderately 'quiet' electrical supply. Older aircraft that were fitted with radial engines generally radiated considerable amounts of ignition noise, which rendered them totally unsuitable for radar trials. Maund also provided three non-commissioned officer (NCO) pilots, Flight Sergeants Shippobotham, Wareham and Slee,[31] with all three having excellent records as test pilots.

The first flight trials, of what was to become known as RDF1R,[32] were undertaken during the autumn of 1936, using the Heyford. Power for the receiver and indicator was drawn from a series of dry cell batteries mounted on the fuselage floor, with the high tension (HT) supply for the CRT being derived from a second-hand Ford ignition coil, driven by a vibrator from a separate 12 volt battery. On the very first flight, with Shippobotham as pilot and Bowen as observer, the Heyford circled the Red Tower transmitter at 2,000–3,000 feet (610–915 metres) and obtained echoes from aircraft in the circuit at Martlesham Heath – a distance of 6¾ miles (11 km). This performance was repeated on subsequent flights, with a maximum range of 12 miles (19 km) being recorded. It is perhaps worth noting, that the performance of this early system was never bettered by any of the wartime radars, at such a low altitude.[33]

Good though the system was, it did have its limitations, which Bowen readily accepted. For the system to work effectively, the airborne receiver required to be synchronised to the ground station's transmitter, and the target aircraft had to be illuminated by it. In situations where the target was behind the transmitter, the receiver would be unable to detect its presence. The system also failed to provide an indication of the target's bearing from the fighter and range measurement was accurate only for so long as the fighter was directly between the ground transmitter and the target aircraft. In all other positions, the range was underestimated. Had it been possible to lay down a grid of RDF1R transmitting stations, a night-fighter would have been able to position itself with a transmitter behind it, before manoeuvring into a stern-chase approach to the target. Potentially, RDF1R offered an effective range of 20 miles (32 km) at low altitude and did not require a sophisticated ground control network. In summary, the system presented a number of performance advantages, that had to set against a number of tactical disadvantages. Bowen, in his book *Radar Days*, expressed his disappointment when RDF1R was cancelled by Watson Watt:

> With hindsight, it is now clear that this was a grave mistake. If RDF 1.5 had been given proper Service trials in the RAF, there would have been several important consequences. In the first place, it would have given them an interim device on which test interceptions could have been carried out at night, two whole years before the outbreak of war. This would have provided pilots and observers with training in the techniques of night interception, something which they did not get until war was declared. Lastly, it would have given us experience of the immense problems, still to be faced, of introducing new and sophisticated equipment into Service use.[34]

Further trials of RDF1R were undertaken during March 1940, with the receiver installed in a Bristol Blenheim, L6622. For these trials the aircraft was based at Manston and used the Foulness Chain Home Low (CHL)[35] station as the illuminating source. Once the RDF1R had locked and synchronised to the CHL transmitter, the Blenheim recorded detection ranges of 4 miles (6.5 km). The system's performance was, however, regarded as 'poor' and the trials were not repeated.

1937

During the early spring of 1937, the Airborne Group obtained a number of American, Western Electric Type 316A 'Giant Acorn' valves, that were capable of generating 20 watts on wavelengths down to one metre (300 MHz). Hibberd incorporated these into the design of a miniature transmitter, operating on 6.7 metres (45 MHz) to match the EMI receiver. With a pulse repetition frequency (PRF)[36] of 1 kilohertz (KHz), a pulse width of 2–3 µsecs, and an output of a few hundred watts, the transmitter was installed in the Heyford during March 1937. During trials the new transmitter proved capable of illuminating the wharfs and cranes in Harwich Docks and shipping approaching the port from the seaward side It did, however, prove impossible to ascertain the range of the ships with any certainty, since the Heyford was not approved for flying over the sea. Nevertheless, the range in the 'sea search' Air-to-surface vessel radar (ASV) mode, was considered to be 3–4 miles (4.8–6.4 km).

The Airborne Group's success with the Heyford as a trials aircraft identified the need for permanent aircraft and crews and the establishment of a properly structured flight test organisation. Bowen was encouraged to investigate a more suitable aircraft, and with the Air Ministry's assistance, he selected the twin-engined Avro Anson, then in service with Coastal Command as a general reconnaissance (GR) bomber. The Anson offered a number of advantages over the Heyford. It was safe and easy to fly, had plenty of cabin space for the stowage of equipment, and accommodation for two pilots and two observers. Being a standard Coastal Command machine, it was equipped to operate over the sea, an important asset, since the Group's remit was extended to cover ASV, in addition to air intercept (AI).

The Air Ministry allocated two Anson Mk.I aircraft, K6260 and K8758, and five pilots and supporting ground crews on detachment from No. 220 Squadron, based at Bircham Newton, in Norfolk. The five pilots comprised, Sergeants Smith, Naish, Argent and Newman, and the detachment commander, Flying Officer D.C. 'Blood Orange' Smith. Aircraft, pilots and ground crews were firmly established at Martlesham Heath by June 1937. Although no formal title was allocated, the unit was frequently referred to after the war as the 'Radar Flight'.

Early trials with the Ansons identified one problem, namely, poorly suppressed engine ignition harnesses, which, on occasions, had the ability to 'drown' the receiver in noise. With the help of RAE, the harnesses were replaced with properly screened and suppressed alternatives, following which, the problem disappeared. Whilst the aircraft were at Farnborough, Bowen set about the task of reducing the operating wavelength to 1 metre. Beginning with the transmitter, Hibberd successfully converted it to push-pull operation, by employing two of the Type 316A valves and reducing the pulse width to 1 µsec and the wavelength to 1 ¼ metres (240 MHz), with the PRF remaining at 1 KHz. This configuration proved to be the absolute limit for the 316As, since any further reduction in wavelength, brought with it a significant reduction in transmitter power. The transmitter power is not known, but Bowen considered it to be less than 1 kW. By inserting a frequency changer section ahead of the EMI chassis, Touch was able to redesign the receiver as a super-heterodyne, with the EMI chassis becoming the intermediate frequency (IF) amplification stage, operating on its frequency of 45 MHz. From then to the end of the War, 45 MHz remained as the standard IF in all British airborne radars.[37]

During August, the Ansons were returned from Farnborough fitted with the necessary bracketry to mount the 1 ¼ metre transmitter, receiver and indicator unit. With the transmitting aerial poked through the cabin escape hatch and the receiving element located within the cabin, K6260 took off from Martlesham Heath on its first flight with the new equipment, on 17 August 1937. With Touch and Wood on board to operate the controls, the set demonstrated a range of 2–3 miles (3.2–4.8 km) against shipping off the Suffolk coast.

Because of a shortage of test equipment to measure the output power of their early radar systems, it became the practice at Bawdsey to mount the experimental equipment in the White Tower and calibrate the range on the nearby water tower at Trimley (today, the village of Trimley St Mary), a distance of a little over 3 miles (4.8 km). During one of these measurements, it was observed that when the wavelength was increased to 1½ metres (200 MHz), a worthwhile improvement in sensitivity was obtained. Thus was born 1½ metre, or 200 MHz radar, which would see continuous service in the UK in Fighter and Coastal Commands until 1942, when it began to be superseded by centimetric technology.[38]

A few days after the successful flight in K6260, Bowen was telephoned by Watson Watt to ask if he would be willing to participate in two days of naval exercises, scheduled to take place the following month (September). This was to be a joint-services event, between ships of the Royal Navy and some forty-eight aircraft of Coastal Command, who would attempt to intercept and report on the 'fleet' as it made its way through the English Channel and the North Sea. The exercise was due to begin during the early hours of Saturday, 4 September, and would last for forty-eight hours. As Bowen later remarked, 'this was too good an opportunity to miss'.[39] With Sergeant Naish, a former Merchant Navy officer, at the controls and with Bowen and Wood operating and observing the radar, K6260 took off from Martlesham Heath in the late afternoon of 3 September, hoping to catch the fleet as it assembled off the Solent. That evening, the Anson's crew found the battleship HMS *Rodney*, the aircraft carrier HMS *Courageous* and the cruiser HMS *Southampton*, with their attendant destroyers, preparing to enter the Straights of Dover. Making several runs over the fleet, Bowen was impressed by the huge echoes that were returned by the large ships. These were far better than anything they had previously experienced.

At day-break the following morning, Naish, Bowen and Wood took off again and headed due east to begin their search. Flying at 3,000 feet (915 metres), and with the radar functioning properly, Naish commenced a box search until, at 0800 hours, they detected a large echo at a range of 5–6 miles (8–9.6 km). Closing to obtain a 'visual', they saw *Courageous* and *Southampton* and their destroyer escort, but not *Rodney*. As the Anson approached, 'all hell broke loose. Signal lights flashed in all directions, guns were fired – no doubt firing blanks – and aircraft started to take off from *Courageous*',[40] all of which were observed by the radar set. This then was the first proof that radar pulses from one aircraft could be reflected from another and received as echoes, and proved conclusively that an AI system could be made to work.

The results of September's sea-search exercise generated a great deal of interest and numerous requests for demonstration flights in K6260. On 10 September, Dr Talbot Harris, the recently appointed head of the Army RDF Group at Bawsdey, was taken for a flight, followed on 14 September, by Squadron Leader Raymond Hart,[41] who was soon to take charge of radar training at Bawdsey. However, the most rewarding from Bowen's viewpoint, was when, on 18 October, Tizard was taken aloft for his first flight in a radar equipped aircraft.

Following the September exercise, the Group was strengthened with the arrival of Robert Hanbury Brown,[42] a scientist recruited by Tizard from Imperial College, London, who joined as a Scientific Officer, followed by two industrial research engineers, Ron Taylor and Paul 'Yagi' Walters. The Technical Assistant's ranks were swelled with the appointment of

Brian 'Chalky' White and R. Mills. Finally, Bill Eastwood, another graduate from London University, was loaned to Bowen from the Army Group to provide additional support with his research into 200 MHz technology.

CONCLUSION

By the end of 1937, and in a period of barely less than three years, radar research in Britain had produced the early warning CH system capable of detecting aircraft formations at a maximum range of 100 miles (160 km) and had received Government funding to begin the construction of a protective chain. In a shorter period, Edward Bowen's Airborne Group at Bawdsey, had reduced the bulk and weight of the CH transmitter and receiver to a point where they could be installed in an aircraft, and brought the wavelength down from 50 metres to a more manageable 1 ½ metres. He and his Group had installed this equipment in a standard RAF aircraft (the Anson) and undertaken a series of flying trials to prove it was capable of detecting air and sea targets. The remaining challenge was to develop the Anson set into practical AI and ASV systems, that were reliable and could be operated and maintained by RAF personnel under service conditions. This work, and the installation of radar in suitable RAF and Royal Navy aircraft, would occupy most of 1938 and 1939.

Notes

1. The League of Nations was established in 1920, as a result of a number of initiatives in the field of international relations, which ended the First World War in 1918. The League's proposals were subsequently written into the Versailles Peace Treaty of 1919, from which most of Europe's troubles subsequently flowed.
2. H.G. Wells *War of the Worlds* and S. Southwold's, *The Gas War of 1940*, published in 1931, foresaw the use of atomic weapons and the destruction of London by air bombardment in the most ghoulish terms.
3. The 1929 Labour Government collapsed in 1931, to be replaced by a Government of National Unity under Labour's Ramsey MacDonald, in which Baldwin served as Lord President. He remained in the post until 1935 when he replaced MacDonald as prime minister.
4. John Terraine, *The Right of the Line* (London: Hodder & Stroughton, 1985), p. 13.
5. *Sound Mirrors on the South Coast*, at www.doramusic.com/soundmirrors.htm.
6. A rank broadly equivalent to an RAF Air Marshal.
7. It was generally agreed that a fighter required a minimum 15 per cent speed advantage over its target. A 25 per cent margin was, however, preferred.
8. Later Air Chief Marshal Sir Robert Brooke-Popham, GCVO, KCB, CMG, DSO, MC, C-in-C Far East, November 1940 to December 1941.
9. Later Viscount Cherwell of Oxford, FRS, and Paymaster General in Churchill's wartime government.
10. MRAF Sir Edward Ellington, GCB, CMG, CBE, served as CAS from May 1933 to September 1937.
11. Later Sir Henry Tizard, GCB, KCB, AFC, FRS.
12. Later Baron Heywood, FRS, of University College, London.
13. Later Baron Blackett, FRS, and the President of the Royal Society.
14. Later Sir Robert Watson Watt, KB, CB, FRS.
15. Later A.F. Wilkins, OBE, MSc.
16. Notes to the author from Phil Judkins dated 24 January 2006.
17. Later Air Chief Marshal Lord Dowding, KCB, CMG, AOC-in-C Fighter Command from July 1936 to November 1940.
18. Average weekly earnings in 1936 were £3.00, whilst today (2005) that average has reached £550 and equals approximately 180 times as much (information supplied by Phil Judkins).
19. It should be noted that until America joined the Second World War late in 1941, radar in Britain was referred to as 'RDF'. It was not until 1942, that the American abbreviation radar – RAdio Detection And Ranging – was formally adopted. For convenience, the term 'radar' will be used throughout this narrative.
20. Later Professor Edward Bowen, CBE, FRS.

21. Later J.E. Airey, MBE.
22. Harold Hartley, *Tizard, Sir Henry Thomas*, the Dictionary of National Biography, 1951–1960.
23. Later Professor Reginald Jones, CB, CBE, FRS, the Assistant Director of Scientific Intelligence in the Secret Intelligence Service (MI6) from 1940–1946 and its Director thereafter.
24. The first IR device, an IR floodlight and detector designed by the German AEG Company. Codenamed *Spänner*, it was introduced to service in 1940.
25. E.G. Bowen, *Radar Days* (Bristol: Adam Hilger, 1987, an imprint of Routledge/Taylor & Francis, LLC), p. 32.
26. *Ibid*, p. 33.
27. Later Dr Gerald Touch, CMG, DPhil, the architect of ASV Mk.II, and the Chief Scientist at GCHQ from 1961–1972.
28. Sidney Jefferson worked for EMI before joining Bawdsey to work on CH receivers, and later as a Senior Scientific Officer, led various groups on radar development. After the War he joined the UK Atomic Energy Authority at Harwell.
29. Later K.A. Wood, OBE, Chief Superintendent of the Blind Landing Experimental Unit, Martlesham Heath.
30. Later Air Vice Marshal A.C. Maund, Air Officer Administration, RAF Middle East Command.
31. Later Wing Commander Slee, and senior pilot of the King's Flight.
32. The ground-based CH early warning radar was designated RDF1 by the Air Ministry, with airborne radar being designated RDF2. Bowen's half and half system was, therefore, designated RDF1R by the Ministry and RDF1.5 by Bowen's team.
33. E.G. Bowen, *op cit*, pp. 36 and 37.
34. *Ibid*, p. 37.
35. CHL was an addition to the CH network that 'filled' low-level holes in the network. Like the later AI radars, CHL worked on the 1½ wavelength (200 MHz).
36. The pulsing performance of radars before and during the Second World War, was quoted in the number of pulses per second (PPS), however, modern term 'pulse repetition frequency (PRF) will be used throughout this book.
37. E.G. Bowen, *op cit*, pp. 39 and 40.
38. It should be noted that 1½ metre AI remained in operational use in the Far East and in training establishments in the UK to the War's end.
39. E.G. Bowen, *op cit*, p. 42.
40. *Ibid*, p. 43.
41. Later Air Marshal Sir Raymond Hart, CB, KBE, MC.
42. Later Professor Robert Hanbury Brown, AC, DSc, FRS.

CHAPTER TWO

The Solution Evolves

1938–1939

The political situation in Europe deteriorated markedly by the beginning of 1938. With his ambition to revoke the Versailles Treaty and restore Germany's dominant position in Europe, Hitler demonstrated his independence by withdrawing from the League of Nations in 1934 and re-militarising the Rhineland in 1936, in the face of negligible protest from France and Britain. Having tested Europe and found it wanting, he turned his attention towards the reunification of the ethnic German populations of Austria, Czechoslovakia and Poland, into the Greater German Reich. Austria was annexed in March 1938, when German troops marched into that country and effectively incorporated it into Germany (Anschluss). Having secured Austria's 6,000,000 Germans, Hitler embarked on the next stage of his conquest, the recovery of 3,000,000 Sudeten Germans in Czechoslovakia. Here he met his first problem. With that country allied to France and Soviet Russia, and with a well trained and properly equipped army, Germany could not mount a successful invasion. Instead, using the threat of war, Hitler encouraged the Sudetens to create internal unrest in the hope of breaking-up the country from within.

Whilst Europe gradually fell apart, Britain, under the premiership of Neville Chamberlain,[1] pursued its policy of rearmament and industrial expansion behind a screen of appeasement. In July 1934, the Cabinet approved the first of a series of 'schemes' that were designed to build an Air Force to match that of Germany and provide a limited reinforcement for the Empire. The responsibility for UK air defence was changed in July 1936, and vested in a new organisation, Fighter Command, under the command of its first AOC-in-C, the recently promoted Air Marshal Sir Hugh Dowding. The new Command gave Dowding operational control of all Regular, Auxiliary Air Force (AAF)[2] and RAF Special Reserve (SR)[3] fighter units, which by September 1938, comprised thirty squadrons,[4] based on seventeen airfields, the CH chain, the Observer Corps and tactical control of AA guns and barrage balloons.[5]

For Fighter Command 1938 was an important year. The first of the eight-gun fighters, the Hawker Hurricane, had entered service with No. 111 Squadron, the previous December, followed by its Supermarine stablemate, the Spitfire, in August and the twin-engined Blenheim Fighter by December. The first five stations of the CH chain around the Thames Estuary were built and operational by August, and were able to track Mr Chamberlain's aircraft on its way to Munich, where the Prime Minister would agree to the handover of the Sudetenland to Germany. Funding was also secured for the expansion of the chain by a further fifteen stations to cover the coast from Portsmouth to the Firth of Forth, with a completion date set for April 1939. Radar training, both technical (radio mechanics and signals officers) and operational, had begun at Bawdsey, with the former being moved to Tangmere and then to Yatesbury, where it would remain as No. 2 Radio School, until the 1950s.

In March 1935, Hitler announced the existence of the *Luftwaffe* to the Foreign Secretary, Sir John Simon, and declared its strength to equal that of the RAF. By the end of 1937, the German aircraft industry had built 5,606 aircraft, to be followed in 1938 by a further 5,235 machines. These enabled the *Luftwaffe* to field a total of 263 *Staffeln*[6] in Germany, of which sixty-seven were fighters and ninety-three bombers with a further three fighter and two bomber *Staffeln*, stationed in Spain to support the Condor Legion.[7]

German involvement in the Spanish Civil War began in July 1936, when the first consignment of 'aid' to the Nationalist forces, comprising aircraft AA (Flak) guns and 'volunteers', under the command of Major Alexander von Scheele, left by sea for Seville. By October, the group had expanded into an expeditionary force (Legion), with formal bomber, fighter, reconnaissance, transport and Flak units and a supporting ground organisation. Although primarily deployed as an aerial 'fire brigade' in support of General Franco's Nationalist Army, the Legion was infamously responsible for a terror raid on the Basque town of Guernica, on 26 April 1937, when twenty-six bombers dropped 45 tons (45.7 tonnes) of bombs, killing at least 250 civilians and destroying, or damaging, some 80 per cent of the housing stock. The bombing of Guernica would go down in popular history, alongside those of Warsaw, Rotterdam and Coventry, as an example of Germany's inhuman use of strategic bombing as a terror weapon that was designed to frighten and eventually to subjugate the nations of Europe.[8] The message was by no means lost on the British Government.

1938

Much of 1938 was taken up with the continued development of the sea search system in the Anson, however, with the ever present threat of war, the need for an effective AI radar assumed greater importance. In order to develop this system, Bowen needed a trials aircraft that better reflected the performance of current fighters. As before, he consulted the Air Ministry as to the requirements for a night-fighter and was advised that the single-engined, Fairey Battle light-bomber possessed the necessary performance and had ample accommodation for equipment and an observer (or two with a bit of a squash) in the rear cockpit. The Battle's power was supplied by the reliable R-R Merlin engine, giving the aircraft a top speed of 240 mph (385 km/hr) at 13,000 feet (3,960 metres) – a vast improvement on the Anson. Bowen accepted the Ministry's recommendation and was duly delivered Battles, K9207 and K9208, along with two experienced pilots, Flying Officer Douglas Rayment[9] and Sergeant Clifford 'Wilbur' Wright.[10]

The arrival of the Battles and the increase in the complement of the Radar Flight, made the arrangement whereby it was a detachment of No. 220 Squadron, no longer tenable, nor was it the proper command for a mere Flying Officer. Consequently, late in 1938, the Flight was renamed 'D' (Performance & Testing) Flight (D per T) of the A&AEE, with Squadron Leader Butler as its first officer commanding (OC), soon to be replaced by Squadron Leader 'Hetty' Hyde from the Establishment's permanent staff.

Watson Watt's attendance and day-to-day involvement with Bawdsey's work declined as he moved up the Air Ministry's hierarchy to become the Director of Communications Development (DCD) at MAP. His place as Superintendent was taken by Jimmy Rowe.

Whilst Watson Watt's technical knowledge and sound advice at Bawdsey was greatly missed, his new position had a positive effect, by enabling him to have a considerable influence on the direction of radar development within the Air Ministry and the armed services. In respect of their characters, Jimmy Rowe and Watson Watt were two very different people. Like his predecessor, Rowe came from an academic and technical background gained at Imperial College and later as the Secretary to the Tizard Committee. He was, however, a better administrator and the effective leader of an organisation that contained many 'prima-donnas'. He was more serious and restrained than Watson Watt, who was shy, warm-hearted and endearing to those who got to know him.[11] Unfortunately, Rowe's 'strong sense of mission' frequently upset his staff and made him, on occasions, 'rather difficult to live with',[12] which would ultimately be one of the contributory reasons for Bowen and Wilkins' departure from Bawsdey.

Before the year was out, two improvements were made to the airborne systems. The first concerned the replacement of the Giant Acorn valves in the transmitter with the Western Electric Type 4304. These increased the output power to somewhere between 1 and 2 kW, with a consequent improvement in the maximum range. The second saw the introduction of the engine driven alternator as the primary electrical power source.

Aero-engines of the period, were capable of accepting a 500 Watt, 24 Volt, direct current (dc) generator, by means of an auxiliary drive shaft, to provide power for an aircraft's electrical systems. Twin-engined aircraft employed a single generator connected to one of the engines, leaving the other spare and available for an alternate use. Investigations were, therefore, initiated for the design of an alternating current (ac) alternator, to fit the drive shaft of the 'spare' engine. Previous attempts by the Air Ministry had produced no concrete results, so taking Watson Watt's advice to try the direct approach, Bowen flew to Sheffield in October to discuss the matter with the Metropolitan-Vickers (Metro-Vick) Company. After landing at Finningley, Bowen and Sergeant Wright removed the generator from the Battle's Merlin engine and took it to the factory's Managing Director, Mr Fletcher, to enquire if Metro-Vick could design an alternator having the same dimensions, pick-up points and spline shaft connection. The problem was given to the chief engineer, Mr Tustin,[13] who left to carry out a number of calculations. On his return, Tustin informed Bowen that it would be possible to build an alternator of the requisite size, which would deliver 800 watts at 80 volts, over an operating frequency of 1,200–2,400 Hz.[14] The weight would be comparable to that of the existing generator and would employ its fixing points and spline shaft. Bowen and Tustin agreed the specification immediately, with the Air Ministry later placing a contract with the company for the supply of eighteen engine-driven alternators. The first model was available and running on the bench within a few weeks, with a performance that fully met the specification. By the end of the following month (November), the first production deliveries had been made, with the remainder being delivered before the outbreak of war. A second contract for 400 was placed with Metro-Vick in October 1939, which, with follow-on orders, would eventually lead to the manufacture of 133,800 units for Britain and her Allies, including the US.

1939

By the beginning of 1939, the Airborne Group's AI radar set was capable of detecting a target at a maximum range of 2–3 miles (3.2–4.8 km), but was unable to give an indication of its bearing (azimuth), elevation, or a defined minimum range. It was also realised that the interpretation of the system's CRT displays, was beyond the capability of the pilot, who was generally fully occupied flying the aircraft. The use of a specialised radar operator, or an observer, was, therefore, necessary. There was also a need to take into account the limitations on the operator, who would be sitting in the rear of an aircraft that was noisy and frequently manoeuvring, whilst trying to interpret the information on his screens. These factors mitigated in favour of simple and easy-to-read displays. The Group therefore examined three display methods:

- The range/azimuth display, later termed 'B-Scan'.
- Phase comparison systems.
- Overlapping beam systems.

The range/azimuth display was temporarily discarded, due to the difficulties likely to be encountered by having to scan mechanically the relatively large 1 ½ metre aerials. It was, however, successfully resurrected and employed with great success on a number of centimetric radars. The phase comparison system, that relied on the principle that two aerials spaced apart, would receive signals that differed in phase due to the position of the target, was also rejected, as the technology was not then available in the 200 MHz band. This left only the overlapping beam system, that had been employed for a number of years in the Lorenz Company's blind approach landing system (Standard Beam Approach – SBA), then in use by the RAF and the *Luftwaffe* (see Figure 1).

Having established the azimuth and elevation method, Touch set about the task of designing a simple display. The easiest means was to provide two CRTs, one displaying azimuth and range information, and the other elevation and range, as shown in Figure 2. From this it will be seen that in both cases, the time-base trace begins with a large return (at the base of the azimuth tube and the left side of the elevation tube), which is the direct reflection of the transmitter pulse. The even larger signal at the end of the time-base trace, is caused by the transmitter pulse striking the ground and being reflected back to the receiver. This characteristic trace was referred to as the 'Christmas Tree Effect', as the pattern bore

Figure 1.   Polar diagrams for 1 ½ metre AI. The polar diagram for the transmitter was not dissimilar in shape to that of a pear and shows that the majority of the radio energy was thrown forward, with little being radiated to the aircraft's immediate port, starboard and rear. The receive azimuth and elevation diagrams provide good coverage forward at about 30–45°, but little directly ahead in line with the aircraft's line of flight.

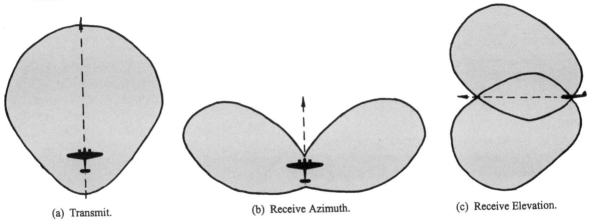

(a) Transmit.   (b) Receive Azimuth.   (c) Receive Elevation.

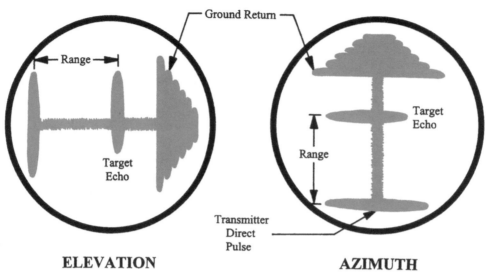

ELEVATION   **AZIMUTH**

Figure 2.   Elevation and azimuth displays for 1½ metre AI. The diagram shows the above/below CRT elevation display on the left and the port/starboard azimuth tube on the right. In this instance the target is shown to be flying above and to port of the fighter at a range of approximately 3 miles (15,850 ft/4.8 km). The ground return appears to have a range of some 3½ miles (5.6 km), which shows the fighter to be flying at approximately 18,500 feet (5,630 metres).

some resemblance to the general shape of a Christmas tree. Any aircraft, therefore, at a range greater than that of the ground return, was concealed by the echoes from the ground return. In metric AI systems (AI Mk.I to Mk.VI), the maximum range of the radar was equal to the height of the night-fighter above the ground. For example, when flying at 15,000 feet (4,570 metres) the early equipments gave a very useful maximum range of approximately 3 miles (4.8 km). However, this reduced to less than 2 miles (3.2 km) when the fighter's altitude was reduced to 10,000 feet (3,050 metres). This restriction, which greatly impeded interceptions at low-levels, was not overcome until the introduction of centimetric techniques in late 1940.

Touch's system provided the observer with an indication of the range, bearing and elevation of a target, relative to his own position, from which he could calculate course corrections for the pilot to effect an interception. In reality, the target aircraft would move with respect to the fighter, which required the observer to maintain a continuous commentary as to its position and the necessary course changes required to intercept it.

Whilst there were four aerials in the receiving system (upper and lower elevation and port and starboard), it was impractical to provide four receivers to 'drive' the signals on the CRTs. After several attempts, Touch devised a two, and later a four-pole, mechanically driven radio frequency (RF) motor switch, which offered each of the four aerials in turn to a common receiver (see Appendix 1). Thereafter, the four-pole aerial motor switch was standardised on all metric AI and ASV radars.

Having realised at an early stage that the system was insufficiently accurate to permit blind-firing, and also taking into account Fighter Command's requirement that all aircraft be positively identified before fire was opened, the Group set about reducing the minimum range to 1,000 feet (305 metres), or less. This range had previously been defined by a series of trials conducted by the RAE at Farnborough, which showed that on moonlit nights the target aircraft could be seen at distances up to 2,000 feet (610 metres). On dark nights, however, visibility was reduced to 1,000 feet (305 metres), or less.[15] The solution to a reduction in the minimum range, lay in adjustments to the system's pulse width.

Flight trials during 1938, demonstrated a minimum range approaching 1,200 feet (365 metres), which appeared to indicate that only a small amount of work was required to reduce this further. Having devised a 'rule of thumb', Bowen calculated that a radar operating on a frequency of 200 MHz and a pulse width of 1 μsec, should be able to achieve a minimum range of 500 feet (150 metres). Unfortunately, a series of practical experiments with 1 μsec pulses showed that the very strong transmitter signals were leaking around and overloading the input stage (otherwise known as the 'front end') of the receiver. To overcome the problem, Touch devised a quenching circuit which inhibited the operation of the receiver during the period of the transmitter pulse. This technique, later referred to as 'Squegging', enabled the receiver to be suppressed under the control of a time-delay circuit in the transmitter. When fully charged, the time-circuit would trigger the transmitter and suppress the receiver.

To establish the minimum range in practice, Touch flew in Anson K6260 with the squegging transmitter rigged for various pulse widths. Using the sea as a target and confirming his height above it from the aircraft's altimeter, Touch was able to demonstrate a minimum range of 500 feet (150 metres), at some cost to the maximum range. A compromise 800 feet (245 metres) was eventually chosen, as being a useful and consistently attainable minimum.[16]

During May 1939, the AI system with its squegging transmitter, was transferred to Battle K9208 and test flown for the first time on 9 June, with Hanbury Brown acting as the radar operator. The target was a Handley Page Harrow bomber/transport. The system is described, as follows, by Robert Hanbury Brown:

The two azimuth aerials were horizontal quarter-wave rods [unipoles] mounted on the opposite sides of the engine of a Fairey Battle and gave overlapping patterns in azimuth. The two elevation aerials were half-wave dipoles mounted one above and one below the wing giving overlapping patterns in elevation. The receiver input was connected to these four aerials in rapid sequence by a rotating mechanical switch and its output was distributed synchronously to the display. After about a dozen flights in the Battle, sometimes using a Hawker Hart, sometimes a Harrow and sometimes a Vickers Virginia as a target, we finally got the complete outfit to work perfectly only three months before war was declared. Flying over Orfordness at a height of 15,000 feet [4,570 metres] on 9 June 1939, I saw echoes from our target [a Harrow] at a maximum range of 12,000 feet (3,660 metres) and in mock interceptions found that the display was reasonably easy to use.[17]

In mid-June, the Battle installation was demonstrated to Air Marshal Dowding at Martlesham Heath. With Flying Officer Smith as pilot and Bowen operating the radar, Dowding was taken on a flight to demonstrate the system's interception capablity. Using the other Battle as a target, Dowding was taken to 15,000 feet (4,570 metres), where Bowen showed off the system's capability by approaching several times at the maximum range and from 30° off the port and starboard dead astern postions. At the end of these manoeuvres,

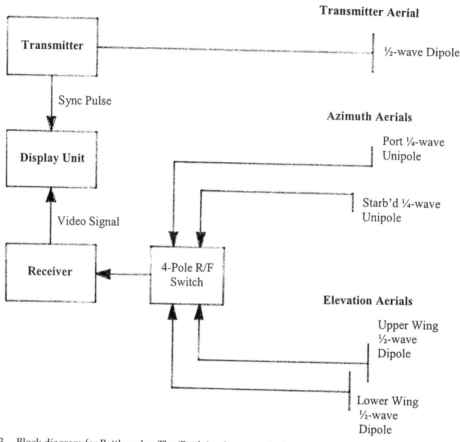

Figure 3. Block diagram for Battle radar. The 'Battle' radar comprised a squegging transmitter that was connected to a simple half-wave transmitting dipole, with the receive elements connected in turn by means of a four-pole R/F switch to the receiver. The video signals from the receiver were synchronised to the outgoing transmitter pulse by a synchronising (sync) pulse, which controlled the timebase on the CRTs on the indicating unit. This relatively crude system suffered from poor minimum range due to the inaccuracy of the synch pulse and the poor rise/fall time of the transmitter pulse. It is not known how the four-pole R/F switch worked. This may have required the operator manually to switch the azimuth and elevation aerials to the receiver. The problem was eventually overcome by the adoption of a motorised RF switch.

with which the AOC-in-C was completely satisfied, he asked for a demonstration of the minimum range performance, as Bowen relates:

> As we made the last interception, he said: 'When we get to minimum range, tell the pilot to hold that postion so that I can see how close we are'. This we did. When we had settled down at minimum range, Stuffy[18] said: 'Now let's take a look'. For the previous 30 or 40 minutes our heads had been under the black cloth shielding the cathode-ray tubes. I whipped the cloth off and Stuffy looked straight ahead. He said: 'Where is it? I can't see it'. I pointed straight up; we were flying almost directly underneath the target. 'My God' said Stuffy 'tell him to move away, we are too close'. We broke away and headed for home. The Commander-in-Chief was clearly pleased with what he had seen.[19]

Following their return to Martlesham Heath, Dowding and Bowen discussed the AOC-in-C's concerns relating to real night-interceptions, which Dowding began by outlining his thoughts on the requirements for a night-fighter:

- The need for good endurance, as night interceptions were likely to be long, drawn out affairs.
- The need for the pilot to retain his night vision by not having to monitor the CRT screens, mitigated against single-seat operation and hence the crew should comprise a pilot and an observer.
- Given that interceptions may well necessitate long patrols over the sea, followed by complex manoeuvres, there was great scope for getting lost, and it might, therefore, be worthwhile to carry a navigator.
- In order that friendly aircraft were not shot down by mistake, the pilot should be required to make a visual identification of his target.
- Finally, the fighter should pack as heavy a punch as was possible, since, if lost, the enemy would be extremely difficult to re-acquire and engage.[20]

On returning to his HQ at Bentley Priory, Dowding advised the Air Ministry of Bowen's progress with AI (RDF2):

I was very much impressed with the potentials of the apparatus (although of course it was initially used only in a lash-up form). The range and approximate position were clearly indicated, and I formed the opinion that an approach in the dark could easily be affected after a small amount of training and practice, provided that a favourable position within range of the target could be obtained.[21]

Later in the document he rightly commented on the system's major limitations, namely, its maximum range being equal to the aircraft's height above the ground and the need to get the minimum range below 900 feet (275 metres).

Although convinced of the necessity for a twin-engined, two/three-seat aircraft, Dowding did not necessarily disregard the use of a single-engined fighter. He considered, but rejected, the Boulton Paul Defiant two-seat fighter, on the grounds that its gunner's position was very cramped. For the time being, the primary requirement was for a twin-engined fighter, which Dowding thought could best be met by the Bristol Beaufighter.[22] However, since it would be some time before this aircraft was in service, he recommended that three or four examples of the 'Battle radar' be built for installation in the Bristol Blenheim, to provide the Air Staff with a factual base from which to draft a specification for an operational night-fighter. As a consequence of the AOC-in-C's recommendations, two 'short-nosed' Blenheim Mk.Is, K7033 and K7034, were delivered to 'D' Flight, with instructions that they be fitted with the Battle radar in the shortest time possible.

Compared to the Battle, the Blenheim was a spacious aircraft that easily accommodated a crew of two or three and the AI equipment. With K7033 fitted with radar and operating as the 'fighter' and K7034 employed as the 'bomber', Bowen began a series of tests to ascertain the system's performance. Whilst the mounting of the elevation aerials on the wings caused little trouble, the location of the azimuth rod elements on the fuselage sides gave poor results. When compared to the smooth contours of the Battle's nose, the Blenheim's comprised a series of short, stubby surfaces made up from a series of small perspex panels, all flanked by two large radial engines, which together produced a multiplicity of reflections. A great deal of effort was expended by Messrs Wood, White and Mills in finding a part of the nose having the necessary smooth contours to provide the overlapping azimuth patterns. A satisfactory solution was eventually achieved, after much swinging of K7033 on Martlesham's compass platform.[23]

The problems with the Blenheim aerial system were brought to the attention of the Air Ministry, who had hoped, following Dowding's report, to have been able to implement a programme for the rapid re-equipment of night-fighters with AI. The problem was raised at the 11 June meeting of the Swinton Committee by Churchill, who demanded that the highest priority be allocated to the development of AI radar:

> In view of the promising nature of the recent trials with the RDF air-to-air sets installed in the Battle at Martlesham and in view of the present weakness of our night defence, it has been decided that Fighter Command should proceed with the development of RDF air-to-air tactics on the highest priority and that we should be prepared at short notice to equip one Blenheim five-gun fighter squadron with this equipment.[24,25]

Authority was therefore granted for the manufacture of the first batch of thirty hand-built AI sets, with six being delivered to Fighter Command and the remainder being held as a 'contingency stock' in the event of war.[26] Eleven aircraft were to be fitted with the necessary cabling and airframe modifications (bracketry), with the first four complete aircraft being delivered to No. 25 Squadron, at Hawkinge, in the shortest time possible. Sufficient transmitters were immediately available thanks to Watson Watt's foresight in late 1938, when, anticipating the demand for AI sets, he had ordered six transmitters from Metro-Vick and six receivers from A.C. Cossor Ltd. Unfortunately, for a variety of reasons, but primarily ones of weight and a reduction in sensitivity, the Cossar receivers were found to be unusable.

By another of those strokes of good fortune that seemed to surround the Airborne Group, Bowen was informed in April or May by Professor Edward Appleton, the Jacksonian Professor of Physics at Cambridge University, under whom Bowen had studied, of the availability of a 45 MHz receiver that was in production by Pye Radio Ltd for the BBC's television service.[27] When approached by Bowen the Company's Technical Director, Mr B.J. Edwards, invited him to visit the Cambridge factory and 'remove' one of the receivers. On his return to Bawdsey, Bowen and Touch quickly recognised they had in their hands a receiver that was smaller, lighter and significantly better than the old EMI model. By placing a 200 MHz mixer stage in front of the Pye receiver, Touch had the makings of a production AI receiver. With the mixer incorporated into the chassis, Pye Radio was awarded an Air Ministry contract on 19 July 1939, for the supply of thirty AI receiver chassis, which the Company duly completed within nine weeks – a very creditable performance.

The Pye receiver was the first equipment to employ the EF50 valve, which, according to Bowen, was to 1½ metre technology, what the resonant cavity magnetron was to centimetric wavelengths. It was later discovered that whilst the EF50 carried a Mullard label, they were in fact manufactured by the Philips Company, at Eindhoven in the Netherlands. Although assured, for the time being, of a plentiful supply, it was necessary to despatch a destroyer to the Netherlands in the spring of 1940 to remove some 25,000 EF50s, together with 250,000 valve bases. Using the valve bases, Mullard's were able to reinstate EF50 production in Britain and thus secured the supply of these very valuable components.

To return to the installation work on the Blenheim. RAE was allocated responsibility for the screening of the aircraft's electrical systems and the fitting of the engine alternators, following which the aircraft were flown to Martlesham for the installation of the AI equipment and the aerial system. By July, the Group had successfully solved the aerial problem and were eagerly awaiting the delivery of the transmitters and receivers, which they were required to have installed in the Blenheims by the beginning of September. Expecting nice shiny, well engineered production-type units, they were appalled to discover that whilst the Air Ministry had been persuaded to include:

> most of the switch-gear, power supplies, control panels and the rest of the components which made a complete assembly. In the Pye contract, there were no racks, brackets and no cable runs – nothing with which to install the complete equipment in an aircraft.[28]

And if this was not enough, apart from the Pye receivers, which were properly engineered, the Metro-Vick transmitters were exact (Chinese) copies of the prototypes, complete with wooden bases for the lecher lines. It was later established that Metro-Vick had been supplied with a 1937 prototype model, complete with wooden insulators, by mistake and hence, they had built the wrong transmitter. These and the Company's transmitter power units, would later prove troublesome. The reasons for the poor design of the early equipments might possibly be blamed on the manufacturers. However, in fairness to them, some of that responsibility might be directed at the Airborne Group, whose limited time prevented them from visiting the factories, to ensure that manufacturing was proceeding properly.

Bowen's problems were further compounded, when the Blenheims delivered to Martlesham Heath were found to be the wrong variant. For whatever reason, the officer in the Air Ministry responsible for arranging the aircraft deliveries, had inadvertently supplied the 'long-nose' Blenheim Mk.IV variant, which had an entirely different nose profile to that of the K7033, a Mk.I. This mistake delayed the installation programme, whilst White and Mills redesigned the azimuth aerials, fabricated new mounting brackets and cable runs and amended a series of drawings.

Faced with the twin problems of incorrect transmitters and the wrong type of aeroplane and a very short timescale, Bowen had little alternative other than to proceed with what he had. When installed at Bawdsey and tested against the Trimley water tower, the system was found to offer an acceptable performance, due primarily to the enhanced sensitivity of the Pye receiver. Once tested, the equipment was transported to Martlesham for installation in the aircraft by a team of five civilian fitters from Farnborough, assisted by members of the Airborne Group.[29] Nevertheless, after expending some considerable effort, the first Blenheim was delivered to No. 25 Squadron at Northolt,[30] on 30 August, with a further five following before the end of September. Three of these aircraft were incorporated into 'A' Flight under Flight Lieutenant J.G. Cave and later under Flying Officer G. Drew.

Shortly before the end of August, Robert Hanbury Brown was transferred to Northolt to work alongside No. 25 Squadron's personnel, to show them how to use the new equipment and sell the concept of night interception 'to the people who were actually going to use it'.[31] Being the only squadron in Fighter Command equipped with a radar-capable night-fighter, and being dangerously close, from the Squadron's viewpoint, to London, No. 25 was frequently visited by a whole host of senior officers. One such visit by Air Chief Marshal Dowding was to cause Hanbury Brown a serious amount of work rearranging the radar 'boxes' in the Blenheim. Arriving at the airfield to begin his day's work (probably on 31 August/1 September), he was confronted by the ground crews busily cleaning the aircraft and the aircrews awaiting the arrival of the AOC-in-C. The atmosphere was extremely tense, but on arriving, Dowding shook hands with Hanbury Brown before proceeding with his inspection of the aircraft. He did not get very far, however, before, in Hanbury Brown's words, he roared:

> Brown, why have you put the AI in the front of those Blenheims? Can't you see that the rear gunner of the enemy bomber will shoot at the light from your transmitting valves, and the AI and its operator will be the first to be hit? Put all the equipment in the back of the aircraft at once.[32]

Brown explained that the transmitter was required to be as close to the transmitter aerial as was possible, in order to ensure the maximum power transfer. Nevertheless, they could shield the light and transfer as much of the equipment as was possible, to the rear of the fuselage. This work, he estimated, would take one week for each aircraft. Dowding, somewhat reluctantly agreed to Brown's proposal, which unfortunately, rendered much of the work carried at Martlesham in the preceding months a complete waste of time!

A few days later, on 3 September, Britain declared war on Germany, following the latter's invasion of Poland on 1 September and her refusal to withdraw. The declaration added a sense of urgency to the installation effort at Northolt and persuaded Bowen to transfer Keith Wood to work alongside Hanbury Brown in the equipment relocation programme. Within a week, a small team of mechanics succeeded in moving the receiver and display unit and the operator's position, to the rear fuselage to provide a darker and quieter enclosure in which to view the screens. The new arrangement worked well in the air, however, one unforeseen problem concerning the TR9D high frequency (HF) radio, did arise. Under normal circumstances, the crew members in the Blenheim (navigator and gunner) communicated with the pilot by means of an intercom, which was an integral part of the radio set. Unfortunately, whilst the radio was transmitting to a ground-based HF station to obtain a navigation 'fix', the intercom was disconnected and the operator could not speak to the pilot. Since this occurred once every fifteen seconds and good interception technique was wholly dependent on the transmission of accurate 'steering' information between operator and pilot, the loss of any communication was likely to effect an interception. With Fighter Command's agreement and the help of the Radio Department of the RAE, the radio was modified to meet Hanbury Brown's requirements for clear and uninterrupted intercom speech.

Having achieved a serviceable radar installation, which had in the meanwhile been designated AI Mk.I by the Air Ministry, and found a solution to the intercom problem, Brown and Wood set about the tasks of training the Squadron's mechanics in the maintenance and testing of the equipment and the operators in the set's capabilities. Maintenance was something of a problem, since, as there were no official manuals, nor specialised test equipment available on the Squadron, Hanbury Brown and Wood were forced to teach the mechanics personally how to set-up their sets and establish procedures to test for, and find faults. Operator training was accomplished by Hanbury Brown and Wood conducting numerous practise interceptions to establish a set procedure and then teaching it to the student operators and their pilots. The first phase of the instruction comprised a detailed introduction to the set's controls and adjustments, followed by an interpretation of the screens to obtain the range and position of the target, and, finally, how to direct the pilot to effect an interception. These lessons were learnt on the ground, before the crews progressed to airborne practice in the Blenheim. The second phase, that of putting their knowledge into practice in the air, took several hundred hours 'intercepting' lone Blenheims that plodded a line from Northolt to Oxford during the hours of daylight. Whilst many of the crews acquired the technique and could carry it out successfully during the day, they were denied the opportunity to try it at night, since the Blenheims were reserved for night patrols with their regular aircrews, albeit on some occasions with Hanbury Brown or Wood acting as the radar operators.

CONCLUSIONS

With the equipment relocation programme completed in October, but with little night flying being undertaken to prove the system under operational conditions and few aircrews being properly trained, AI could only be regarded as a trials and development system during the early months of the Second World War. Nevertheless, sufficient progress had been made by Bowen's Airborne Group to establish the basic requirements of an AI set and produce a number of prototypes to prove the technology and build the first production batch. The credit for this achievement rightly belongs to Edward Bowen, for his leadership of the Airborne Group and to the scientists, engineers, technical assistants and service pilots, who made up the Group in its formative years. Credit is also due to Henry Tizard, who first identified the need for an airborne radar set and acted as Bowen's mentor, and to Air Chief Marshal Sir Hugh Dowding who sponsored the development and helped define its military requirements. Without these men of the highest calibre, airborne radar in Britain would not have progressed at the rate it did, nor would an AI set have been available at the time when the country needed it most.

By the autumn of 1939, therefore, Britain was equipped with its first radar equipped night-fighter, that was set to play its part in the overall defence of the country, against a *Luftwaffe* that was well trained and motivated to conduct a strategic bombing campaign against the British Isles for the second time in barely twenty years.

Notes

1. Chamberlain replaced Stanley Baldwin as Conservative leader and Prime Minister in May 1937.
2. The Auxiliary Air Force was created by MRAF Lord Trenchard in 1925, to establish a Citizen Air Force, based along the lines of the Territorial Army. The majority of AAF squadrons converted to the fighter role in 1935.
3. The Special Reserve, created alongside the AAF in 1925, was manned by officers having previous flying experience. In 1936, a number of its squadrons were transferred to the AAF and converted to fighters.
4. Three with Hawker Furies, six with Gloster Gladiators, nine with Gloster Gauntlets, eight with Hawker Demons, four with Hurricanes.
5. Michael Gething, *Sky Guardians* (London: Arms & Armour Press, 1993), pp. 74 and 75.
6. A *Staffel* was the lowest flying element in the *Luftwaffe*, with a nominal strength of nine aircraft and, therefore, broadly comparable to an RAF Flight.
7. E.R. Hooton, *Phoenix Triumphant*, Appendix 6 (Brockhampton Press, 1999), p. 279.
8. *Ibid*, pp. 131 and 132.
9. Later Squadron Leader Douglas Rayment.
10. Later Flying Officer Clifford Wright.
11. J.A. Radcliffe, *Robert Alexander Watson Watt*, The Dictionary of National Biography.
12. Dr E.H. Putley in Colin Latham and Anne Stubbs, *Pioneers of Radar*, (Thrupp: Sutton Publishing, 1999), p. 33.
13. Later Professor Tustin, Professor of Electrical Engineering at Imperial College.
14. This range was a consequence of the need to provide the necessary power over the speed (revolution) range of the engine.
15. E.G. Bowen, *op cit*, pp. 67 and 68.
16. *Ibid*, pp. 67–9.
17. R. Hanbury Brown, *Boffin* (Bristol: Adam Hilger, 1991), p. 28.
18. Dowding's nickname throughout the RAF was 'Stuffy'.
19. E.G. Bowen, *op cit*, p. 70.

20. *Ibid*, pp. 71 and 72.
21. Air Chief Marshal Dowding in a letter to the Air Ministry, dated 10 July 1939, and quoted in *The Official Signals History of the Second World War, Volume 5*, pp. 112 and 113 and filed as AIR10/5485 in the National Archives, Kew. Hereafter referred to as *The Signals History, Volume 5*.
22. The prototype Beaufighter, R2052, flew at Bristol's Filton airfield on 17 July 1939.
23. It should be noted that aerial systems and their polar diagrams were first tested and mapped on the ground.
24. An extract from the 11 June 1939 meeting of the Swinton Committee, quoted in *The Signals History, Volume 5*, p. 5.
25. It should be noted that the five-gun Blenheim fighter refers to the Mk.If fitted with a four-gun ventral tray, plus the single gun mounted in the dorsal turret.
26. The exact number of sets in this contract is difficult to ascertain. Bowen in his book *Radar Days*, quotes thirty, whilst *AI* in *The Signals History* states twenty-one. The author is inclined to Bowen's figure as much of the subsequent equipment orders relate to the thirty figure.
27. The BBC television service was opened on 2 November 1936 and closed on the outbreak of war.
28. E.G. Bowen, *op cit*, pp. 78 and 79.
29. By the summer of 1939, the Airborne Group comprised some twenty-odd personnel, including the lady cleaner.
30. No. 25 Squadron moved from Hawkinge to Northolt on 22 August 1939.
31. R. Hanbury Brown, *op cit*, p. 34.
32. *Ibid*, p. 34.

CHAPTER THREE

Building the Solution

September 1939–April 1940

The opening months of the Second World War proved to be something of an anti-climax from Fighter Command's viewpoint. Attacks by the *Luftwaffe* on Britain's towns and cities failed to materialise because the Country's eastern seaboard was flanked and protected by French, Belgian and Dutch territory, which the enemy could not easily cross. Poland, on whose account France and Britain had gone to war, was quickly overrun from the west by co-ordinated German ground and air forces that successfully employed the *Blitzkrieg* (Lightning War) technique for the first time. By 9 September 1939, German armoured forces (*Panzers*) had taken the Polish capital, Warsaw, followed nine days later by an invasion from the east by Soviet troops.[1] Caught between two fronts and poorly equipped to fight a modern war, the Polish army and air force were gradually worn down, before collapsing completely on 16 October, when all fighting ceased.

In a repeat of their actions in the First World War, the British Government resurrected the BEF to bolster the French Army and transported it to the Continent, where it took up a position between the French 1st and 9th Armies along the Belgian border. To support the Army and a poor French Air Force, the Air Staff formed an Advanced Air Striking Force (AASF) from the Fairey Battle squadrons of Bomber Command's No. 1 Group, supplemented by two Blenheim squadrons from its No. 2 Group and four of Hurricanes from Fighter Command, and transferred them to France during September.[2]

On the outbreak of war, Fighter Command had thirty-four squadrons available in the UK, against a requirement for forty-six, which Dowding considered necessary to meet his commitments to protect coastal convoys, the Home Fleet at Scapa Flow and the UK mainland. Of these, seventeen were equipped with Hurricanes, twelve with Spitfires, six with Blenheim fighters, two with Gladiators and one with Gloster Gauntlets.[3] Aware that the Blenheim squadrons would stand little chance of survival when faced by the Messerschmitt Bf 109 and Bf 110, for it had a performance that was barely equal to some of the *Luftwaffe*'s bombers and inferior to most, Fighter Command relegated the Blenheim to the night-fighter role in October 1939. If the situation of the fighter force gave Dowding cause for concern, that of the CH network was in much better shape. The twenty-station CH scheme was very nearly completed, with eighteen completed and passing 'plots' to Fighter Command's Filter Room at Stanmore by September 1939. Work was also in hand to build twenty-four CHL stations, many of which would be collocated with a CH station, and have these installed and operational in the shortest possible time. By the time of the Battle of Britain in August 1940, there would be thirty-two operational radar sites, each accommodating a CH and CHL station and twenty-five stand-alone CH sites, covering the sea approaches from Strumble Head in South Wales to the Orkney Islands in the north of Scotland.[4]

Whilst its units reorganised and replaced the losses sustained in the Polish campaign, the *Luftwaffe* began a series of armed reconnaissance operations in the North Sea and around the coastal waters of Scotland, to establish the movements of British shipping and, when the opportunities presented themselves, to attack maritime targets. These operations were undertaken by regular bomber units (*Kampfgeschwader* – KG) and dedicated coastal patrol and minelaying squadrons (*Küstenfliegergruppe* – KfGr), operating in close co-operation with the German Navy (*Kriegsmarine*). In mid-October, the *Luftwaffe* carried out its first bombing operations over British soil, when nine Junkers Ju 88As of the 1st *Gruppe*[5] of KG 30 (I./KG 30) bombed shipping in the Firth of Forth on 16 October, followed by two raids on 17 October against warships in Scapa Flow. Two Ju 88s from the Firth of Forth operation were shot down into the Firth by Scottish-based Spitfires and one Ju 88 from the Scapa raids was brought down by AA fire and crashed on the island of Hoy. In both cases there were no casualties, but the old, unseaworthy battleship, HMS *Iron Duke*, was hit at Scapa and had to be beached to prevent its sinking.

Throughout the autumn and winter of 1939/1940, a period that became known as the 'Phoney War', or the 'Sitzkrieg', the German High Command (*Oberkommando der Wehrmacht* – OKW) undertook the detailed planning for the invasion of France and the Low Countries, that was scheduled to take place in the spring of 1940. However, in April 1940, at the instigation of the First Lord of the Admiralty, Winston Churchill,[6] the territorial waters of neutral Norway were mined to impede shipping carrying Swedish iron ore to ports in northern Germany. Threatened with the disconnection of these vital supplies, and attracted by the potential of basing U-Boats and aircraft in the northern waters, the *Kriegsmarine*'s C-in-C, *Gross Admiral* Raeder, who had for some time contemplated an intervention in Scandinavian waters, urged Hitler to mount a seaborne invasion of Norway and Denmark. On 8 April, German airborne and naval forces struck at Norway and Denmark, overrunning both in quick succession before defeating an Anglo-French counter-invasion force and securing both countries.

With its northern flank secured, German *panzer* units, ably supported by *Luftwaffe* tactical aircraft, invaded France, Holland and Belgium on 10 May. Outflanking the defences of the Maginot Line, German armoured troops broke the Allied lines through the Ardennes forest and reached the Meuse river by 12 May, crossing it the following day. With the French Army in disarray and slow to react, the Germans reached St Quentin on 18 May and stopped on the Channel coast at Abbeville on 20 May. Having split the Allied armies in two, the BEF, which had moved forward into Belgium, began falling back alongside the French and Belgians towards the coast. With the ports of Boulogne and Calais in German hands by 24 May, the BEF was driven back to Dunkirk, from where, by the courage and determination of the Royal Navy and the 'Little Boats', some 338,000 troops (224,000 British) were taken off the beaches and returned to England.[7]

With the BEF evacuated from Dunkirk and out of the battle, Italy, sensing an opportunity, declared war on Britain and France on 11 June and invaded her neighbour forthwith. Paris fell on 14 June and the Premier, Marshal Pétain, concluded an armistice that took France out of the War on 22 June. With Denmark, Norway, France, Belgium and Holland defeated and German troops occupying the Continental coast from the North Cape to the Spanish border, Britain and her Empire stood alone.

Whilst the Phoney War and the subsequent invasion of Western Europe may not be regarded as one of the most illustrious deployments in the annals of British arms, it did, nevertheless, provide the Country and the RAF with one significant facility – time. During this period Fighter Command, despite the haemorrhaging of its aircraft to France, was able to expand its operational squadrons by seventeen, to fifty-two, by July 1940: twenty-five of Hurricanes, nineteen of Spitfires, six of Blenheim fighters and two of Defiants.[8] Work had also begun on establishing an operational training organisation to improve pilot training and provide them with some experience on the aircraft they would fly in combat. From the spring of 1940, the first of the Operational Training Units (OTU) was created to train fighter crews. Technical training was not neglected and was as important as that for aircrews. By April 1940, the number of Schools of Technical Training had doubled compared to April 1939, and airmen whose training had previously taken several years, had it reduced to several months, without any apparent loss of efficiency.

One of Fighter Command's principal requirements was the need to secure a supply of fighter aircraft – the other being an extension of the CH and CHL networks to cover the west and northern coasts. On 14 May 1940, the Air Ministry's research and production departments were detached to form MAP, under the direction of the Canadian newspaper magnate, Lord Beaverbrook, who later assumed the responsibility for development and manufacture of radio/radar equipment. Prior to Beaverbrook's arrival in the new Ministry, the production of fighter aircraft had begun to rise steadily as bottlenecks in the aircraft industry were overcome and production was concentrated on five specific types: the Hurricane, Spitfire, Blenheim, Whitley and Wellington. During June, the industry delivered 446 fighters, followed by 496 in July and 476 in August.[9] These numbers were supplemented by aircraft repaired by the RAF's repair depots and the Civilian Repair Organisation which had been established from within the aircraft manufacturers in 1938 and transferred to MAP in 1940. By the time of the Battle of Britain, some 65 per cent of aircraft delivered to the RAF came from the aircraft factories and the remaining 35 per cent from the repair organisations.[10]

Autumn 1939

The declaration of war on 3 September had an immediate impact on 'D' Flight and the A&AEE. Concerned at the supposed vulnerability of Bawdsey and Martlesham Heath to air attack, the Air Ministry decided that in the interest of safety, it would be prudent to move the A&AEE to a site more remote from the coast. During September a reorganisation of the research establishments saw 'D' Flight detached from the A&AEE and renamed the Special Duty Flight ('SD' Flight) and the Bawdsey Research Station retitled the Air Ministry Research Establishment (AMRE). With the A&AEE destined to be moved to Boscombe Down in Wiltshire, Rowe set about the task of identifying a suitable site for AMRE and its 'SD' Flight. His choice fell on the campus of Dundee University, where Watson-Watt had studied as an undergraduate, with the 'SD' Flight and its aircraft being accommodated at the nearby civil airfield at Perth. A special train was chartered to ferry the heavy laboratory equipment north, with the smaller items being flown in the Flight's aircraft and the staff making their own way by road. By this time (early September), the 'SD' Flight comprised ten experimental aircraft, in addition to twenty or so of the Blenheims destined for No. 25 Squadron, many of which were partly fitted out with their AI installations. The Flight's pilots flew their own machines, whilst ferry crews handled the No. 25 Squadron aircraft.

Sadly, one of the latter failed to reach Perth, with the aircraft and its crew being presumed lost over the Irish Sea.

Although abandoned as a research site, Bawdsey remained as a part of the east coast CH network. As events transpired, it was not on the *Luftwaffe's* target list and was never intentionally attacked, although a few stray bombs did fall within the station's perimeter later in the war. Nearby Orford was, however, subjected to a raid during 1942, when thirteen people died and a number of houses were damaged.

Perth's primary wartime role was the training of *ab initio* pilots, with whom the 'SD' Flight was expected to share the airfield's limited facilities. Negotiations between Bowen and the airfield manager produced an agreement whereby one of the site's two hangars was divided between the two parties, whilst he and the manager shared the latter's office. This agreement enabled the Airborne Group to continue the Blenheim AI fitting programme inside the hangar, whilst those already completed, or awaiting modification, were parked on the airfield, where their delicate equipment was exposed to the elements. The operating conditions for the scientific staff were somewhat rudimentary, compared to the facilities at Bawdsey and Martlesham, and could only be achieved by setting up the laboratory space in a partitioned corner of the hangar. Similarly, the Flight's air and ground crews were in poor shape, with much of the aircraft servicing having to be conducted in the open, whilst the pilots and observers occupied a small room in the airport building.

Despite these not inconsiderable drawbacks, work continued on the modifications to the Blenheims which, thanks in no small part to everyone's tireless efforts, was completed during October. Further work on the installation of ASV in Lockheed Hudsons and Short Sunderland flying boats, followed by an Admiralty request for the installation of 'forward looking ASV' (ASV Mk.II) in Fleet Air Arm Swordfish and Walrus aircraft, deferred all serious research work. For the time being, therefore, the Airborne Group was only able to undertake installation tasks.

The use of the AI equipped Blenheims by No. 25 Squadron's 'C' Flight, highlighted a number of deficiencies in the quality of the installation work undertaken at Perth and a series of operational limitations. During September, complaints directed at Fighter Command, showed that aircraft were being delivered without the necessary fittings that held the equipment in place, and identified an interference problem between the AI and the aircraft's TR9D HF R/T set. The latter was so serious that the Blenheims were forced to fly part of an interception with their R/T switched off, which precluded proper communication with the ground controller and reduced still further the possibility of a contact with the enemy. A technical investigation found that it was not possible to eliminate the interference generated by the AI equipment; however, when the HF radio was replaced by the superior very high frequency (VHF) R/T, the fault became manageable. Since it was Fighter Command's policy to replace all HF radios with VHF sets, including those in the Blenheim fighter, the problem resolved itself over a period of time. Nevertheless, this first instance, of what today would be known as 'Electro-Magnetic Compatibility' (EMC), indicated that a more thorough testing of the AI set with the aircraft's systems was required.

No. 25 Squadron then raised two further problems that required Fighter Command's attention. First, the art of successful night interception required a considerable amount of skill and co-operation between pilot and RDF operator[11] (R/O), which were only acquired after a great deal of training and operational experience – something that had not been well considered by Fighter Command, or AMRE. Second, night flying showed that whilst the

CH radar was capable of dealing with large formations of aircraft, its ability to detect and track single aircraft, was very limited, and non existent when they passed behind the chain.

To address these two problems, a detachment of No. 25 Squadron was formed at Martlesham Heath on 16 September, as a trials unit to conduct experiments into night interception techniques with the Bawdsey CH station and to teach night-fighter crews the correct procedures when engaging the enemy at night. The latter required the crew to approach the interception of an enemy aircraft in much the same way as a hunter would stalk game, that is by stealth and with patience. In the methodology devised by the Martlesham Detachment, it was the R/O's responsibility to detect the enemy aircraft and then compute its position in three dimensions in his mind, before directing his pilot by a series of course corrections to a position where he could obtain a 'visual' interception and identification of the enemy. From this point on it was the pilot's responsibility to close with the enemy and place himself in the most advantageous position (usually behind and below to the left or the right) before opening fire. The essence of this type of interception was 'teamwork', which was why, somewhat belatedly, night-fighter crews were credited with a kill and not just the pilot; decorations would eventually be awarded in the same way.

At the beginning of the war, R/Os were recruited from the ranks of air-gunners, many of whom proved temperamentally unsuited to the operation of sensitive electronic equipment. Nevertheless, a number did go on to become good operators. To overcome the shortage, the Air Ministry very sensibly revoked the rule whereby only those individuals having uncorrected good eyesight were allowed to train as aircrew. This enabled a number of schoolmasters (who could think and reason in three dimensions) to be recruited into the R/O's ranks to provide much needed, albeit, bespectacled operator material. Ground radio and wireless mechanics were also recruited by this method.

Teamwork also involved the ground controller, whose skill in directing the night-fighter within range of its AI, was critical to the success of a night interception. Trials with the Martlesham Detachment and the Bawdsey CH station, showed that AI Mk.I had insufficient range to undertake an independent search for the enemy, and would, henceforth, require the assistance of a ground controller. It was also discovered that the CH station could not provide the requisite accuracy to place the night-fighter within AI range of an individual bomber. This was due in part to the poor discrimination of the CH set, which was designed to detect and track large bomber formations, and the extended communications chain that placed the tracking information on the plotting table (henceforth, the plot) of the Sector Controller who was handling the interception. Experiments were conducted during the autumn to see whether the RDF controller, located in a CH station, could himself direct the operation and communicate directly with the night-fighter. This technique, termed 'All RDF Interception', enjoyed some success, but proved something of a hit-and-miss affair, that required considerable guesswork and imagination on the controller's part. Nevertheless, the basic work on All RDF Interception was to lay the foundations for the introduction of Ground Control of Interception (GCI) radar in late-1940, which would itself greatly improve the interception success rate at a critical point in the country's defence.

The CH trials by the Martlesham Detachment also uncovered the principal shortcoming of the early metric sets – poor low-level performance. Attempts to apply All RDF Interception techniques to the night interception of low-level minelaying aircraft proved unsuccessful, since the low level at which they were flying reduced the Blenheim's search range to 5,000 feet (1,525 metres), or less. The failure against the minelayers was reported to the Air Staff, who in turn passed the problem to AMRE to investigate and report back. AMRE undertook a further series of experiments, but these confirmed the earlier report. In a written reply to the CAS, Tizard commented:

> The Director of Communications Development [Watson-Watt] and I have been having discussions about the interception of low flying aircraft at night, and I think you might like to know our conclusions. We agree that the AI apparatus now fitted to our machines is quite unsuited to the purpose. It [AI Mk.I] was not designed to meet these conditions and we do not think that you ought to rely on it in the least.[12]

Maintenance was yet another problem. Due to the pressures exerted on the Airborne Group to get AI Mk.I into service as quickly as possible, the system had not been engineered in a proper manner. Consequently, when the system failed, the RAF's radio and wireless mechanics were insufficiently trained to cope with faults on Acorn valves and CRTs. The problem was compounded by the lack of servicing manuals and test equipment. However, by dint of some imaginative thinking on the part of the Air Ministry, a number of radio amateurs (radio hams) were recruited to provide a ready trained source of radar mechanics. This, along with the establishment of schools for the training of radar mechanics, and the improved design of later marks of AI, ultimately solved the maintenance problem.

Another aid to the design and improved reliability of the AI set was discovered during the autumn of 1939. On a visit to London, Bowen was visited by representatives of Imperial Chemical Industries (ICI), who wished to show him samples of a new material that had been developed at their Warrington Factory. This material, a plastic, offered similar properties to rubber, but also displayed good RF characteristics, which they thought might prove useful as an insulation for the co-axial cables used in aircraft radio equipment. Samples of the new material were sent to Perth for analysis, where they quickly fulfilled ICI's expectations. Unbeknown to ICI, the material was ideally suited to the manufacture of RF co-axial cable, since it had all the flexibility properties required for aircraft installations. Named 'Alkathene' by the Company, but better known by its chemical name of 'Polythene', the new material was put into limited production for a pilot batch of cable to fulfil the Airborne Group's immediate requirements. Although not easy to produce, Polythene went on to become one of the standard materials employed by the radar and radio industry for the manufacture of cables and other dielectric devices.

Fortunately for Bowen and his team, the Air Ministry quickly realised that Perth was not the ideal site on which to conduct radar research and the installation of sensitive electronic equipment. By the middle of October, the Group was informed they would shortly be packing their bags and moving into better accommodation at No. 32 Maintenance Unit (MU), near St Athan in South Wales. No. 32 MU's primary task was the repair and maintenance of RAF aircraft and their engines, to which was added the installation of AI and ASV equipment in fighters and maritime aircraft. On arriving at St Athan, Bowen set about teaching the RAF wireless fitters the skills of radar installation and testing, in order that the Airborne Group and its staff might return to their primary function of research and development. It was also the Air Ministry's intention that the Group should remain at St Athan long enough to pass on their skills to the RAF, following which, they would be moved to a more permanent station to continue their research work.

Unlike Perth, St Athan was a large pre-war airfield, comprising a good main runway, with Bowen's training school to the north and several C-Type hangars and No. 32 MU's workshops to the south. A well appointed mess and barrack accommodation housed several thousand recruits undergoing their trade training. In addition to the MU, the station hosted a navigation school, a fighter training pool and a transport training unit.

The main body of the Airborne Group, which comprised some twenty-five civilian staff and forty RAF officers and airmen from the SD Flight, left Perth on 5 November. Although unaware of their arrival, the station commander, Group Captain Lucking, and his Chief Engineering Officer, Wing Commander Fowler, promptly made one of the C-Type hangars available to Bowen – a number that was destined to rise to three to accommodate the Group, its flying crews and the fitting parties. The area inside the hangar was quickly partitioned off, using a number of conveniently discarded Heyford wings, to provide laboratory and storage space. The conditions inside the hangar were far from ideal, since for most of the day the doors were open to the elements, which caused great gusts of ice cold air to be sucked through the building.

In his book, *Radar Days*, Bowen places the blame for the conditions at St Athan, under which he and his staff had to work during the winter of 1939–40, squarely on the shoulders of Watson-Watt and Jimmy Rowe:

> We might have got by with those canvas partitions if it were summer, but it was already November and for several months there was no source of heat whatsoever. For a large part of the time our people wore greatcoats and gloves inside this improvised laboratory space. . . . The conditions were simply appalling. . . . Here was one of the most sophisticated of defence developments being introduced to the Royal Air Force, and it was being done under conditions which would have produced a riot in a prison farm. I often wonder how much better the night defence of Great Britain would have been in the latter half of 1940 if we had been provided with better facilities and greater manpower during those critical days at Perth and St Athan. It is easy to be wise after the event, but I blame it all on that precipitate flight from Bawdsey; responsibility for this must be placed firmly on the shoulders of Rowe and Watson Watt. . . . As was already abundantly clear, Bawdsey and Martlesham had not been singled out for special treatment by the enemy and, once this was obvious, I will never understand why we were not allowed to go back.[13]

In stark contrast to Bawdsey and Martlesham, St Athan was bombed a few weeks after their arrival by a single Junkers Ju 88, that struck the main runway with a single bomb which failed to explode.

Despite the conditions, the fitting parties settled down to the routine of aircraft install-ations. The various equipments that comprised the AI and ASV systems were delivered to St Athan by the principal manufacturers, Pye Radio and E.K. Cole Ltd, whence they were bench tested, accepted and installed in the aircraft, complete with cables, wiring looms and locally made fittings and mounting brackets. When this work had been completed to the satisfaction of those concerned, the equipment was ground-run prior to a comprehensive air-test by pilots from the SD Flight, with one of the civilian personnel checking the radar – usually Messrs Mills, White or Wood. With the test flights successfully completed, the aircraft were flown directly to their Fighter, or Coastal Command unit.

Given the desperate need for night-fighters, progress was slow whilst the fitting parties learned their trade. By the middle of November twenty-one Blenheims were passing through No. 32 MU's workshops, of which four were awaiting delivery to No. 25 Squadron's detachment at Martlesham (two for operational research and two for testing by Fighter

Command) and three (L4825, L4829 and L4837) to No. 600 Squadron at Hornchurch. How-ever, matters gradually improved with seventeen Blenheims being delivered during the first full month of the installation programme, December, rising to eighteen in January 1940 and a very creditable 100 during July. These figures, however, do not reflect the overall radar installation effort, which also included work on the ASV fitting programme for Coastal Command.[14]

1940

Despite the best efforts of Bowen and his team at Perth and St Athan, further criticism of AI Mk.I, citing its poor minimum range, where the target echoes were lost in the transmitter pulse on the CRTs, continued to be reported to the Air Staff. The extent of this problem caused the AOC-in-C Fighter Command to condemn AI Mk.I as unfit for operational use and fit only for training purposes.[15] Nevertheless, and aware of its shortcomings, the Air Ministry had already instituted the design of a 'production-ised' version of AI Mk.I to be designated AI Mk.II. Initial discussions during October 1939, established a requirement for 300 sets to be delivered to the RAF by Christmas. Work on the Mk.II system began the same month, with E.K. Cole being assigned the task of developing an improved version of the Mk.I's transmitter, which retained the Acorn valves. Pye Radio was given the task of designing a better receiver timebase circuit that would incorporate a number of improve-ments, including suppression circuits for the local oscillator. It was hoped that with the introduction of transmitter suppression, the system's minimum range might be improved. Subsequent reports of minimum ranges as low as 400 feet (122 metres) under laboratory conditions, were later claimed, but never realised in practice. The continued need to meet a Fighter Command requirement for a maximum range of 10 miles (16 km) was also pursued, but was never achieved, since the Blenheim could not fly at 53,000 feet (16,155 metres).[16]

During January, Bowen reported a number of difficulties with the first aircraft installation of AI Mk.II. Whilst the suppression stage in the transmitter worked well and cancelled the transmitter returns, when test flown in a Blenheim, the minimum range was found to have increased to 1,000 feet (305 metres), whilst the maximum range was 14,000 feet (4,270 metres) on a large target. Attempts at reducing the minimum range failed to get lower than 800 feet (245 metres), but the cause was established as being an unexpectedly high capacitance between the valve filaments in the transmitter and ground (chassis). Realising the capacit-ance problem was a significant failure in the new set, Bowen recommended the manu-facturing programme be suspended whilst a suitable solution was sought. In the meanwhile, the six sets that had been delivered from the manufacturers, were installed in five Blenheim Is of Nos 25 and 600 Squadrons during February, for testing and evaluation purposes.

Following No. 25 Squadron's pioneering work with the introduction of AI, Fighter Com-mand authorised the formation of five more Blenheim night-fighter squadrons, Nos 23, 29, 219, 600 and 604, in October 1939, who began their night-flying training that same month. No. 25 Squadron's 'C' Flight at Martlesham Heath began operational night-fighter patrols over the North Sea during November, with the objective of intercepting any enemy aircraft attempting to raid London, whilst No. 600 (City of London) Squadron, AAF based at Manston and equipped with Blenheim Ifs, received three AI-Blenheim IVs in November 1939, with which to patrol the Thames Estuary in search of minelaying Heinkel He 115s.

During February 1940, the question of the deployment of AI Mk.II in the Blenheim was raised in a memorandum from the Assistant Chief of the Air Staff (ACAS) (Radio) to the

AOC-in-C Fighter Command.[17] In this document the ACAS accepted the fact that whilst the Mk.II system had proved inadequate in intercepting low flying aircraft, the equipment could be of value when operating at heights above 5,000 feet (1,525 metres). Dowding must have concurred, or been persuaded by his junior, for authority was given for the installation of AI Mk.II in three Blenheims each of Nos 23, 25, 29, 219, 600 and 604 Squadrons. However, it was stressed that these Blenheims were to be used solely for training purposes in order not to impede their operational commitments. The squadrons were warned of the Mk.II's short-comings, but were advised that the employment of an AI set could provide valuable experience for the crews and stand them in good stead when an operational version was available for service. Conversely, Fighter Command insisted that the aircraft were not to be exempted from their present operational duties, and as such, little or no valuable training appears to have accrued from this exercise!

In order that senior officers could better understand the theories of night-interception and the conditions under which the crews had to operate, Bowen conducted a series of lectures at Fighter Command's HQ at Bentley Priory, at Stanmore, Middlesex. Beginning with the classic stern-chase, in which the fighter identifies the bomber, and tries, by a series of manoeuvres, to close to a position directly behind the enemy aircraft, Bowen extrapolated a number of facts. First, the intercepting fighter required a speed advantage over the bomber, of some 20–25 per cent, otherwise there was little prospect of the raider being caught. Taken simply, this meant that if the bomber was flying at 250 mph (400 km/hr), the fighter had to be capable of at least 300 mph (480 km/hr). It should be noted that the Blenheim If, when fully laden, could manage 280 mph (450 km/hr) at 15,000 feet (4,570 metres), whilst the *Luftwaffe's* medium bombers, the Dornier Do 17Z, the Heinkel He 111H/P and the Junkers Ju 88A, were all capable of 250 mph at similar heights. Second, in order to stand a reasonable chance of completing the interception successfully, the fighter had to reside within a cone of some 40–50° behind the enemy aircraft and be on a course that was not more than 30° from that of the target. Given the inaccuracy of the CH stations at that time, this was a formidable problem that would only be resolved with the introduction of GCI towards the end of the year.[18]

Before leaving the Mk.II story, it is perhaps worthwhile mentioning two experimental versions. The first, AIH – 'H' for high power, or high powered, was an attempt to increase the power output by introducing Micropup valves in the transmitter and a telegraph system between the R/O and pilot. A Blenheim I was fitted with the new transmitter and test flown at St Athan during the first week of April, with encouraging results. The Blenheim was then despatched to Manston for further testing, whilst a second was prepared by No. 32 MU. The system was further improved by the fitting of 'new' aerials, possibly installed in the vertical plane, during the third week of April. These indicated an increase in range, provided good azimuth indications and an improved front-to-back ratio. Flight trials against a Wellington confirmed the designer's expectations, leading to the conversion of four more Blenheims at St Athan. With the closure of the Airborne Group at St Athan on 30 April and the transfer of its staff to Worth Matravers, of which more anon, all trace of AIH has been lost in AMRE's correspondence. Correspondence with Professor Hanbury Brown confirms that further development of AIH ceased, for what reason we do not know, but perhaps because of the introduction of the superior AI Mk.IV during the summer of 1940.

The second derivative, AIL – 'L' for locking timebase, comprised a Mk.II system that was connected to aerials that were possibly canted upwards to give greater illumination of the upper elevation area. Tests during the second week of April, showed an increase in maximum range, but a poor minimum of 3,000–5,000 feet (915–1,525 metres), that led to the formal cancellation of the system in June 1940.

By the winter of 1940, the number of Government and service agencies[19] involved in the design, testing, introduction and operation of AI radar, served to induce a sense of confusion into the delivery programme. This lack of co-ordination was recognised by, amongst others, the Air Staff, Bowen, Tizard and Watson-Watt, who pressed the Air Ministry for the appointment of an overall authority to link the efforts of those attempting to solve the night-interception problem. Subsequently, a committee under the chairmanship of the Deputy Chief of the Air Staff (DCAS), Air Marshal Sir Richard Peirse,[20] was established to co-ordinate all matters relating to night-interception and the initiation of such action as was necessary to that end. The monitoring of technical developments in the interception field, I-R, the detection of engine ignitions and searchlight fighter control, were also included in its terms of reference. Entitled the 'Night Interception Committee' – the 'Interception Committee' from July 1940 and the 'Air Interception Committee' from July 1941 – the new body met for the first time on 14 March to review the progress on AI, fighter control and developments in related fields. Discussions ranged over a wide variety of subjects, but the Committee recognised that AI still offered the best and most promising solution to the night-interception problem and gave it their full support.

The meeting also recognised the need for a specialised unit, operating under the direction and with the authority of Fighter Command, that was charged with the operational testing of AI-related equipment and experiments in night-interception techniques. This unit, initially named the 'Night Interception Unit', but later changed to its more familiar title of the 'Fighter Interception Unit' (FIU), was formed at Tangmere, Sussex, on 10 April, under the command of Wing Commander G.P. Chamberlain.[21] Its preliminary establishment was ten officers and ninety airmen. Such were the Air Ministry's concerns that FIU should begin its work at the earliest opportunity, the CAS directed its equipment and manning should rate the highest priority equal to that accorded to operational fighter squadrons. The Unit's preliminary aircraft strength comprised five, possibly six, Blenheims, with their AI provided by No. 32 MU. Accommodation was provided for the attached scientific staff, along with laboratory and workshop facilities for in-house equipment modifications and trials. FIU's role was threefold:

- First, to provide a test-bed facility for the trialling of AI radar and associated equipments under operational conditions, in order that any shortcomings be identified and corrected prior to that equipment being committed to full scale production.
- Second, an examination of the equipment's 'maintainability' to ensure its servicing was within the capability of the average wartime radio mechanic and to establish day-to-day maintenance routines and calibration procedures.
- Third, to conduct examinations of interception techniques based on the enemy's operating methods, with the results being passed to Fighter Command and promulgated amongst the squadrons.

These tasks were undertaken by the Unit's scientists, engineers, pilots and civilian and service observers, who condensed their assessments and ideas into a series of monthly reports to Fighter Command and engineered tested modifications for equipment and aircraft. By these means, FIU was able to ensure a steady flow of tried and tested AI systems to the night-fighter squadrons and provide a pool of operational and technical expertise, unrivalled outside of the UK.

During this period Robert Hanbury Brown was posted to Tangmere to assist FIU in the testing of AI in the short-nosed Blenheim Mk.I, and provides a sketch of Peter Chamberlain and the difficult working relationship between the civilian scientist and RAF officers:

It [FIU] was commanded by an extraordinarily active officer, Wing Commander G.P. Chamberlain, who fired out decisions like a machine gun, worked flat out 20 hours a day and expected everyone else to do the same …. Together with one of our original team from Bawdsey, Brian White, I helped FIU put all the various prototypes of metre-wave AI through their paces. It was a busy time.

Although I held an honorary Commission as a flight lieutenant in the RAFVR I never wore my uniform except when flying over enemy territory. … To wear a uniform is to define your position in the military hierarchy which makes it more difficult to talk freely and critically to people of higher rank. On the other hand being a civilian in a military unit during war can be uncomfortable. … Officers are anxious to preserve their authority and don't like untidy civilians shambling about, while civilians are anxious to preserve their independence and don't like being bossed about.

I don't think Peter Chamberlain liked untidy civilians cluttering up his immaculate Unit any more than I liked him trying to tell me what to do. Nevertheless we got on very well, and I would rather fly with him than anyone else I know.[22]

Chamberlain's deputy at FIU's formation was Flying Officer Glyn 'Jumbo' Ashfield,[23] who could, and frequently did, put the stately Blenheim through a series of aerobatic manoeuvres unimagined by its designers. His apparent speciality was high speed, low level flying under electricity cables.[24] Other known FIU aircrews in its early days were pilots: Flight Lieutenant Anthony Miller (Flight Commander), an Auxiliary officer from No. 600 Squadron,[25] Flight Lieutenant Robert Ker-Ramsey from No. 25 Squadron,[26] Pilot Officer John White from No. 3 Squadron,[27] Sergeant Dicky Ryalls,[28] R/Os Sergeant Edgar le Conte,[29] Sergeant Reginald Leyland,[30] and observers Sergeant George Dixon[31] and Sergeant Geoffrey Morris.[32]

FIU's preliminary education was imparted by Bowen, who travelled to Tangmere on 20 April to brief Wing Commander Chamberlain and his staff on the principles and practice of AI and night-interception. According to Bowen, Chamberlain and his team rapidly acquired the basic techniques and quickly brought their experience to bear on the operational aspects of radar testing and evaluation.[33]

Whilst FIU were establishing themselves at Tangmere, the operational testing of AI Mk.II was continued on Fighter Command's behalf by No. 600 Squadron at Martlesham. The Squadron's report concluded the system's minimum range to be 1,000 feet (305 metres) and its maximum somewhere between 5,000–6,000 feet (1,525–1,830 metres), using a Blenheim as the target aircraft – a very poor result. The tests were carried out under ideal conditions, with one of the Airborne Group's members acting as observer and making the necessary in-flight adjustments – an advantage that would not normally be available to an operational crew. It also appears that the trials were carried out in daylight, since *The Signals History* states that:

the maximum range quoted did not give a pilot the same advantages as the equivalent amount of daylight'. Continuing, the History comments that 'the effective field of AI vision was restricted to a fairly narrow cone directly in front of the aircraft and unless the fighter positioned himself some 6,000–1,000 feet [1,830–305 metres] behind the [enemy] aircraft, and flying in the same direction, at roughly the same height, he had very little chance of completing the interception.[34]

Once outside this narrow cone, the system's direction finding qualities were poor, with those contacts at a wider angle being impossible to hold. Further, ambiguous returns due to irregularities in the field strength of the Blenheim's radiation pattern (polar diagram), made it possible for the R/O to see a target that was in front of, and above the fighter, when in fact it was behind and below. These irregularities, therefore, ensured that interceptions with the Mk.II system would be few and far between. In reality, as Bowen had concluded after the first few sets had been delivered early in the year, AI Mk.II felt far short of the performance expected of an operational radar set. Like AI Mk.I before it, AI Mk.II was a failure as an operational AI set and was fit only for training and instructional purposes.

CONCLUSIONS

The premature move of AMRE from Bawdsey to Dundee in September 1939, and the subsequent move of the Airborne Group to St Athan, at a time when the RAF was in desperate need of AI and ASV radars, can only be described as an unmitigated disaster to the radar research programme. Nevertheless, the pioneering work undertaken by No. 25 Squadron during this period, and its detached flight at Martlesham, yielded valuable information on AI Mk.I that confirmed its poor minimum range and condemned the set as an operational system.

Its replacement, AI Mk.II, introduced early in 1940, fared little better. Poor management control during manufacture served to impede the delivery of a properly designed set, and testing quickly confirmed its principal shortcomings: poor low-level cover and poor minimum range, that would, unfortunately, render it fit only for research and training purposes.

Whilst both sets failed to deliver a viable system to Fighter Command, the testing and deployment of AI Mk.I and Mk.II did demonstrate their potential as an interception tool and encouraged the Air Staff, and more particularly Dowding, to continue the research programme. Other shortcomings associated with AI generally were also identified: a lack of qualified servicing mechanics and technical documentation, poor reliability, a lack of performance on the part of the Blenheim fighter and most importantly, the need for a better ground control system.

This period was not, however, all doom and gloom. The need for an organisation to co-ordinate all matters relating to night interception was recognised by the Air Ministry, with the establishment of the Night Interception Committee in March 1940 and the FIU the following month, with the latter being responsible for the operational testing of AI equipment in night-fighters. Both organisations would eventually go on to provide valuable contributions to the night-fighter community and air-interception tactics and techniques. The first steps in the development of night-fighter control and the raising of a requirement for a GCI radar, were also undertaken during this period. However, the overriding requirement was the delivery of an AI set that was capable of being deployed in quantity by Fighter Command, a requirement that was to be partially fulfilled by AMRE's next product, AI Mk.III.

Notes

1. With the Soviet Union keen to avoid war at any cost, but wishing to expand her sphere of influence in the east and the Baltic, and Germany wishing to secure her eastern borders when she invaded Poland, Russia and Germany concluded a non-aggression pact on 24 August 1939.
2. John Terraine, *The Right of the Line* (London: Hodder & Stoughton), pp. 96 and 97, and Denis Richards, *The Royal Air Force 1939–1945, Volume 1, The Fight at Odds* (London: HMSO, 1993), p. 63.
3. Chaz Bowyer, *Fighter Command 1936–1968* (London: J.M. Dent & Sons Ltd, 1980), pp. 188 and 189.
4. Winston Ramsey [Ed.], *The Blitz Then & Now, Volume 1* (London: Battle of Britain Prints Ltd, 1987), pp. 126–8 and Jack Gough, *Watching the Skies* (London: HMSO, 1993), p. 8.
5. A *Luftwaffe* bomber *Gruppe* comprised thirty aircraft and was commanded by a captain (*Hauptmann*), or a major, who carried the title of *Kommandeur*. Each *Kampfgeschwader* comprised three operational *Gruppen* and one training *Gruppe*.
6. Winston Churchill entered Chamberlain's Cabinet as First Lord of the Admiralty on 3 September 1939.
7. Mark Hichens, *The Troubled Century, British & World History 1914–1993* (Bishop Auckland: Pentland Press, 1994), pp. 151–4.
8. Chaz Bowyer, *op cit*, pp. 189 and 190.
9. Denis Richards, *op cit*, pp. 152–4.
10. Michael Armitage, *The Royal Air Force, An Illustrated History* (London: Brockhampton Press, 1996), p. 101.
11. The term 'radar operator' is generally used to describe the crew-member who operated the AI radar and will be so used throughout this narrative until the title 'Navigator' was accepted later in the War.
12. *The Signals History, Volume 5*, p. 9.
13. Edward Bowen, *op cit*, p. 93 and 94.
14. Whilst St Athan was turning out AI Blenheims it was also modifying Hudsons and RAF parties at Pembroke Dock and Helensburgh were installing ASV in Sunderlands and Catalinas.
15. Memo from AOC-in-C Fighter Command, dated 10 January 1940 and quoted in *The Signals History, Volume 5*, p. 119.
16. Using 1½ metre radar, which restricted the maximum range to the height of the fighter above the ground, this figure translated into a height of nearly 53,000 (10 × 1,760 × 3) feet which no aircraft could achieve. Equally, it was only with the introduction of centimetric radar that ranges approaching 10 miles were achieved!
17. *The Signals History, Volume 5*, p. 120.
18. E.G. Bowen, *op cit*, pp. 122 and 123.
19. AMRE, the Airborne Group, No. 32 MU St Athan, the Air Ministry, RAE Farnborough, MAP, HQ Fighter Command, the manufacturers and the squadrons, to name but a few.
20. This post was renamed the Vice Chief of the Air Staff from 22 April 1940.
21. Later Air Vice Marshal Peter Chamberlain, CB, OBE.
22. R. Hanbury Brown, *op cit*, p. 56.
23. Squadron Leader Ashfield, DFC, AFC, was killed in a flying accident in December 1942 when he commanded a flight of No. 157 Squadron.
24. E.G. Bowen, *op cit*, p. 124.
25. Later Group Captain A.G. Miller, DFC, who commanded No. 17 Squadron during the Battle of Britain and No. 600 in 1942, before retiring from the RAF in 1946.
26. Later Squadron Leader Ker-Ramsey, MBE.
27. Later Group Captain J.W. White, MBE.
28. Later Squadron Leader D.L. Ryalls, shot down and killed as a flight commander with No. 219 Squadron in December 1944.
29. Later Wing Commander E.F. le Conte, OBE.
30. Later Flying Officer R.H. Leyland.
31. Later Flying Officer G. Dixon.
32. Later Wing Commander G.E. Morris.
33. E.G. Bowen, *op cit*, p. 124.
34. *The Signals History, Volume 5*, p. 121.

CHAPTER FOUR

Deploying the Solution
April–July 1940

Following the conquest of France, Norway, Denmark and the Low Countries, the *Luftwaffe*, along with the other arms of the German forces, halted its military operations whilst it replenished its losses and regrouped. On the outbreak of war, the *Luftwaffe*'s aircraft were concentrated within the boundaries of the Reich to defend Germany and provide a jumping-off point for the invasion of Poland. By the end of June 1940, with Europe secure and Germany's opponents temporarily held in check, the *Luftwaffe* set about a reorganisation in preparation for the invasion of Great Britain. *Luftflotte* 2[1] and *Luftflotte* 3, that had previously been responsible for providing the air support during the *Blitzkrieg* on France and the Low Countries, were brought forward from its bases in western Germany and repositioned in France, Belgium and Holland, whilst a third, smaller unit, *Luftflotte* 5, was established to control *Luftwaffe* elements in Norway and Denmark.

As far as Great Britain was concerned, the principal threat came from *Luftflotte* 2 that was based in Holland and Belgium and headquartered in Brussels, under the command of *Generalfeldmarshall* Albert Kesselring, and *Luftflotte* 3, based in northern France at Saint-Cloud, near Paris, under *Generalfeldmarshall* Hugo Sperrle. *Luftflotte* 2 was allocated the responsibility for attacking targets in eastern England, whilst *Luftflotte* 3 covered western Britain. The secondary threat, that from *Generaloberst* Hans-Jurgen Stumpff's *Luftflotte* 5, headquartered at Oslo, was centred on operations over northern England and Scotland, in conjunction with *Luftflotte* 2.[2] By 20 July, following a significant airfield construction programme that would provide the *Luftwaffe* with a number of all-weather runways, the three *Luftflotten* were able to field 864 serviceable medium-range bombers (Heinkel He 111, Dornier Do 17 and Junkers Ju 88), 248 dive-bombers (Junkers Ju 87), 656 single-engined fighters (Messerschmitt Bf 109) and 280 twin-engined fighters (Messerschmitt Bf 110).[3] To oppose this force Fighter Command had twenty-five squadrons of Hurricanes, nineteen of Spitfires, two of Defiants and six of Blenheim fighters.[4] The Blenheims being, 'of doubtful value during the day against escorted bombers',[5] were prudently allocated to night-defence.

April 1940

The location of the Airborne Group at St Athan had, to an extent, separated them from the main stream of radar research being conducted at AMRE, Dundee. Accordingly, the assistance of the scientists at Dundee was called upon to design the next system, AI Mk.III. For this design, the Mk.II's transmitter was discarded and replaced by a similar one taken

Figure 4. Fighter Command Groups 1941–1945. It should be noted that No. 9 Group was disbanded on 15 April 1944 and its area of responsibility was transferred to No. 12 Group.

from ASV Mk.I, then in production for Coastal Command, that employed Micropup valves and a change in operating frequency.[6] Twenty AI Mk.III sets were ordered for installation in Blenheim Ifs, for which subsequent tests showed an improvement in maximum range to 17,000 feet (5,180 metres), but little in the way of reduced minimum range. The system still suffered from squint at certain angles of elevation and an excessive pulse width. The latter problem manifested itself in the approach to the target, where the time interval between the transmitter pulse and the returning target echo was so small, that the echo was swamped in the tail of the transmitter pulse. This phenomenon occurred at target range of 1,000 feet (305 metres) or more, and before the pilot had made visual contact with the target. The overall effect, was for the target to disappear off the screens when it was outside visual range.[7]

The new set's shortcomings had been communicated to HQ Fighter Command (probably by FIU, as was their duty), which led, as Hanbury Brown described it, to the 'great minimum range controversy' that would itself lead to the creation 'of unnecessary friction ... between the airborne group and the main body of AMRE' and the 'alienation' of Bowen. In the early days of radar development, Bowen had been advised by the Air Ministry that a minimum range of 1,000 feet (305 metres) was adequate, however, when AI Mk.III was tested it produced ranges between 800 and 1,500 feet (245–460 metres) depending on the adjustment of the receiver. Bowen and Hanbury Brown were of the opinion that this could be reduced to 800 feet by means 'of a simple modification to the receiver', which they were reluctant to introduce to the production lines, as this would impact on the delivery schedule. Bowen and Hanbury Brown were, therefore, astounded to hear that Jimmy Rowe and Dr W.B. Lewis,[8] Rowe's deputy, had spoken with Fighter Command's Research Section, headed by Harold Larnder, who, according to Hanbury Brown 'presumably'[9] persuaded them that the in-adequate minimum range was a significant defect that required their immediate attention. Indeed, so important was this work to AMRE, that Lewis took personal responsibility for the investigation, whilst, at the same time, placing a development contract on EMI to find alternative solutions. Lewis' actions, nevertheless, had the unfortunate effect of precipitating Bowen's departure from AMRE, as Robert Hanbury Brown explains:[10]

> Why Lewis started this work without discussing the whole problem with the airborne group, and in particularly with Taffy Bowen, I don't know; no doubt it was partly due to the separation between Dundee and St Athan and partly to the fact that Rowe and Lewis never got on with Taffy. I suppose Lewis accepted what he was told by Larnder and thought it was time someone else, other than Taffy, had a look at what he thought was an important problem. Whatever the reason, he infuriated Taffy and that was the last straw in the already strained relations between Taffy and the top brass of AMRE. In effect Taffy, who had been one of its brightest stars was lost to AMRE, but not to radar as will be seen later.[11]

Lewis believed the minimum range could be restored by shortening the tail of the trans-mitter pulse, which he proposed to accomplish by connecting a second transmitter across the tuned circuit of the first, to act as a damper. By this means, a minimum range figure of 600 feet (183 metres) was expected. An alternative approach, originating from within AMRE, comprised a series of minor modifications to the transmitter and more serious alterations to the receiver, by which means it was hoped the minimum range could be reduced to 800 feet (245 metres). The latter approach was designated AI Mk.IIIA, whilst Lewis' double trans-mitter system was designated AI Mk.IIIB.[12]

In order to prove the theory, Lewis ordered that a number of the Mk.III sets be delivered to Dundee, where a young Oxford undergraduate and electronics engineer, Edmund 'Ted' Cooke-Yarborough,[13] was given the responsibility for conducting a series of practical experiments. The sets were duly delivered to Dundee by Mr P.L. 'Yagi' Waters (an expert in aerial design, hence the 'Yagi'), who proved extremely useful in explaining the Mk.III system to Cooke-Yarborough. The first flight trials of the Mk.IIIB transmitter were flown in a Blenheim, piloted by Sergeant Button, from Leuchars, sometime before the end of April 1940. With Lewis and Cooke-Yarborough acting as observers, the results proved inconclusive due to their not being able to independently measure the range to the target, when the tail of the transmitter pulse disappeared from the CRT trace. Further trials with AI Mk.IIIB were deferred when the SD Flight was moved to Christchurch, following AMRE's transfer to Worth Matravers, near Swanage, in May.[14]

By 26 May, three Blenheims, L6835, L6836 and L6838, had been fitted with AI Mk.IIIA. Preliminary testing in L6835[15] demonstrated a marginal reduction in minimum range to 950 feet (290 metres) and a maximum of 8,500 feet (2,590 metres), whilst the picture quality on the CRTs was considered to be very clear and readable, and a great improvement on the Mk.II. On the down side, the D/F indications were assessed as 'less than satisfactory', as was the level of interference with the crew's intercom, which frequently interrupted conversations between the pilot and the R/O.

Similar tests with L6836 gave near identical results, with the exception of the maximum range, which was reduced to 5,000 feet (1,525 metres). The reason for the reduction is not known, as the installation in L6836 was similar to those in the other two aircraft, excepting for the transmitter, which was modified to give a longer pulse width. It was hoped that increasing the pulse width would bring about a corresponding increase in maximum range, at the expense of a 100 feet (30 metres) increase in minimum range. Further trials with the two systems were conducted by FIU during June, which confirmed a further reduction in the minimum range to 900 feet (275 metres) against a Blenheim target aircraft. However, one significant shortcoming, that of poor system reliability, relating to the burning-out of transformers in the power unit, the short life of the CRTs in the display unit and deficiencies in the aerial system, were also confirmed by FIU.

These problems had already been identified by AMRE before the FIU trials, having been the subject of a meeting held at Farnborough on 17 May, which was attended by representatives from the Air Ministry, HQ Fighter Command, Messrs E.K. Cole and Pye Radio, and chaired by the RAE's Group Captain de Burgh. The evidence was examined in great detail, with the burning-out of the power supply transformers being attributed to the need to draw large amounts of power from the transmitter, in order to overcome a poorly designed coupler in the aerial system. The failure of the CRTs being due to shortcomings in the indicating unit's timebase circuitry that produced a high beam current, which dramatically reduced the tube's life. The solution to the former was the introduction of a starter switch in the transmitter and the complete redesign of the timebase circuitry, both of which would delay production and service introduction. In the meanwhile, something had to be done to prevent the Mk.III development shuddering to a halt. A compromise was devised, whereby ten Mk.II or Mk.III equipments, that did not incorporate the timebase or transmitter modifications, might be produced from a number of Mk.II transmitters and receivers being held at St Athan. These could be delivered immediately and since there was little difference in the Mk.II and Mk.III receivers (the Mk.III had an additional control), testing could proceed. Fortunately, ten Blenheims that could take either type of receiver, were available at St Athan, where the fitting-out could take place.

The symptoms of aerial unreliability, first identified in L6838, were defined by FIU as a reversal of the azimuth indications on the indicating unit at angles of elevation or depression, greater than 45°. An air test on 28 May, flown by Wing Commander Chamberlain, with

Hanbury Brown acting as observer, examined the ratio of the azimuth signals when the target was dead ahead and when it was at other angles. Their examination showed that:

> when the angle of elevation, or depression, of the target relative to the Blenheim [fighter] was less than about 45 degrees its azimuth was shown correctly, but at angles greater than 45 degrees the display of azimuth was not only wrong in magnitude, but was sometimes wrong in sense. Above an angle of about 60 degrees a target on the left of the fighter was actually shown as being on the right and vice versa.[16]

Like the transmitter and timebase problems before it, the aerial reversal phenomenon had the potential to delay the clearance of the system by FIU and delay the introduction of AI Mk.III to the night-fighter squadrons. After a short period of 'head scratching' Hanbury Brown was able to attribute the failure to the fact that the layout of the aerial system in L6838, a short-nosed Blenheim Mk.I, had been designed for the long-nosed Blenheim Mk.IV. As far as he was able to fathom, the reversal was due to the azimuth aerial on L6838 being enclosed by a pseudo cavity made by the nose and the engine nacelle. At high angles of elevation or depression, the returning signals from the target were coming over the aircraft's nose and being reflected off the engine cowling and thereby interfering with the direct signal from the target.

In order to overcome the situation, Hanbury Brown had the azimuth aerial elements moved outwards to a position on the outboard side of the engine cowling, as shown in Figure 5. This arrangement was air tested by Hanbury Brown on 5 June, but showed little improvement over the previous arrangement. With pressure being applied by Tizard and others to get the system into service as quickly as possible, the Airborne Group tried out a number of horizontally polarised aerial systems, none of which worked. It slowly dawned on Hanbury Brown that the only thing they had not tried was to change the polarisation from horizontal to vertical. 'This completely cured the trouble', and provided the designers with 'remarkably neat' aerials 'which could be mounted far away from the engines where the troublesome reflections from the engine cowlings could not reach them'.[17] Hanbury Brown later discovered that horizontal polarisation had originally been chosen for ASV radar, where the 'clutter' from the sea returns was small, and since the early AI sets shared much of their technology with ASV, AI was consequently designed with a horizontal aerial system.

The new vertically polarised installation fitted to Blenheim L4846, comprised a simple half-wave dipole with reflector on each wing, outboard of the engine nacelle, acting as the azimuth elements and a quarter-wave unipole, with reflector, above and below the port outer wing (moved to the starboard outer wing in later versions) and a transmitting dipole, with director, mounted on the nose. Flight trials beginning on 1 July, with Hanbury Brown once again acting as observer, confirmed the elimination of the azimuth reversal, alongside a reduction in the ground returns and aerodynamic drag. With production underway, Fighter Command decreed that a programme of retrospective modifications to upgrade the Mk.III's aerial system to the vertical plane, was too great to contemplate. However, all 1½ metre equipments from AI Mk.IV onwards, were built with vertically polarised systems, that were, in most respects, identical to that of L4846.

Whilst Hanbury Brown and others were trying to resolve the technical issues with AI Mk.III, the members of the Night Interception Committee were endeavouring to prioritise the manufacture and fitting of airborne radar systems. At the Committee's fourth meeting on 2 May, they reached the conclusion that the threat presented to the country by the

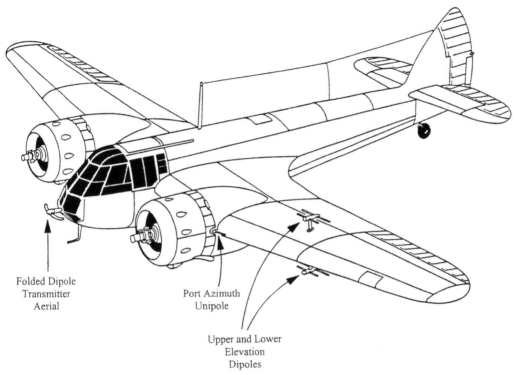

Folded Dipole
Transmitter
Aerial

Port Azimuth
Unipole

Upper and Lower
Elevation
Dipoles

Figure 5. Aerial system for AI Mk.III on Blenheim If. In this instance the Blenheim's dorsal turret has been removed to save some 800 lb (360 kg) of weight and improve drag. The aircraft also needed just two crewmen – a pilot and R/O.

Luftwaffe's bomber fleets, posed more of a problem than that of the *Kriegsmarine*. Consequently, in their opinion, AI production should be accorded a higher priority than that for ASV. In order to meet this commitment, eighty of the 140 transmitters being built for ASV production were transferred to the AI programme, with, in addition, another seventy being ordered from E.K.Cole. This was to meet a requirement for the equipping of sixty Blenheims with AI Mk.IIIA and a further forty with Mk.IIIB, despite the problems that were being experienced with the equipment. Furthermore, the responsibility for the engineering design of the Blenheim's AI installation should be transferred to the Radio Department of the RAE, Farborough.[18,19]

The Night Interception Committee's decision was not greeted enthusiastically in all quarters of the Air Ministry, one of whose most important senior officers, the Director of Signals (Air) (DSigs[Air]), concluded in a report towards the end of May:

> that he had come to the conclusion that AI in any form that it existed at the time of writing, was unsuitable for use with service squadrons. He stressed that the greatest danger with AI was that its premature introduction to service use in an unreliable and unfinished form, would create prejudice against it in the minds of pilots, who were notoriously conservative.[20]

DSigs(Air)'s comments on AI were valid, since it had long been recognised that when the system could clearly demonstrate a moderate level of reliability, its introduction into the

RAF would not be impeded by him, 'not even for a single day'.[21] Continuing, he highlighted the current claims for AI's performance, which he thought exaggerated, and the unrealistic dates for the manufacture of equipment and the aircraft fitting-out programme. In conclusion, the Director proposed that sufficient time and effort should be allocated to the execution of proper service trials, with a view to ensuring that AI fulfilled its operational requirements before it was handed over to the squadrons – something that Fighter Command and the FIU were, even then, trying to achieve.

The theme of reliability was continued in a paper written by Tizard for the fifth meeting of the Night Interception Committee on 23 May, in which he recommended that FIU should first master the techniques of night interception by a series of experiments and then fly these by day until they had been mastered. Until then, he considered that 'service personnel were trying to run before they could walk'.[22] Like the DSigs(Air), Tizard also thought that greater attention should be paid to perfecting the system's operation and reliability, at the expense of reducing the minimum range. He nevertheless agreed that whilst the basic principles of AI's design and construction were correct, it was let down by its engineering, which he thought to be of an 'inferior standard'. This may well have been one of the causes of the reliability problem, and something which the RAE's Radio Department had the ability to fix. As for the manufacturing companies, Tizard believed they did a poor job, which in turn had induced a number of failures. He also criticised a series of shortcomings in the design of the controls, which made the system difficult to use in the air.[23]

By 26 July, the fitting parties at No. 32 MU had delivered seventy Blenheims equipped with AI Mk.III,[24] whose reliability was still the cause for some concern in the Air Ministry and at RAE. A report by RAE concluded the design of the receiver was fundamentally flawed on account of its being based on a commercial television chassis, whose mechanical rigidity and component tolerances were below the standard required for airborne equipment. At this late stage, and with production in full swing, it was not considered feasible to redesign the receiver and consequently a high degree of technical expertise would be required to maintain the system in good working order. As may be appreciated, the RAF was not exactly awash with highly skilled radar mechanics during the summer of 1940! The report also highlighted the problems associated with the horizontally polarised aerial system. Overall, RAE assessed the Mk.III installation in the Blenheim as being 'partially reliable', and accorded it a performance rating of 'variable'. In terms of improvements, they considered the design of a less critical aerial system to be paramount (their tests highlighted deficiencies in the transmission cabling and connectors and problems with aerial matching). However, in their opinion, only a complete redesign of the system that incorporated a separate modulator, would produce a reliable AI set. To this end, they endorsed the award of AMRE contract to EMI to develop the modulator design.[25] In the same vein, Hanbury Brown and Cooke-Yarborough were instructed (probably by Lewis) to cease further work on AI Mk.III and concentrate on its successor, AI Mk.IV.

During the spring of 1940, authority had been given by the Air Staff and HQ Fighter Command for the re-equipment of the six designated night-fighter squadrons (Nos 23, 25, 29, 219, 600 and 604), each with an establishment of twelve Blenheim If fighters, with AI Mk.III. By the end of June, thirty-one aircraft had been delivered, rising to seventy, as has already been described, by the end of July and 140 at the end of October, with most receiving the Mk.IIIA version. In a move to overcome the radio interference problem and give improved air-to-ground communications, most of the Blenheims were equipped with VHF R/T sets. Due to the greater complexity of the Mk.IIIB system and the consequent increase in manufacturing and installation time, this equipment did not see as rapid an introduction to service as the Mk.IIIA variant. The development of AI Mk.IIIB was finally abandoned in June,[26] when comparative trials with the new EMI pulse modulator showed that, 'whilst both systems could provide minimum ranges below 500 feet (150 metres), the more complex EMI system could "see" aircraft at substantially greater maximum ranges'.[27]

Following the delivery of the Blenheims, a number of difficulties arose concerning the maintenance of the equipment and the quality of the operators. To begin with, each squadron was allocated three, sometime four, untrained radio mechanics, who possessed little or no test equipment with which to carry out routine maintenance. In a similar vein, such was the rapid expansion of AI, that some units lacked R/Os having the necessary aptitude and training. Before AI training became standardised, many ground crew in the wireless and radio branches volunteered to fly as operators, with little in the way of screening to assess their suitability for the role. To add to the problem, the regular pilots and air gunners disliked having a third crew member on board the Blenheim, as they got in the way when it came to abandoning the aircraft in an emergency. However, the need for a separate observer to monitor the radar was thought to be necessary, since the gunner's night sight, and hence his ability to search the sky for the enemy, was destroyed after a prolonged period staring into the CRTs. Despite an earlier Air Ministry decision to employ bespectacled schoolmasters in the role, there was simply not enough of them to fill Fighter Command's needs. However, with the creation of the Operational Training Unit (OTU) system in the spring of 1940 and the establishment of No. 54 OTU at Church Fenton in November 1940, a standardised method for the assessment and training of night-fighter pilots and R/Os was introduced for Fighter Command.[28] In the meanwhile, and until the system began to deliver trained aircrews from late December, the squadrons had to do the best they could with the material available to them.

The lack of a proper maintenance regime and poorly trained R/Os was compounded by two other issues: the inadequate performance of the Blenheim and the lack of a coherent system of ground control. The Blenheim If was an adaptation of the standard Blenheim bomber, which entered RAF service as a light-bomber with Bomber Command in March 1937. The need to replace the Hawker Demon in the two-seat fighter role in the late 1930s, brought about the employment of the Blenheim as a two-seat, long-range, day-fighter, for Fighter Command. In its new guise, the Blenheim's bomb-bay doors were removed to make way for a ventral pack that held four 0.303 inch (7.69 mm) American Browning machine-guns, complete with 2,000 rounds of ammunition (500 rounds per gun). The gun-packs, built in the Southern Railways workshops at Ashford, Kent, were attached to the bomb beams, and the resulting fighter was known as the If. The dorsal turret and its 0.303 inch Vickers K machine-gun was retained, but later removed to save weight and maintenance effort. The Blenheim was a thoroughly conventional, all-metal, stress skinned, cantilever monoplane. Powered by two 840 hp Bristol Mercury VIII radial engines, the Blenheim had a top speed of 279 mph (450 km/hr) at 15,000 feet (4,570 metres), a service ceiling of 30,000 feet (9,145 metres) and an endurance of five hours[29] – figures that were reduced when some 600 lb (270 kg) of AI equipment was added in 1940. Blenheim If's were introduced to Fighter Command from December 1938.

As a night-flying aeroplane, the Blenheim provided a stable platform on which to learn the skills and tactics required in night-fighting, but as a night-fighter, it was barely adequate, with a performance that did not provide the requisite 20–25 per cent speed advantage (see also p. 35) and was, by the standards of the day, lightly armed. One pilot, Squadron Leader

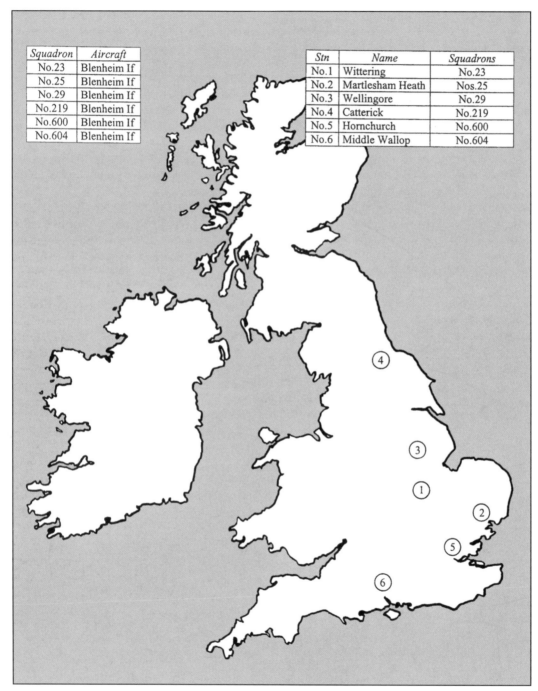

Squadron	Aircraft
No.23	Blenheim If
No.25	Blenheim If
No.29	Blenheim If
No.219	Blenheim If
No.600	Blenheim If
No.604	Blenheim If

Stn	Name	Squadrons
No.1	Wittering	No.23
No.2	Martlesham Heath	Nos.25
No.3	Wellingore	No.29
No.4	Catterick	No.219
No.5	Hornchurch	No.600
No.6	Middle Wallop	No.604

Figure 6.   Blenheim night-fighter Squadrons at August 1940. By August 1940 all six Blenheim squadrons were equipped, or equipping with AI Mk.III. The diagram also shows that in the aftermath of the fall of France and the Low Countries how the defence was primarily orientated to protect against attacks from the east and south-east.

Jeremy Howard Williams, DFC, who trained on Blenheims at No. 54 OTU and flew them with 604 Squadron, whilst accepting the aircraft's design was somewhat antiquated, thought:

> The Blenheim's instruments, apart from the standard blind-flying panel, are not so much grouped in the cockpit as thrown together haphazardly whilst its cockpit has a multitude of small windows separated by metal framework like a greenhouse. These panels have a disconcerting tendency at night to pick up reflections from the cockpit lighting, so that the eye is constantly distracted by lights seemingly coming from out-side the aircraft.[30]

Despite these shortcoming, the Blenheim was the only night-fighter available in reasonable numbers between the early summer and the late autumn of 1940.

The lack of an accurate ground control system for night-fighting was the other issue affecting the squadrons in the summer of 1940. During daylight hours, the sector control centres, using data supplied by the CH stations, were capable of directing fighters with an accuracy of between 3 and 5 miles (4.8–8 km), which was more than sufficient for the day-fighter boys to intercept a bomber, or fighter formation. However, at night, with an aircraft's search limit restricted to something less than 1½ miles (2.4 km), the CH station's performance was wholly inadequate. This, allied with the Blenheim's poor performance, made the interception of the enemy's bombers at night dependent, to a large extent, on luck.

On the night of 18/19 June, a Blenheim of No. 29 Squadron, operating out of Digby, Lincs, was vectored on to the track of an enemy bomber. Indications on the AI enabled the R/O to guide his pilot into a position where he could see the exhaust flames from the bomber's engines. However, before he could close the range, he was spotted and fired upon by one of the bomber's gunners, following which the enemy aircraft rapidly drew ahead and escaped.[31]

Similarly, the R/O of a Blenheim on night patrol, flown by the CO of the Wittering-based No. 23 Squadron, Squadron Leader Bicknell, reported they were closing rapidly on an unidentified aircraft. Having already been appraised of the presence of a 'hostile' in their vicinity, the pilot executed a stall-turn that brought the Blenheim into the target's slip-stream, whereupon he identified it as hostile. However, before he could obtain a good shooting position, the Blenheim was observed and the enemy rapidly drew away and escaped.[32]

A third incident concerned a report submitted by Aircraftsman 1st Class (AC1) E.L. Brine, who was flying as an R/O in a Blenheim (squadron and pilot not known), operating under the direction of an FIU controller at Tangmere, on the night of 14/15 July. The crew were alerted by the controller that an enemy bomber was being tracked by the Poling CH station, some 10 miles (16 km) to the south of Selsey Bill. After being given a course to steer, Brine's AI set detected an aircraft flying well below the fighter. Directing his pilot to lose altitude and increase speed, the aircraft descended and the range closed very slowly to 6,000–7,000 feet (1,830–2,135 metres). Brine's pilot executed two turns to starboard (20° and 10°) and a corrective turn to port (10°), bringing the target into the dead-ahead position, with both aircraft at the same altitude. The range at this juncture was 6,000 feet. AC1 Brine continues:

> Our own lack of altitude hampered operations somewhat, but the response was very clear and of excellent amplitude, despite being in the ground return most of the time. It was continued but we were unable to approach nearer than 4,000 feet [1,220 metres]

owing to a lack of speed and eventually the range commenced to increase again very slowly. The pursuit was eventually abandoned when approximately over the French coast. It is respectfully pointed out, that in the opinion of the writer, that only the lack of speed prevented a successful and comparatively easy interception, as AI indications were excellent and continued to be maintained until the abandonment of the pursuit'.[33]

These examples, which were reported to the Night Interception Committee, illustrate three facts: first, the need for greater performance from a better night-fighter, second, that if properly maintained and operated, AI Mk.III could perform well, and, three, despite the improbability of their being detected, *Luftwaffe* bomber crews kept a sharp lookout at night.

All was not, however, doom and gloom in the early summer of 1940, for on the night of 22/23 July, the everyday routine of exercise and trials by FIU and the controllers at Poling CH station, was finally to be rewarded. On this night, an FIU crew, comprising Flying Officer Ashfield, pilot, with Pilot Officer Geoffrey Morris[34] observing and Sergeant Reginald Leyland[35] operating the AI set, flying a Blenheim If, were on patrol over the English Channel to the south of Bognor Regis, Sussex:

Our aircraft was patrolling at 10,000 feet [3,050 metres] as Raid No. 9 composed of six aircraft at 6,000 feet [1,830 metres], appeared. As the sub-controller at Poling was providing good information on this raid, the FIU controller directed the Blenheim to intercept it, but the raid turned south-east and as there was little chance of an interception, the Blenheim was vectored to Selsey Bill. The raid again turned north, the Blenheim was ordered to vector 180 degrees and to switch on AI. At approximately two minutes from crossing the coast after the first vector of 180 degrees, the AI operator reported a contact, but below. Height was lost and after four corrections, each of 5 degrees to starboard, a bandit was reported at a range of 5,000 feet [1,525 metres], the height of the fighter being 5,000–6,000 feet [1,525–1,830 metres]. Shortly afterwards, the observer reported 'bandit to port'. Through the side window the pilot saw the enemy silhouetted against the moon at an angle of about 45 degrees and about 200 feet [60 metres] above the Blenheim. By the silhouette it was ascertained to be a Dornier 17. Distance was closed, and our aircraft took up a position about 200 feet below and dead-astern. The bandit's track was approximately 150 degrees and our aircraft could hold him at the first gate. The backsight was brought into position, and though the pilot could not see his foresight, he closed to approximately 400 feet [120 metres] and opened fire. A grand fireworks display was observed from the tracer and incendiary striking home. No return fire was noticed and the firing button was pressed to an approximate range of 100 feet [30 metres]. The enemy aircraft lurched to starboard and the nose dropped. Our pilot attempted to follow, pressing the firing button, when the whole of the cockpit perspex was covered in oil and, from the sensations experienced, the crew found themselves on their backs. By the use of instruments, the aircraft was brought back to a level keel at 700 feet [215 metres], but had lost complete contact with the enemy aircraft. Our aircraft called for a home bearing, which was received from Poling, and having flown on 30 degrees for six or seven minutes, the aircraft crossed the coast somewhere near Littlehampton. As our aircraft crossed the coast, our observer reported a big blaze astern, and the glow was seen on the water. The observer pin-pointed the fire on the water about 5 miles [8 km] south of Bognor. Enemy casualty, one Dornier unconfirmed – our casualties, nil.[36]

Ashfield's claim that night, was later confirmed as one enemy aircraft destroyed; however, the identity of the aircraft and its crew has never been properly established. A number of contemporary accounts state it to be a Do 17Z of the 2nd *Staffel* of *Kampfgeschwader* 3 (2./KG 3), but this is not known for certain, nor is the final fate of the crew, who were almost certainly killed in the crash or drowned. The action on the night of 22/23 July, therefore, was the first occasion, when a radar equipped night-fighter intercepted an enemy aircraft and destroyed it. Given the importance of the event, and in particularly its impact on the morale of the night-fighter squadrons, it is strange no decorations were awarded to Ashfield or his crew, although Ashfield was decorated with the Air Force Cross (AFC) at the end of his tour with FIU in March 1941.

The third significant problem affecting night interception was the inability of the CH network and the fighter control organisation to provide precise directions to the night-fighter crews. Whilst seconded to No. 25 Squadron at Northolt in the autumn of 1939, Robert Hanbury Brown undertook a trial using a small force of Blenheims to act as the 'enemy', who were then flown against Northolt's Sector HQ, whilst a number of No. 25 Squadron's Blenheims, under the command of the Sector Controller, attempted to intercept them. Due to the proximity of the London Balloon Barrage, the trial was held during the hours of daylight, with Hanbury Brown monitoring the operation from the ground. The tracks of the 'enemy' force were plotted on the sector map and the controller scrambled the defenders and gave them a course to steer to intercept the bombers. As the tracks changed, the controller strove desperately to bring the fighters within AI range, but failed at each attempt.

Later that day in the Officer's Mess, Hanbury Brown got into conversation with a Squadron Leader, who explained that the 'enemy force' had flown right across the sector without being challenged. The Squadron Leader then roared with laughter, before explaining that the Sector Controller had spent most of the morning (so presumably the trial took place in the afternoon) trying to intercept drifting barrage balloons travelling at 15 mph (24 km/hr) and he did not, therefore, stand much chance against a bomber doing 200 mph (320 km/hr)! This fact led the Squadron Leader to the conclusion that the chance of intercepting bombers 'was vanishingly small by day, and zero at night'.[37]

From Hanbury Brown's viewpoint, something, therefore, had to be 'done about the control of night-fighters by radar from the ground', otherwise, their very existence was threatened.[38] Given that the maximum range of AI at that time was less than 3 miles (4.8 km), and it could never exceed the height of the aircraft above the ground, the minimum requirement for the control system was to be able to place a fighter within 2 miles (3.2 km) of its target. To achieve anything less than this was a complete waste of time. However, his experience with the training of night-fighter crews at Northolt had taught him that a successful interception depended as much on being able to fly the fighter accurately, with regards to its speed, altitude and course, as it did by having a good ground control system. The successful night-fighter crew, therefore, had to acquire the stealth and patience of the hunter.[39]

Another factor identified by Hanbury Brown was navigation, which he considered had not been properly addressed. Whilst a number of exercises had been conducted before the war to study the control aspects of day-fighting,[40] no one had undertaken any study that would show how a fighter might be vectored onto a bomber at night, and how that fighter might navigate itself to and from its patrol area, under blackout conditions.

On the completion of his tour with No. 25 Squadron, Hanbury Brown compiled a long memorandum that laid out his thoughts and observations on night-fighter control.[41] In this document, he outlined the aforementioned factors necessary for a successful interception, which Bowen forwarded to Rowe at Dundee, with the following conclusion:

> that the solution to the problem was to build a special radar with a narrow rotating beam, which he [Bowen] called a 'Radio Lighthouse'. Such a radar would have a display like a map with the radar station at the centre and aircraft would appear as short arcs (Plan Position Indicator, PPI).[42]

This system perfectly describes GCI radar.

CONCLUSIONS

Whilst the results with AI during the summer of 1940 were encouraging, there was, nevertheless, a pressing need for a high performance fighter, fitted with an equally improved and more reliable radar, to replace the Blenheim/AI Mk.III combination. However, the introduction of such a fighter would come to nought unless, and until, it was directed by an effective ground control system and the night-fighter crews were trained to a higher standard in terms of their flying and interception skills. The solution to the first requirement, was, fortunately, to be completed during the early months of the summer of 1940, but would not provide the country with the required level of defence, until the introduction of GCI in December of the same year. In the meanwhile, the British people were destined to undergo a testing trial by day and then by night.

Notes

1. Air Fleet, a *Luftwaffe* tactical unit containing fighters, bombers and reconnaissance aircraft, comparable in size to a small RAF Command, or US Air Command.
2. Alfred Price [1], *The Luftwaffe Data Book* (London: Greenhill Books, 1997), pp. 21 and 31.
3. Alfred Price [2], *The Blitz On Britain, 1939–1945* (London: Ian Allen, 1977), p. 39.
4. Chaz Bowyer, *op cit*, pp. 189 and 190.
5. Alfred Price [2], *op cit*, p. 76.
6. ASV and AI radars operated on different frequencies.
7. *AI* in *The Signals History, Volume 5*, p. 12.
8. Dr W.B. Lewis, CBE, FRS, joined AMRE from the Cavendish Laboratory where he specialised in radio technology.
9. R. Hanbury Brown, *op cit*, p. 60.
10. *Ibid*, pp. 59 and 60.
11. *Ibid*, p. 61.
12. *Ibid*, p. 60 and *AI* in *The Signals History, Volume 5*, p. 122.
13. Later E. Cooke-Yarborough, CBE.
14. The information on the flight trials of AI Mk.III is drawn from documents and correspondence in the AVIA7, 13 and 15 series of files held in the National Archives, Kew, and detailed in the bibliography.
15. L6835 had previously been used for testing AI Mk.II and was regarded as something of a 'lash-up' and was perhaps not the best aircraft in which to test the new set.
16. R.Hanbury Brown, *op cit*, p. 58.
17. *Ibid*, p. 58.
18. Prior to the development of radar, the RAE's Radio Department was responsible for the system design of all wireless and radio sets in RAF and Fleet Air Arm aircraft.
19. *The Signals History, Volume 5*, p. 123.
20. *Ibid*, p. 123.
21. *Ibid*, p. 123.
22. *Ibid*, p. 123.

23. *Ibid*, p. 123.
24. AVIA15/136, Memo from the Chief Superintendent RAE to DCD, dated 26 July 1940.
25. According to Ted Cooke-Yarborough, EMI had patented the inductor pulse modulator in December 1939. However, there was no formal contract with AMRE, or the Air Ministry until April 1940.
26. It is not known if AI Mk.IIIA was abandoned at the same time.
27. Ted Cooke-Yarborough, when correcting the Author's first draft.
28. Ray Sturtivant, John Hamlin and James J. Halley, *Royal Air Force Flying Training & Support Units* (Tunbridge Wells: Air Britain [Historians] Ltd, 1997), p. 241.
29. The manufacturer's data as quoted in Chaz Bowyer [2] *Bristol Blenheim* (Shepperton: Ian Allen Ltd, 1984), p. 124.
30. Jeremy Howard Williams, *Night Intruder* (London: David & Charles, 1976) pp. 27 and 28.
31. *The Signals History, Volume 5*, p. 125.
32. *Ibid*, p. 125.
33. *Ibid*, p. 125.
34. Later Wing Commander G.E. Morris.
35. Later Flying Officer R.H. Leyland.
36. Combat report of FIU Blenheim for the night of 22/23 July 1940, quoted in *The Signals History, Volume 5*, p. 125.
37. R. Hanbury Brown, *op cit*, p. 39.
38. *Ibid*, p. 39.
39. *Ibid*, pp. 39 and 40.
40. The most important of these exercises was that undertaken in August 1936, at Biggin Hill, and devised by Henry Tizard, to understand the problems of fighter control.
41. *A Suggestion for Fighter Control by RDF*, by Robert Hanbury Brown, dated 24 November 1939.
42. R. Hanbury Brown, *op cit*, p. 43.

CHAPTER FIVE

The Best That Could Be Done

July–November 1940

Until it acquired bases in France, Belgium, Holland and Norway, in the summer of 1940, the *Luftwaffe*'s bombers were restricted to long-range armed reconnaissance and meteorological sorties over the North Sea and the Scottish naval bases of Rosyth and Scapa Flow. Bomber attacks against towns and cities on the British mainland were banned by the personal order of Hitler, for he still cherished the hope that Britain could be persuaded to sue for peace, given the right conditions. He did, however, give permission for attacks against naval vessels and the commencement of submarine warfare by the U-Boat arm of the *Kriegsmarine*. Mining operations against merchant shipping in the Thames and Humber Estuaries and off the East Coast, were also authorised and carried out on the *Kriegsmarine*'s behalf by the *Luftwaffe*. Using the new, parachute deployed, magnetic *Luftmine* (aerial mine), a weapon that would later be adapted as a dual sea mine and blast bomb (codenamed *Monika*) to devastating effect, shipping losses around the coastal waters of Britain began to increase.

Throughout his political career, it had been Hitler's intention not to invade Britain, but to reach some form of compromise, whereby she would retain her Empire and international trade, whilst Germany was given a free hand in Europe, and later, in Soviet Russia. Therefore, at the beginning of July 1940, the German Chancellor was expecting the British Government to sue for peace and accept his 'generous' offer. Hitler waited in vain. On 10 May, the day the German Army (the *Heer*) invaded Holland, Winston Churchill replaced Neville Chamberlain as Prime Minister and changed Britain's attitude towards Germany. On 2 July, Hitler directed the OKW to begin its preliminary planning for the invasion of Britain; Operation *Sealion*. Having heard nothing further from the British Government by 16 July, he ordered the planning to proceed and to be completed by the middle of August. Three days later, Hitler's final appeal to the British people was rejected on their behalf by the Cabinet and military planning was accelerated. The *Luftwaffe*'s part in the forthcoming invasion was simple:

'The English Air Force must be so reduced, morally and physically, that it is unable to deliver any significant attack against an invasion across the Channel'.[1]

On 21 July, Hitler confirmed his intention to invade Britain, a decision that was communicated by *Reichsmarshall* Goering to his senior *Luftflotten* commanders the same day, during a conference at his country seat, Karinhall. The invasion proper was scheduled to begin in 'a week or so', in the meanwhile, operations were to be stepped up with small-scale attacks on shipping in the English Channel and probing flights to ascertain the British defences.

In order that the invasion should succeed it was necessary that the *Luftwaffe* achieve complete air superority over the British Isles, by means of the total defeat of the RAF. The close proximity of the Continental airfields enabled the *Luftwaffe* to expand its offensive role to include strategic raids against the RAF's airfields and the country's industrial centres. However, the defeat of the RAF and the intimidation of the population was recognised by the Germans as requiring a considerable resource. On this occasion, unlike in previous campaigns, the responsibility for a successful invasion would rest firmly on the shoulders of the *Luftwaffe* and its commanders. Therefore, for the first time in history, the *Luftwaffe* would be responsible for the conduct of an independent campaign, whose objectives were:

- to neutralise, by bombing, the RAF's ability to defend the country, and
- to extend that bombing to Britain's towns and cities, in order to interfere with industrial output and, more importantly, to reduce civilian morale to such an extent that the country's leaders would be forced by public opinion, to sue for peace.

From the very beginning of the Battle of Britain (which for the sake of brevity, falls outside the scope of this book) a small force of German bombers conducted small-scale night raids against aircraft factories, dock installations and other valuable targets. These raids achieved little in the way of material damage, but frequently caused loss of life in the residential areas that closely surrounded these targets. The *Luftflotten*'s experience in the day-battle showed that when operating against a determined defence that was equipped with sufficient numbers of high-performance fighter aircraft and pre-warned of their approach by radar, they could expect heavy casualties. However, when flown at night, their bombers were almost invulnerable to interception and, if flown above 12,000 feet (3,660 metres), from illumination by searchlight. Cloudy nights were also much favoured by the bomber crews, since they reduced the searchlight's effectiveness to virtually 'zero'.

Bombing by night did, nevertheless, raise a number of problems for the *Luftwaffe*, the chief amongst which was navigation. Despite being located very close to Great Britain, in terms of flying distance, it was anticipated that some difficulty would be experienced by the crews in identifying their targets on dark nights and in the face of a total blackout. To aid them in their navigation and provide an accurate bomb-release point, the *Luftwaffe*'s Signals Staff developed a series of VHF beam systems that worked with the bomber's *Lorenz*[2] airfield approach receiver. The first, codenamed *Knickebein*, employed two VHF transmitters located in occupied Europe, whose beams could be swung until they intersected over a given target. Although relatively accurate,[3] *Knickebein* was only capable of marking a town or a city, since the spread of its beam at long-range, mitigated against it being employed against smaller, high values targets (gasworks, factories, etc.). By the late summer of 1940, five *Knickebein* transmitters were operational in Europe.

The second system, designated *X-Gerät* (X-System), utilised the approach beam of *Knickebein* which intersected three fine cross beams for automatic bomb release. Tests with the system demonstrated a practical range of 180 miles (290 km) from the transmitter. Unlike *Knickebein*, *X-Gerät* required the fitting of two additional radio receivers and associated aerial masts to the bomber,[4] making *X-Gerät* aircraft readily identifiable to the RAF's intelligence officers. The system also required a high degree of crew training and co-operation to achieve accurate results and, consequently, was only issued to three specialised units: *Kampfgruppe* 100 (KGr 100), III./KG 26 and II./KG 55. *X-Gerät* was introduced to service over Britain by KGr 100 in December 1939.

The final, and most accurate and sophisticated system, *Y-Gerät* (Y-System), could be linked to a bomber's autopilot to provide automatic tracking of the navigation beams. First deployed in September 1940, *Y-Gerät* employed a conventional approach beam, along which the bomber could fly and synchronise to. As the aircraft approached the target area it received separate signals from another ground station, which it re-transmitted to enable a ground controller to calculate the bomber's distance from the transmitter. D/F indications within these signals enabled the controller to keep the bomber in the Approach Beam, and by means of a VHF link to tell the observer when to release his bombs. For its day, the system was very accurate, being capable of hitting a power station sized target at a range of 120 miles (190 km). Like *X-Gerät*, *Y-Gerät* required specially equipped aircraft and properly trained crews and was only operated by a single unit, III./KG 26.

The 'beam bombers' existence and purpose were known to Air Intelligence, who provided ground-based radio counter-measurers (RCM) transmitters under the control of No. 80 Wing, to jam the systems. They were also prized targets for the night-fighter squadrons, some of whom took special measures to identify and destroy them.

The *Luftwaffe*'s reversion to night-bombing began towards the end of the Battle of Britain, on the night of 28/29 August, and for three nights thereafter, 160 aircraft from Sperrle's *Luftflotte* 3 bombed Liverpool, loosing only seven aircraft (a loss rate of barely 1 per cent). On 7 September, the enemy carried out a daylight attack on London and followed it up with a night raid by 318 bombers. With little to fear from the defences, and with the fires from the daylight attacks clearly visible, the *Luftwaffe* bombed the city, more or less unopposed, causing nine conflagrations,[5] nineteen major fires,[6] forty serious fires,[7] and 1,000 smaller fires. In addition to the damage, which was widespread, 430 Londoners lost their lives and a further 1,600 were seriously injured.[8] No enemy aircraft were lost to night interception, or AA fire.

The air defences over this period, comprised six squadrons of Blenheim Ifs and two of Defiants, No. 141 (Biggin Hill and Gatwick) and No. 264 (detachment at Northolt), which were transferred to night-fighting following their poor showing during the day battle, in addition to a number of single-seat Spitfires and Hurricanes that flew night patrols. As Hanbury Brown had predicted:

> The instructions from the ground on the whereabouts of the raiders were so imprecise as to be virtually useless, with the result that although scores of sorties were flown each night, successes were few and largely a matter of luck.[9]

With the fighters being tracked by their Identification, Friend or Foe (IFF) transmissions (codenamed *Pipsqueak*) and the bombers by the CH stations, and both plotted on a board in the Sector Operations Room (Ops Room), the Sector Controller was able to steer the fighters towards their quarry. The *Pipsqueak* system was in turn supported by those searchlight batteries that were fitted with gun-laying (GL) radar, who were able to point their beams in the general direction of the enemy, where, it was hoped, the night-fighters would be able to see them and attack. This 'lash-up' system, although somewhat primitive and crude by later standards, nevertheless, proved to be the correct approach, since a number of the Blenheim R/Os gained contacts. Successes, however, were few and far between, due to the Blenheim's lack of performance and firepower. The searchlights also failed to illuminate the bombers sufficiently for the Blenheim's pilot to see the enemy aircraft, as one pilot, Pilot Officer R.C. Haine of No. 600 Squadron recalled:[10]

> It might seem a simple matter for night fighter crews to see the bombers that had been illuminated by searchlights, but this was not the case. If the raiders came on bright moonlight nights, which was unusual during this time, the beams of the 90 cm searchlights were not visible at heights much above 10,000 feet [3,050 metres]. If the searchlights were actually on the enemy bomber the latter could be seen from some way away, but only if the fighter was beneath the bomber and could see its illuminated underside; if the fighter was higher than the bomber, the latter remained invisible to the fighter pilot. If there was any haze or cloud it tended to diffuse the beams, so that there was no clear intersection to be seen.[11,12]

The pattern of bombing continued throughout the autumn months, with heavy raids[13] on London (35), Birmingham (10), Coventry (2), Wolverhampton (1), Liverpool (4), Southampton (4), Bristol (3), Plymouth (1), Portsmouth (1), Manchester (2) and Sheffield (2), many of which were led by the pathfinders of KGr 100. Of these raids, that against Coventry on the night of 14/15 November was perhaps the worst. Beginning at 1920 hours and ending at 0535 hours the following day, 104 aircraft from *Luftflotte* 2 and 304 from *Luftflotte* 3, led by KGr 100, dropped a heavy load of bombs, including a large proportion of incendiaries, that destroyed much of the commercial, industrial and residential areas of the city, halted war production, and killed 506 people and seriously injured 432. Although the tonnage of bombs was no different from that dropped on London on an average night, the proportional scale of destruction was much greater, as Coventry was a much smaller city.

Devastating though the bombing certainly was, there was some hope for the future. In September, Fighter Command introduced the first versions of the Bristol Beaufighter, equipped with the improved AI Mk.IV, to service with the night-fighter squadrons and by the year's end the first GCI station was commissioned at Durrington. These three factors, together with improved training and AI maintenance, were to lift the defence to a new level and signal the decline of the *Luftflotten* in the West.

July 1940

In an effort to bolster the Airborne Group's work, AMRE brought in more scientists and engineers to reinforce Bowen and his team. Much of the team's efforts had earlier been depleted nursing the Mk.III system through its birth pangs, and by the diversion of personnel to the development of ASV, in the light of the worsening situation in the Atlantic. Help had, however, been provided by RAE, whose staff helped with the more practical aspects of AI systems engineering.

Bowen's disagreements with Rowe and Lewis' appointment as Rowe's deputy, brought matters to a head concerning the relationship between the senior management at Dundee and the Airborne Group. In late July, Bowen was nominated by Tizard to join his technical mission to the US, as its representative on radar matters. Bowen accepted Sir Henry's offer and left AMRE in August to prepare for his trip to America, where he was instrumental in the development of the Radiation Laboratories at the Massachusetts Institute of Technology (MIT) in November 1940.

Bowen's contribution to the evolution of airborne radar in Great Britain, cannot be over emphasised. It was he who nursed the technology from its birth in 1936, to the manufacture, in 1940, of a practical system that could be operated and maintained by service personnel – a very significant achievement, when one considers the technical complexity of packaging a radar set into an aircraft's fuselage. Like the other Government scientists of his day, Bowen never received the public recognition he so richly deserved. Nevertheless, at the war's the Government saw fit to reward him as a Commander of the British Empire (CBE)[14] for his contribution to the air defence of the UK, but by then he had left these shores and settled in Australia.

In fairness to Lewis, his far-sighted decision to divert some of the country's radio industry resources into the radar field, was to pay handsome dividends in the near future and impact on the development of AI throughout the remainder of the war. By this means, EMI was brought into the programme to develop AI Mk.III's replacement, AI Mk.IV, with the placement of a contract on the Company on 17 May 1940. Whilst EMI's contract was being finalised, AMRE's senior management were casting around for a new site, as Bowen relates:

As the work of the airborne group at St Athan drew to a close [it was being taken over by RAE], yet another move was contemplated. From every point of view, the sensible thing for us to do was go back to Bawdsey and Martlesham, where many of us still had roots and where the facilities were vastly superior to anything we had experienced during the first few months of the war; but the bureaucrats did not see it that way. What happened was really decided by events at Dundee, where things reached crisis point. The fiction that Dundee was a good place for radar research could no longer be maintained and another move simply had to be made. In describing these events in his book *One Story of Radar*, Rowe said: '... it is difficult to find good in the Dundee era ....'. If however we were seemingly bankrupt it is not to be supposed that grand work was not done elsewhere'. He was presumably referring to the field work done away from Dundee, even the work at St Athan. He went on to say that it: '... was the last place to choose for radar research and all concerned with selecting it or not opposing the Dundee site, including myself, had learned a lesson not to be forgotten'.[15]

As Bowen agrees, 'this was a generous admission',[16] but had little apparent meaning, as AMRE, with Rowe at the helm, apparently failed to learn their lesson by relocating the Establishment to Worth Matravers on the Dorset coast, a few miles to the east of Swanage and some 60 miles (96 km) from the enemy on the Cherbourg Peninsula. The Swanage site 'consisted of a series of unfurnished wooden huts with no facilities, without heating and surrounded by a sea of wood'.[17] Overall, a location totally inadequate for the housing of scientific staff and for the testing of delicate electronic equipment.

The problem was further compounded by a second decision to move the SD Flight to the nearby civil airfield at Christchurch, near Bournemouth:

To anyone with the least experience of flying, it was totally unsuitable for the work in hand .... Christchurch had a single grass runway, too short for safety. It was surrounded on all sides by domestic residences which prevented any possibility of extending the runway. At the western end was a tennis court which was quickly demolished by a Hurricane which failed to clear the wire surround .... The building available at Christchurch to shelter both the air and ground crews for about twenty aircraft consisted of a 20 × 20 foot [6 × 6 metre] wooden hut. There was no place at all for the testing of new radar equipment going into aircraft. Compared with Bawdsey and Martlesham, it was a ludicrous choice.[18]

Coincident with the move to Swanage, AMRE was transferred from the Air Ministry to the MAP, where it was renamed the Telecommunications Research Establishment (TRE), a title it was to retain until the war's end. In spite of these moves and changes, the replacement design for AI Mk.III somehow managed to move forwards.

Aware that much of the Mk.III system was based on sound engineering principles, the EMI team, led by Alan Blumlein[19] and Eric White, placed their modulator in the centre of the new system. This device, which EMI had patented in December 1939, replicated the circuitry of Lewis' double-transmitter concept (as in AI Mk.IIIB), to dampen the tail of the transmitter pulse by means of clever circuitry that interacted with the transmitter's Micropup valves. By this means, the modulator was able to produce a sharp pre-pulse, which triggered the Micropup valves in the transmitter and shut-down the receiver.[20] It had the additional advantage of employing the Mk.III's transmitter, slightly modified, in AI Mk.IV and saving considerable time and effort in its development.

The aerial system was a replica of that installed in Blenheim L4846 (see Figure 7), with the transmitter and receive aerials connected by means of polyethylene insulated co-axial cables. The new system employed the display unit taken from AI Mk.III, with the elevation display on the left and the azimuth on the right (see Figure 2).

The new system, with Hanbury Brown and Eric White observing and operating, undertook its first flight at Tangmere in an FIU Blenheim between the 30 June and 1 July and proved an immediate success. It did, however, suffer from one problem, namely, instability in the echoes on the CRTs, which made it very difficult for the R/O to read off the elevation and azimuth indications. The system was removed from the Blenheim, modified and flown again from Christchurch, before being returned to Tangmere for a further series of trials between the 17 and 26 July. The modifications cleared the echo instability, enabling FIU to re-test the system and communicate their satisfaction to Fighter Command. The new set was found to have a greater maximum and a lower minimum range than AI Mk.III, as a direct consequence of the use of the vertically polarised aerials. They also confirmed the eradication of the VHF radio interference problem, and the restoration of proper air-to-ground R/T communication.

Figure 7. Aerial elements for AI MkIV.

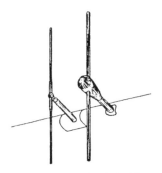

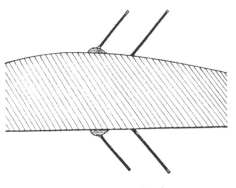

Transmitter Aerial
Folded Dipole with
Director

Starboard Azimuth Aerial
(Part of Aerial System Type 25)

Upper and Lower Unipoles
with Reflectors
(Part of Aerial System Type 25)

Addressing the 11th Meeting of the Interception Committee, Dowding said he was proposing to recommend that development of the Mk.IV equipment be continued with a view to its being fitted as standard in the night-fighter versions of the Bristol Beaufighter. His recommendation was accepted, and from 10 September, the installation of the Mk.IV system in Blenheim fighters was stopped, with all the effort being diverted towards the equipping of Beaufighters.[21] In the meanwhile, the night defence of Great Britain would be in the somewhat shaky hands of the Blenheim!

Throughout the spring and summer of 1940, the inexperienced night-fighter crews spent many hours searching the skies over England in their underpowered machines, for an enemy, which at first, they could not see. Much of their time was, of necessity, spent learning the basic elements of their craft by day and practising it by night. The path to their mastery of that craft was, however, impeded by the poor serviceability and ambiguous indications from the AI and an enemy that frequently flew below 7,500 feet (2,290 metres), which significantly reduced the set's range, and made them difficult to detect. These factors were further complicated by the lack of a decent ground control system.

Getting airborne, flying the course and landing safely were something of a trial, as Flying Officer Roderick 'Rory' Chisholm,[22] who did not find night flying particularly easy, explains on the night of his first operational patrol:

Thus I and my crew found ourselves on the night of August 16th as No. 5 on the programme, with I believed, little prospect of flying. Several patrols were sent off, and then two aircraft were reported as unserviceable for the rest of the night. We were next off. It was getting late and there were only a few more hours of darkness. I thought there would probably be no more flying, and so I lay down and tried to sleep. I must have slept because the ringing of the telephone made me come to with a start. Orders were passed for a patrol on a certain line near Bristol.

This was it. I stumbled out into the darkness. I felt scared but stoical …. I hurried as best I could towards the aircraft, towards the torches of the ground crew who were waiting. We flew for two and a half hours in beautiful weather, came back and landed successfully at the third attempt.

The patrol was a mixture of nervous tension and great exhilaration. Bristol was being blitzed 15,000 feet [2,415 metres] below me. I saw bombs explode and anti-aircraft fire and plenty of fires, but since I was never sure of myself, of the attitude of the aircraft or of my own position, this patrol was of no potential value to our night defence.[23]

The Blenheims flew with a crew of three: pilot, air-gunner and radar operator, with the conditions for the latter in the rear fuselage being very uncomfortable. Although the height of summer, the crews were forced to wear a combination of service issue clothing and items of their own invention to stay warm at the higher altitudes of their patrol. First a pair of pyjamas, over which one thick silk and wool aircrew vest, a roll-neck pullover and serge battledress. To keep the feet warm, two pairs of cast-off ladies silk stockings, followed by two pairs of thick woollen socks. Having got all items properly 'installed' the completed apparel was covered by fur-lined leather trousers and jacket (Goon Skins) and flying-boots.[24]

Whilst the gunner had a properly designed station in the turret, the operator was forced to kneel down on the floor of the fuselage to operate the radar. With nothing to do until commanded to switch-on the set (flash the weapon) and contribute to the patrol, many fell asleep swathed in flying suit and blankets, much to gunner Sergeant Jimmy Rawnsley's disgust[25]:

Then came the coded instruction from the ground to switch on the radar set. 'Flash your weapon' the Controller said. So there was something ahead of us! I stooped down again and looked at the magician. He was still asleep. Reaching out, I prodded him with my flying-boot. 'Hi ... Abracadabra!' I exclaimed. 'It's time to do your stuff.' He stirred and unwound himself, and thrust the blankets into an untidy heap on the floor in front of the seat, and knelt on them. A familiar humming came over the intercom as he switched on .... 'Have you got anything?' John [Cunningham] asked, eager but patient. The only answer the operator made was a series of muffled grunts. 'It it's very indistinct' the operator muttered. 'I don't think'. The muttering died away into incoherence. 'Well keep trying' John said ... no response came from the kneeling figure amidships. I wondered if he was praying. We went blundering on impotently into the darkness, but it was no good: we found nothing at all.[26]

Despite these problems, when the circumstances were favourable, the defence did achieve some success. Following Ashfield's success on the night of 22/23 July, a few more enemy raiders were detected, but none was converted into a successful interception. This spell was broken on the night of 17/18 August, when Blenheim L6741 of No. 29 Squadron, operating out of Tern Hill, and flown by Pilot Officer Richard Rhodes, with Sergeant William 'Sticks' Gregory,[27] as his gunner, was vectored onto an enemy aircraft, flying some 15 miles (24 km) to the south of Chester. At 0228 hours, Rhodes made out the lights of an aircraft, which, upon closing, he identified as an He 111. For the next two hours, Rhodes and Gregory patiently stalked the Heinkel, by which time they had crossed from the west to the east of England, and passed out into the North Sea near Spurn Head. When Rhodes had closed the range to 400 yards (365 metres), he opened fire with the Brownings until the ammunition boxes were empty. Coming abeam of the raider, Rhodes gave Gregory the opportunity to add the fire from his Vickers to the Heinkel's discomfort. Severely damaged by the Blenheim's guns, the Heinkel circled slowly in the night sky, whilst gradually losing height, until it landed gently on the sea and sank.

Rhodes' success was followed a few nights later by another 29 Squadron crew, comprising Pilot Officer Bob Braham[28] as pilot, Sergeant Albert Wilsdon, gunner, and Aircraftsman 2nd Class (AC2) Norman Jacobson operating the AI set. On the night of 24/25th, the crew, in Blenheim L1463, were on patrol on a dark, but clear night at 10,000 feet (3,050 metres) over the Humber Estuary, when they were vectored onto a raider that was illuminated by search-lights and travelling north at 160 knots[29] (180 mph or 290 km/hr). Braham gradually closed the range, but in his eagerness to engage the enemy he opened fire at too great a range (a common mistake in those times), and before he had positively identified the target. Seeing his tracer falling short, he closed further and identified the aircraft as a Dornier Do 17. Unbeknown to Braham and Wilsdon, a Hurricane night-fighter had also attached itself to the Dornier, before diving on it and causing Wilsdon to let-off a burst from the Blenheim's turret. Shouting over the intercom for Wilsdon to cease firing, Braham was now rapidly overhauling his opponent:

I was well within range of the Dornier which the searchlights were still industriously illuminating for us, but unfortunately my closing speed was now so fast that I only had time for a very short burst, as I vainly tried to slow my aircraft down to stay behind him. However, I had the satisfaction of seeing him blossom smoke and could see sparks as my bullets bit into him. As I couldn't keep behind him, I gradually overtook him on his starboard side as closely as possible to give Wilsdon a crack with his gun. This would enable him to fire bursts into the enemy's most vulnerable area, the cockpit. The German gunners must either have been blind, or injured, for I received no return fire as we slowly flew past. Wilsdon was firing long bursts from his gun. I saw the flames at the same time as I heard him whoop with joy 'Got him'. The Dornier slowly peeled off to port, its dive steepening as it headed for the ground with flames trailing astern. We circled, waiting for the crash on the ground, and in a highly elated mood headed back for base.[30]

Braham and his crew were subsequently awarded one Do 17Z 'destroyed'. Sadly, Rhodes and Jacobson did not long survive their combats, being posted as 'missing', along with their gunner, Sergeant Gouldstone, on the night of 25/26 August. They are believed to have been in action with an enemy aircraft over Wainfleet in Blenheim L1330, before crashing into the sea. Rhodes was aged only nineteen and Jacobson eighteen, when they were lost.

The beginning of September, was to see another squadron, No. 25, open its account, when, on the night of the 4/5th, New Zealander, Pilot Officer Michael Herrick,[31] and his R/O, Sergeant John Pugh,[32] intercepted an He 111H-3 of *Stab* I./KG 1[33] that was illuminated by searchlights, whilst bombing Tilbury, Essex. Herrick's fire disabled the Heinkel's port engine, forcing the crew to jettison their bomb-load and abandon the aircraft at very low level, before it crashed at Rendlesham, near Eyke, Suffolk. The loss of this particular aircraft must have been keenly felt by KG 1, since it was commanded by the *Gruppe's Kommandeur*, Major Maier, who, along with *Oberleutnant*, Graf von Rittberg, *Oberfeldwebel*[34] Stockert and *Unteroffizier* Bendig, was amongst those killed. The only survivor, *Oberleutnant* Biebrach, became a prisoner-of-war (PoW).

Herrick's second victory came later that month, on the night of 14/15 October, when, with Pilot Officer Archibald Brown[35] as R/O, he intercepted an He 111H-4 of I./KG 4, flown by *Oberleutnant* Kell, that was illuminated by searchlights over London at a height of 10,000 feet (3,050 metres), and shot it down at Down Hall, to the north-east of Harlow, Essex. Kell and another member of the crew, *Feldwebel* Hobe, escaped by parachute to become PoWs, but the remainder of the crew, comprising *Unteroffiziers* Muller-Wernscheid and Topfer, were killed.

The third Blenheim success that month fell to an Auxiliary squadron, No. 600 (City of London), on the night of the 15/16 October, when an Australian, Flight Lieutenant Charles Pritchard,[36] with Pilot Officer Henry Jacobs as gunner, destroyed a Junkers Ju 88, that crashed into the sea off Bexhill, Sussex, shortly after midnight.

The British air defences at the beginning of the Battle of Britain, in August 1940, comprised the coastal early warning CH network, backed-up by the visual plotters of the Observer Corps (later the Royal Observer Corps – ROC) These two networks provided an adequate supply of information to the various sector controllers, who in turn managed the day-battle. However, as has already been mentioned, during the periods of darkness, interception was a near impossibility. This situation was well understood by Dowding, but until such time as a GCI network was established and reliable AI sets were available in quantity, he was virtually powerless to prevent the enemy from roaming the night skies more or less, unmolested. Such was the lack of an effective night-fighter, that during September, not one enemy aircraft was destroyed by AI interception. This state of affairs prompted the Minister of Aircraft Production, Lord Beaverbrook, to suggest that someone in the Air Ministry be tasked with an investigation of the night defences. The CAS, MRAF Sir Cyril Newall,[37] agreed to the formation of a committee, under the chairmanship of MRAF Sir John Salmond,[38] to advise on the 'preparation of night-fighters'[39] and their operations.

The 'Night Defence Committee', otherwise known as the 'Salmond Committee', which included such notable officers as Air Chief Marshal Sir Wilfred Freeman,[40] Air Marshal Sir William Sholto Douglas,[41] Air Marshal Sir Arthur Tedder[42] and Air Chief Marshal Sir Philip Joubert de la Ferte,[43] met for the first time during September to discuss the subject and report to the CAS. The report was discussed at a meeting on 1 October, which was attended by the CAS and the Secretary of State for Air, Sir Archibold Sinclair, and recommended several courses of action, namely:

- The acceleration of the production of AI Mk.IV equipped Beaufighters.
- The control of night-interceptions from coastal CH stations.
- The production of radio aids for night-fighters: AI beacons and Lorenz blind landing aids.
- The introduction of specialised night-fighter OTUs.[44]

In relation to the acceleration of Beaufighter production, there was little that could be done, that had not already been done, as even then the first machines were entering squadron service. No doubt under pressure from the Committee to bolster the night-fighter force, Dowding reluctantly transferred three Hurricane squadrons, No. 85 (at Church Fenton), No. 87 (Exeter, Hullavington and Bibury) and No. 151 (Digby and Wittering) to the so-called 'Cat's Eye' night-fighter role during October. The Hurricane Mk.I was ideally suited to night-fighting, where its stout construction, ease of handling on the ground and in the air, good all-round performance and stability as a gun-platform, was greatly appreciated.

By June 1940, it had been clearly established at AMRE and with Fighter Command's Stanmore Research Station and its Signals Staff, that the design of a new radar set based on 1½ metre technology, incorporating the PPI display taken from CHL, was the right approach for the direction of night-fighters. A specification, for what was now formally called 'GCI', was agreed sometime in June or July. The principal requirements, stipulated:[45]

- A range of 50 miles (80 km), with coverage upwards of 5,000 feet (1,525 metres).
- The measurement of height and bearing.
- Freedom from vertical gaps.
- An IFF capability.
- The provision of a PPI display giving continuous indications of the position, speed and direction of flight of the fighter and its target.

The matter was discussed and approved at the next meeting of the Air Interception Committee on 18 July, mainly thanks to the persuasive powers of Wing Commander Raymond Hart[46] from HQ Fighter Command, and thereafter issued to AMRE. It is strange, therefore, that the Salmond Committee were not aware that the specification for GCI had been released and the prototype set was literally being assembled during September. This knowledge should, therefore, have made their requests for further trials at coastal CH stations redundant.[47]

AI beacons were radar transponders, operating in the frequency AI band 188–198 MHz, which could be interrogated by the night-fighter's AI set and respond with a coded sequence that would identify the airfield on which they were located. The beacon response was shown on the AI's azimuth display and was selected by means of a rotary switch on the set.

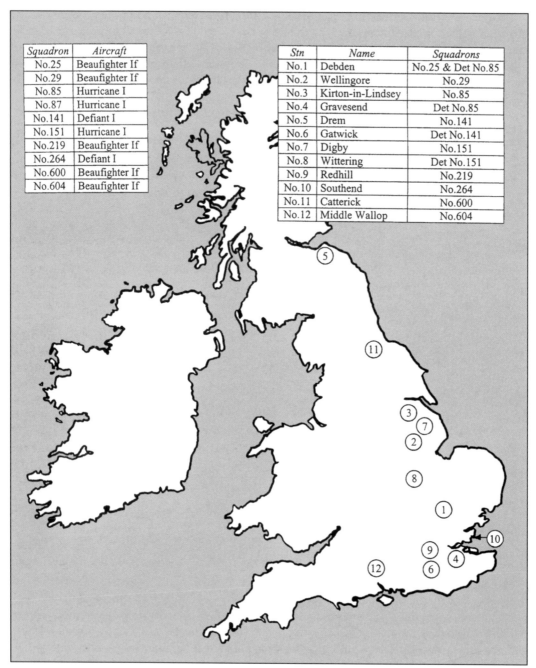

Squadron	Aircraft
No.25	Beaufighter If
No.29	Beaufighter If
No.85	Hurricane I
No.87	Hurricane I
No.141	Defiant I
No.151	Hurricane I
No.219	Beaufighter If
No.264	Defiant I
No.600	Beaufighter If
No.604	Beaufighter If

Stn	Name	Squadrons
No.1	Debden	No.25 & Det No.85
No.2	Wellingore	No.29
No.3	Kirton-in-Lindsey	No.85
No.4	Gravesend	Det No.85
No.5	Drem	No.141
No.6	Gatwick	Det No.141
No.7	Digby	No.151
No.8	Wittering	Det No.151
No.9	Redhill	No.219
No.10	Southend	No.264
No.11	Catterick	No.600
No.12	Middle Wallop	No.604

Figure 8.   Night-fighter Squadrons at October 1940. The diagram reflects the significant changes that Dowding had achieved since August 1940, by the conversion of a number of AI fighter squadrons to the Beaufighter and the introduction of five cat's eyes squadrons equipped with the Hurricane or Defiant.

Fighter Command's insistence that night-fighter crews be able to clearly identify friendly aircraft from those of the enemy, required the new set to be able to respond to the coded pulses of IFF Mk.I and Mk.II. This required the night-fighter to carry a small interrogator which was triggered by the R/O to broaden the target echo of friendly aircraft on the azimuth and elevation displays. Mk.II IFF was installed on all fighter, bomber and coastal aircraft beginning in July 1940 and ending in February the following year.[48]

Salmond's suggestion concerning the establishment of night-fighter OTUs was agreed by the Air Staff, beginning with No. 54 (Night) OTU, which was formed at Church Fenton, near York, on 25 November 1940, with Blenheims, Douglas (DB-7) Havocs and Defiants. A number of the Blenheims were later replaced by Beaufighters, whilst the Defiant training was passed to No. 60 (Night) OTU in June 1941. Thereafter, No. 54 taught twin-engined crews, leaving No. 60 to concentrate on single-seat night-fighters.[49]

On 12 August, in preparation for the delivery of FIU's first Beaufighter, Flying Officer Ashfield flew to Bristol's factory at Filton to receive instruction on the handling of the Beaufighter, with the first machine, R2059, a Mk.If, being delivered to Tangmere the same day. Following a raid on Tangmere on 16 August, FIU moved to the civil airfield at Shoreham, Sussex. FIU's first operational patrol with R2059, was flown by Flying Officer Ashfield[50] on the night of 5/6 September without any success, due to equipment failure. By the 20 September, FIU was in possession of three Beaufighters, R2055, R2059, and R2078, minus a fourth which was wrecked at Tangmere on the 9 September, when its canopy detached and it crashed whilst executing an emergency landing. The three serviceable Beaufighters were despatched to St Athan for the installation of AI Mk.IV, with R2055 being the first equipped, followed by R2078 and R2059. Testing of AI Mk.IV in the Beaufighter continued throughout the rest of September and into October.[51,52]

On 2 September, four Blenheim fighter squadrons, No. 25 (at North Weald), No. 29 (Wellingore), No. 219 (Catterick) and No. 604 (Middle Wallop), received their first Beaufighter If, followed six days later by a fifth squadron, No. 600 (Hornchurch). The sixth Blenheim squadron, No. 23, ceased night-fighter operations in October, before transitioning to night-intruder operations, minus AI, with Beaufighters and Havocs. It was these five units that were to bear the brunt of the Winter Blitz, which followed the end of the Battle of Britain and continued to the end of May 1941.

Armed with four 20 mm cannon and six 0.303 inch (7.69 mm) Browning machine-guns, the Beaufighter was a vast improvement on the Blenheim. Although down in performance,[53] compared to that which was promised earlier in the year, the machine was, nevertheless, greeted with a great deal of enthusiasm by the air crews. One of whom, Pilot Officer Jeremy Howard-Williams,[54] recorded his impressions:

The cockpit is well laid out as the Blenheim's was bad, and everything lies under the hand which has to use it. On take-off, the left hand opens the throttles (right throttle a little in advance of the left, to counteract swing) and merely has to move forward a few inches to raise the undercarriage lever and can then immediately return to the throttles to make sure they don't creep back at this critical stage of a flight. Switches are thoughtfully arranged so that they are split into small groups; this means that the pilot can identify each switch at night by feel. The trim controls are easily reached by the right hand, and each one works in the natural direction of the control. Perhaps the biggest bonus of all is the excellent view both of outside and inside [an important aspect, particularly when a lot of instrument flying is done]. The Blenheim's instruments, apart from the standard blind-flying panel, are not so much grouped in the cockpit as thrown together haphazardly. But someone has studied the placing of the ancillary instruments and controls on the Beau, so that the engines dials are together and easily checked, minor gauges not so frequently required are a little out of the way but still visible when wanted and close by their appropriate controls when appropriate.

The Blenheim cockpit has a multitude of small windows separated by metal framework like a greenhouse. These panels have a disconcerting tendency at night to pick up reflections from the cockpit coming from outside the aircraft. The Beaufighter's windshield, which suffered from a similar disadvantage in its earlier version, is now [February 1942] one large perspex moulding surrounding a bullet-proof centre portion. There are few metal members to block the view, and clear-vision panels at each side and overhead can be opened to give draught-free access to the outside. Cockpit lighting is good. The aircraft has a good turn of speed – more than 300 mph [483 km/hr] but is heavy and not particularly manoeuvrable, a quality it does not necessarily need for night fighting. Its worst point, from both the pilot's and navigator's angle, is the lack of heating. ... The heater can only produce a small volume of hot air close by the [pilot's] right heel, ... whilst the RO's heater merely sends a dribble of lukewarm air to dissipate among the myriad cold draughts whistling down the fuselage. The only other drawback is a certain instability in the pitching plane which manifests itself as a tendency to tighten up in a turn, so that positive forward pressure has to be applied to the control column to maintain a steady turn.[55]

Unlike the Blenheim, the Beaufighter was not equipped with a gun turret, as gunner, Sergeant Rawnsley discovered:

There she stood, sturdy, powerful, fearsome, surrounded by an enthusiastic crowd. ... Pilots, engineers, fitters, riggers, armourers and signals mechanics were in attendance. ... For the gunners, however, there was a shattering disappointment. Where the turret should have been there was nothing but a plain, moulded dome of perspex. But where were the four free guns in the turret in the back that could fire forwards and upwards into the belly of an enemy bomber? There was not even a single free gun with which we could foster our delusion of usefulness.[56]

Faced with the prospect of a transfer to bomber or coastal squadrons, where experienced gunners were in short supply, the Wireless Operator/Air Gunners (WOp/AGs) of Fighter Command were offered the opportunity of retraining as R/Os. Many took up this option and went on to forge successful careers on Beaufighters and later on Mosquitoes, whilst others, who, for a variety of reasons, failed to make the grade, or preferred a posting, were siphoned-off to the other Commands, where many subsequently perished in the night skies over Germany and in the Atlantic.

The first successful night interception with AI Mk.IV occurred shortly after the infamous raid on Coventry, on the night of 19/20 November, during a raid by 356 aircraft from *Luftflotten* 2 and 3, on Birmingham. Led by the 'firelighters' of KGr 100 and II./KG 55, the enemy struck at targets 'associated with the aircraft and aero-engine industry' ... 'that were clearly visible to the bomber crews'.[57] One of *Luftflotte* 3's aircraft despatched that night, a Ju 88A-5 of 3./KG 54, took-off from its airfield at Evreux at 2345 hours, with instructions to bomb factories in the Birmingham area. The crew comprised, *Unteroffizer* Sondermeier, pilot, *Gefreiter*[58] Seuss, radio operator, *Unteroffizer* Liebermann, observer, and *Flieger*[59] May, gunner.

Waiting for the raid was Beaufighter If, R2098, of No. 604 Squadron, crewed by Flight Lieutenant John Cunningham[60] and R/O, Sergeant John Philipson,[61] patrolling the night sky over Oxfordshire. Cunningham and Philipson had earlier been unlucky when they established contact with a number of raiders flying at 18,000 feet (5,485 metres), but failed to acquire them visually, due to cloud cover. Returning to their patrol line in the hope of better luck, the pair made contact with an aircraft in the Brize Norton area and steered towards it. Under Philipson's guidance, Cunningham made visual contact with the enemy and confirmed it as a Ju 88. Closing to 200 yards (180 metres), he opened fire with a six to seven second burst with all six machine-guns and the four cannon. The effect on the Junkers was devastating; flashes from the exploding ammunition enclosed the aircraft, mortally wounding observer Liebermann and starting a fire in the starboard engine cowling. The fire quickly took hold, forcing pilot Sondermeier to close the throttle on the damaged engine, feather the propeller, which reduced the fire, and jettison the bomb load of two 250 kg (550 lb) and four 50 kg (110 lb) bombs. Turning south for home, the aircraft, by now down to 6,000 feet (1,830 metres), and with the starboard engine well alight once again, was abandoned by the crew to crash into soft farmland, near East Wittering, Sussex, at 0135 hours. Sondermeier and Seuss escaped by parachute to become PoWs, but May drifted out to sea, where he perished in the icy waters of the English Channel.

Cunningham did not claim the Junkers as destroyed, reporting the combat as 'inconclusive', until the interrogation of the crew confirmed they were fired on from the air shortly after reaching Birmingham, which clearly identified his Beaufighter as the culprit. Shortly after Cunningham and Philipson were formally credited with the victory – theirs and Fighter Command's first in the Beaufighter.

John Cunningham's victory in November 1940, marked the nadir in Fighter Command's fortunes in its battle against the night-bomber. Although the fight was by no means over, nor won, the Command's performance began a steady improvement from December, with the introduction of the first GCI stations, increased Beaufighter deliveries and the creation of further cat's-eye squadrons. Improvements in night-flying training and radar maintenance, together served to bring about a gradual strengthening of the defences and a consequent increase in the numbers of enemy aircraft destroyed.

CONCLUSIONS

Although John Cunningham's victory on the night of 19/20 November, may be regarded as the end of the low point in Fighter Command's fortunes in the night battle, the elements that were the cause of its poor performance during the autumn of 1940 had been identified, namely:

- The lack of a GCI network for the inland tracking of enemy bombers.
- A shortage of AI Beaufighters.
- The lack of sufficient cat's-eyes squadrons as an interim solution pending the availability of AI-equipped fighters.
- The lack of proper night-flying training and radar equipment maintenance.

From December 1940 onwards the correction of these faults would bring about a gradual strengthening of the night defences and a consequent increase in the number of enemy aircraft destroyed.

Notes

1. Instruction to the *Luftwaffe* quoted in Alfred Price [2], *Blitz on Britain, 1939–1945* (Ian Allen, 1977), p. 37.
2. The *Lorenz* beam approach system, introduced into *Luftwaffe* service in 1933, was also sold by the Lorenz Company to airlines in America and the RAF, where it was known as 'Standard Beam Approach' (SBA) – Ken Wakefield, *Pfadfinder* (Tempus Publishing, 1999), p. 5.
3. At its maximum range of 150–180 miles (240–290 km), *Knickebein's* beam was 1 mile (1.6 km) in width.
4. The X-Gerät system was fitted to Heinkel He 111H-1x, H-2x and H-3x aircraft and was readily identifiable by the three masts on the aircraft's upper fuselage.
5. A conflagration was defined as a fire that required more than 100 pumps to bring it under control.
6. A major fire was defined as one requiring more than thirty pumps.
7. A serious fire was one that required between eleven and thirteen pumps.
8. Alfred Price [2], *op cit*, pp. 92 and 93.
9. *Ibid, op cit*, p. 95.
10. *Ibid*, pp. 95 and 96.
11. *Ibid*, p. 97.
12. Generally, an aircraft was tracked by two or more searchlight beams.
13. A heavy raid was defined by the *Luftwaffe* as one in which more than 100 metric tonnes (98½ tons) of bombs were dropped on the target.
14. The CBE in 1945 was a significant award for a public servant.
15. E.G. Bowen, *op cit*, pp. 136 and 137.
16. *Ibid*, p. 137.
17. *Ibid*, p. 137.
18. *Ibid*, pp. 137 and 138.
19. A.D. Blumlein was an engineer of considerable talent and accomplishment, who, by the time he was inducted into the radar secret, held patents for co-axial submarine cable systems, stereo gramophones, radio aerials, television broadcasting, cathode ray tubes and power systems.
20. E.B. Cooke-Yarborough, in corrections of the author's first draft.
21. *The Signals History, Volume 5*, p. 128.
22. Later Air Commodore R.A. Chisholm, CBE, DSO, DFC*, and CO of FIU in 1943.
23. Roderick Chisholm, *Cover of Darkness*, (Morley: Elmfield Press, 1953), pp. 40 and 41.
24. C.F. Rawnsley and R. Wright, *Night Fighter* (Morley: Elmfield Press, 1976), p. 37.
25. Later Squadron Leader C.F. Rawnsley, DSO, DFC, DFM, of 604 Squadron, who with his pilot, Group Captain John Cunningham, DSO**, DFC*, went on to forge one of the most successful night-fighter teams of the war.
26. C.F. Rawnsley and R. Wright, *op cit*, pp. 41 and 42.
27. Later Wing Commander W.J. Gregory, DSO, DFC, DFM, whose partnership with Wing Commander Bob Braham, created another very successful night-fighter team. Gregory acquired the nickname 'sticks' because he played the drums in a pre-war band.
28. Later Group Captain J.R.D. Braham, DSO, DFC, AFC, who later commanded No. 141 Squadron on intruder patrols over Germany and became one of the highest scoring Allied night-pilots of the war.
29. The speed quoted in nautical mph, where 1 knot approximately equals 1.1 mph.
30. Bob Braham, *Scramble* (London: William Kimber, 1985), pp. 48 and 49.
31. Later Flight Lieutenant M.J. Herrick, DFC*, who commanded No. 15 Squadron in March 1943, and was a flight commander with No. 302 (Polish) Squadron. He was killed in action on 16 June 1944, when flying in company with Wing Commander Braham.
32. Later Flight Lieutenant J.S. Pugh, DFC.
33. Each *Luftwaffe Gruppe* had a staff, or *Stab* section, that contained the senior officers of the *Gruppe*.
34. A *Luftwaffe* rank broadly equating to an RAF Flight Sergeant.
35. Later Flight Lieutenant A.W. Brown.
36. Later Wing Commander C.A. Pritchard, DFC, and the CO of 600 Squadron in December 1940.
37. MRAF Sir Cyril Newall, GCB, CMG, CBE, AM, was CAS from September 1937 to October 1940.

38. MRAF Sir John Salmond was CAS from January 1930 to May 1933.
39. *The Signals History, Volume 5*, pp. 128 and 129.
40. Air Chief Marshal Sir William Freeman, KCB, DSO, MC, replaced Dowding as Air Member for Research & Development and would later be instrumental in the development of the de Havilland Mosquito.
41. Air Marshal Sir William Sholto Douglas, KCB, MC, DFC, was destined to replace Dowding in November 1940, as the AOC-in-C Fighter Command.
42. Air Marshal Sir Arthur Tedder, GCB, was the Director General of Research & Development at the MAP, and technically subordinated to Beaverbrook.
43. Air Chief Marshal Sir Philip Joubert de la Ferte, KCB, was ACAS (Air Operations).
44. *The Signals History, Volume 5*, p. 129.
45. *The Signals History, Volume 5*, p. 184.
46. Later Air Marshal Sir Raymond Hart.
47. *The Signals History, Volume 5*, p. 185.
48. Ian White [2], *The Origins & Development of Allied IFF During World War Two* (Bournemouth University: September 1998), p. 8.
49. Sturtivant, Hamlin and Halley, *Royal Air Force Flying Training & Support Units* (Tunbridge Wells: Air Britain [Historians] Ltd), p. 241.
50. Flying Officer Ashfield was promoted to acting Flight Lieutenant on the 8 September 1940.
51. The Operational Record Book of the FIU, pp. 6–9, filed as AIR9/27 in the National Archives. Hereafter, the *FIU ORB*.
52. It is also possible that the Beaufighter prototype, R2052, was delivered to FIU in September 1940 for testing, as a reference to the 'Beaufighter prototype' is recorded for the 28 September. *FIU ORB*, p. 9 refers.
53. Because they were unable to employ the Hercules VI radial engine of 1,675 hp, Bristol's design team were forced to substitute the lower powered Hercules Mk.III of 1,270 hp. With these engines the prototype, R2052, demonstrated a maximum speed of 335 mph (536 km/hr) at 16,800 feet (5,120 metres),whilst the fully equipped R2054 achieved only 309 mph (495 km/hr).
54. Later Squadron Leader J. Howard-Williams, DFC, who served with 604 Squadron and FIU, flying Blenheims, Beaufighters and Mosquitoes.
55. Jeremy Howard-Williams [1], *Night Intruder* (David & Charles, 1976), pp. 27 and 28.
56. C.F. Rawnsley and R.Wright, *Night Fighter* (Morley: Elmfield Books, 1976), pp. 46–8.
57. Winston Ramsey [2], *The Blitz Then & Now, Volume 2* (London: Battle of Britain Prints, 1988), p. 282.
58. A rank broadly equating to an RAF AC1.
59. A rank broadly equating to an RAF AC2.
60. Later Group Captain John Cunningham, DSO**, DFC*, who commanded Nos 604 and 85 Squadrons during the war and headed the de Havilland Company's Flight Test Department as Chief Test Pilot and subsequently, as a director.
61. Later Warrant Officer J. Philipson, DFC, killed in a flying accident on the 5 January 1943.

CHAPTER SIX

The Winter Blitz is Ended

December 1940–May 1941

During the long nights of December 1940, the *Luftwaffe* stepped up the pace of its bombing campaign and extended it to include raiding over the whole of the British Isles. During that month alone, the long-range bomber force flew 3,979 sorties and sixty-six long-range night intruder patrols. As might be gauged from a force of this size, there was considerable loss of life amongst the civilian population, particularly in their homes and at their places of work. Many public buildings, and the utilities (gas, water, electricity, sewage and telephones) that sustained them, were destroyed or seriously damaged. Road, rail and the inland waterway networks frequently suffered dislocation, or interruption, which reduced war production and the distribution of food and essential supplies. The situation might have been far worse, had it not been for the poor weather conditions which precluded flying during the third week of the month, and an unofficial truce covering the three nights over Christmas, when neither Air Force ventured out.

In response, the defence offered little in the way of opposition to the night raiders, with German records showing that accidents of one type or another accounted for the majority of the fifty-three long-range bombers lost during December.[1] In reply, the night defences claimed the destruction of only four enemy aircraft, three by visual means,[2] and one to the peculiar Long Aerial Mine (LAM)[3] carrying Harrow bombers of No. 93 Squadron. The guns claimed a further ten, for the expenditure of 70,000 rounds of heavy AA ammunition. Fighter Command lost five night-fighters in the same period.

In December, the first hand-built GCI station was tested and commissioned at Durrington, near Worthing, Sussex, and was quickly followed by a further five by the end of January, to provide an inland belt extending from the Humber Estuary to the west of Portland in Dorset. Improved deliveries of AI Mk.IV equipped Beaufighters provided sufficient aircraft to convert five of the Blenheim squadrons, whilst additional squadrons were formed from Defiant and Hurricane day-fighters in the cat's-eyes night-fighter role. With the *Luftwaffe* wholly converted to the night-bombing role and thus providing the day-fighters with fewer targets, Fighter Command, under the leadership of its new AOC-in-C, Air Marshal Sir William Sholto Douglas, went over to the offensive towards the end of the old year. In line with this thinking, and needing some way of causing the enemy pain, No. 23 Squadron was stood down from night-fighting and transferred its Blenheims, and later Douglas Havocs, to intruder patrols over occupied Europe.

By January 1941, the first twin-engined night-fighter OTU, No. 54 at Church Fenton, equipped with AI Mk.III Blenheims, began turning out crews for the expanding Beaufighter force. In line with improved flying training and the introduction of a variety of ground radio equipments, including radar, the Air Staff initiated an expansion of the technical trades within the RAF. Those recruits destined to serve as R/Os and requiring a higher level of academic attainment than for other RAF trades, were filtered-out of general trade training at No. 2 Signals School at Yatesbury, Wiltshire. At Yatesbury they were given a grounding in radar and basic instruction on AI, before being posted to a squadron, where they arrived having never been in an aeroplane. This arrangement proved inadequate to fulfil the needs of an expanding night-fighter force; consequently, a second radio school, No. 3[4] was established at Prestwick in December 1940, for the sole purpose of training R/Os and giving them flying experience, before posting to No. 54 OTU to team-up with a pilot and undergo further training to qualify as a night-fighter crew.

From January 1941, onwards, the combined effect of the increasing numbers of Beaufighters, improved ground control, greater aircrew efficiency and improved equipment serviceability, began to take a toll on the enemy, but did not reduce its effectiveness as a bomber force. The raids against Britain's industrial targets, its docks and the civilian population, continued unabated throughout the winter and spring of 1941, and on into the early summer, when relief, in the form of *Luftwaffe* redeployments in preparation for the invasion of Soviet Russia, brought the Winter Blitz to an end.

On 25 November 1940, Air Chief Marshal Sir Hugh Dowding stood down as the AOC-in-C Fighter Command in favour of Air Marshal Sir William Sholto Douglas. Although the victor of the Battle of Britain, Dowding suffered criticism from certain quarters of the Air Staff for his handling of the night battle, and when Air Chief Marshal Sir Charles Portal replaced MRAF Sir Cyril Newall as CAS in October 1940, Dowding's time at Fighter Command was limited. The new AOC-in-C, 'a man who was more happy in attack than defence',[5] brought a more aggressive stance to his new job. With abundant numbers of under-employed day-fighters at his disposal, he sacked his No. 11 Group commander, Air Vice Marshal Sir Keith Park,[6] replaced him with his arch critic, Air Vice Marshal Trafford Leigh-Mallory,[7] and went over to the offensive.

In relation to night defence, Sholto Douglas' first action as AOC-in-C was to tackle the twin problems of confidence in AI interceptions and the serviceability of the equipment. A note written by the Director of Air Tactics on 26 November 1940, highlights these weaknesses:

> Although the AI operators at FIU are specially selected, those in units earmarked for night-fighting are not. It is of the greatest importance that the AI operator should be keen, intelligent and of a patient and painstaking disposition. He is, after all, the brains of the aircraft until the moment that the pilot actually sees the silhouette of the enemy aeroplane and opens fire. But for the present, insufficient attention has been given to the carrying out of interceptions in daylight. This is essential in order to convince the pilots and AI operators that AI really works, and to allow them to see that the blips on the tubes correspond to the movements of the target aeroplane.[8]

and with reference to servicing:

> It is evident that in certain aircraft the AI has never been properly calibrated. For example, during a daylight test it was found that two Beaufighters showed the target aircraft as being dead ahead according to the tubes, when it was actually some 40 degrees on either bow. Lack of knowledge of a fault of this nature would account for

several failures to interception at night, when as the attacking aircraft closed range, the target appeared to suddenly flick away and disappear from the tubes. Apparently no testing gear, as yet, has been provided to discover faults such as those described, or to check AI sets for functioning immediately before flight.[9]

This state of affairs was not missed on the politicians, for whom the night defence of Britain was a pressing issue. Commenting on the Director of Air Tactics' report, the Secretary of State for Air, Sir Archibald Sinclair,[10] proposed the introduction of civilian Scientific Officers to liaise with the squadrons to rectify the problem. In turn, the AOC-in-C recommended that each AI night-fighter squadron, five at December 1940 and growing, should have a Scientific Officer posted-in to smooth the introduction of the new equipment.[11] As sensible as this suggestion might appear, AMRE had insufficient staff to provide liaison officers for each unit. However, and perhaps unbeknown to these senior officers, action had already been taken to expand the establishment of each night-fighter squadron to include a Special Signals Section. This section, under the command of a rapidly trained reserve officer,[12] was (eventually) provided with test equipment, documentation and wireless mechanics, to service and maintain the AI equipment. The first Special Signals Sections were created in the autumn of 1940 and became a permanent feature of every night-fighter squadron to the war's end and beyond.

Prior to his departure from Fighter Command, Dowding, through his staff and the Command's wireless monitoring service (the Y-Service), had identified KGr 100 as the principal user of the beam-bombing systems (*X and Y-Gerät*). It was established that the pathfinders frequently approached the coast at dusk, to position themselves over the target. If these aircraft could be intercepted and destroyed, the accuracy of the following raid would, therefore, be disrupted and its bombing accuracy reduced. Thanks to the efforts of the Y-Service, who were able to detect the bearing of the *Y-Gerät*'s approach beam and provide D/F fixes on the radio transmissions from KGr 100's aircraft, the defence was forewarned of their approach and the route they would take. However, when placing fighters across their known route and searching for them against the afterglow of the setting sun, the defence failed to make any impact on the *Gruppe*, nor stem the assault, until Beaufighters were introduced in serious numbers.

In his first Night Interception Report as AOC-in-C, dated 8 December, Douglas addressed the need to increase the number of night interceptions and the overall number of night-fighters. Following an approach by General Sir Frederick Pile, the General Officer Commanding (GOC) Anti-Aircraft Command, Douglas agreed to the relocation of searchlights and their GL sets, to work in groups of three, to track enemy aircraft on clear nights and provide indications of their route to the defending fighters. Much of this work, which was aimed primarily at the cat's-eyes force, was rendered obsolete in the New Year, with the availability of GCI stations in the South-East (of which, more later).

Douglas' efforts to increase the number of night-fighter squadrons and improve their airfields had a modest effect. By early December, Fighter Command's night-fighter defences comprised five AI squadrons operating a mixture of Blenheims and Beaufighters, and five of cat's-eyes flying Hurricanes and Defiants. His overall objective was to increase these to twenty squadrons, with at least twelve being equipped with AI Beaufighters and/or Havocs. With the prospect of few, if any, more fighters, Douglas reshuffled the 'pack' to make best use of his resources. No. 87 Squadron operating Hurricane flights from Colerne and Charmy Down, was brought together at Charmy Down, Wilts, with a detachment based

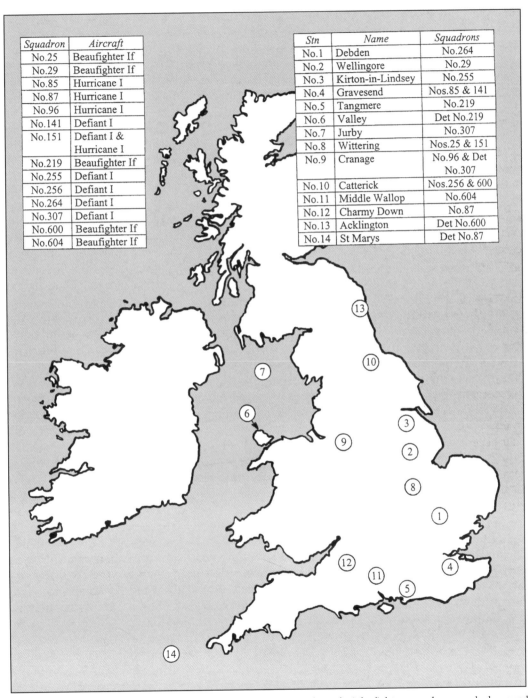

Squadron	Aircraft
No.25	Beaufighter If
No.29	Beaufighter If
No.85	Hurricane I
No.87	Hurricane I
No.96	Hurricane I
No.141	Defiant I
No.151	Defiant I & Hurricane I
No.219	Beaufighter If
No.255	Defiant I
No.256	Defiant I
No.264	Defiant I
No.307	Defiant I
No.600	Beaufighter If
No.604	Beaufighter If

Stn	Name	Squadrons
No.1	Debden	No.264
No.2	Wellingore	No.29
No.3	Kirton-in-Lindsey	No.255
No.4	Gravesend	Nos.85 & 141
No.5	Tangmere	No.219
No.6	Valley	Det No.219
No.7	Jurby	No.307
No.8	Wittering	Nos.25 & 151
No.9	Cranage	No.96 & Det No.307
No.10	Catterick	Nos.256 & 600
No.11	Middle Wallop	No.604
No.12	Charmy Down	No.87
No.13	Acklington	Det No.600
No.14	St Marys	Det No.87

Figure 9. Night-fighter Squadrons at December 1940. The number of night-fighter squadrons peaked around December 1940, by which time Dowding had been replaced by Sholto Douglas and the Blenheim had more or less been totally replaced by the Beaufighter. Note how the defence had been strengthened in the west to counter raids by *Lufflotte* 3 against the West Midlands, Northern Ireland and Glasgow Clydeside. Further, significantly more than half the night defence comprised cat's eyes fighters.

at St Mary's. A second Hurricane squadron, No. 85, based at Debden, Essex, and Kirton-in-Lindsay, was moved south to operate alongside the Defiant equipped, No. 141 Squadron, at Gravesend, Kent, with 264's Defiants also moving from Debden to Gravesend. The Defiants of No. 151 Squadron[13] based at Digby, Lincs, moved south to Wittering and Bramcote, to enable No. 29 to bring its Wittering flight together with the remainder of the Squadron's Blenheims at Digby. Fighter Command's night-fighter deployment at the end of December 1940, is shown at Figure 9.

Whilst pressing the Air Ministry for more fighters, Douglas also sought to create more, and better, night-fighter airfields. The Ministry responded positively on 9 December, by agreeing to provided twelve such airfields and equip them with Standard Beam Approach (SBA) systems and to begin SBA courses that month for Beaufighter crews at Watchfield.

In an effort to put more fighters into the night sky, Fighter Command introduced 'Fighter Night' patrols during December to cover a number of the more important target areas. This system, originally designed and constructed during Dowding's leadership, was introduced on 12 December, replacing an earlier scheme, which utilised patrols by Hampden bombers tasked with observing and reporting the route of enemy raiders. The Fighter Night concept was based on a layered approach, whereby six single-engined fighters were stepped at 2,000 feet (610 metres) intervals, between 10,000 and 20,000 feet (3,050–6,095 metres) on the approach to a city, with their patrol lines marked by flares on the ground, and flown by night-fighter squadrons and by selected pilots from day-fighter units. The system's significant drawback, was the need for the pilots to be in visual touch with the ground and to be able to see the marker flares. On cloudy nights the patrols were cancelled. Each patrol had a planned duration of one hour, with some pilots flying two, or more patrols on any one night. During the periods of Fighter Night patrols, the AA guns ceased firing and searchlights were doused in the patrol areas. It was hoped the patrols would offer a better prospect of engaging the enemy and release the Hampdens to return to their proper role of bombing the enemy.[14]

December also saw the introduction of offensive patrols operating against the enemy's Continental airfields, in an attempt to harass and disrupt the *Kampfgruppen*. Codenamed *Intruder*, these operations were conducted under the operational control of No. 2 Group, Bomber Command, from its HQ at Wyton. However, the need to include night-fighters in these operations necessitated the co-operation of Fighter Command, therefore, from December 1940, No. 23 Squadron was allocated the role of long-range night-intruder. On account of the secrecy of its radar equipment, the AI sets were removed from the aircraft, and since they were destined to spend long hours 'stooging' around over German airfields, properly qualified navigators were posted into the Squadron. To increase their time over the targets, the Squadron was moved to Wattisham, near the Suffolk coast. For the intruder role, the aircraft retained their four-gun pack and carried small bombs. No. 23 Squadron flew its first intruder sortie on the night of 21/22 December.[15]

1941

The first week of the New Year opened with North-West Europe dominated by an anti-cyclone located over the North Sea and a ridge of high pressure extending over the south-west of the British Isles. These not unseasonal conditions soon had England in the grip of biting north-easterly winds, that brought heavy falls of snow along the East Coast. The bright, clear night skies caused the temperature to drop rapidly to well below freezing.

Despite there being a number of nights (thirteen) on which the *Luftflotten* were unable to operate, due to the poor weather conditions, the people of Great Britain, and more especially the long suffering citizens of London, were to experience another month of serious bombing. Attacks were directed at dock installations at the ports of Avonmouth, Cardiff, Plymouth/ Devonport, Portsmouth, Southampton and Swansea and the principal cities of Manchester, Birmingham and Bristol, in addition to four heavy strikes on London and its suburbs and a single precision raid on the R-R factory at Derby. In addition to the inland raids, IX *Fliegerkorps* were extremely busy in laying mines in the coastal shipping lanes. The raids on the docks and port facilities, were all part of Germany's strategy of increasing the pressure on the country by escalating the effect of the emerging U-Boat blockade.

Records for the month of January credit the night-bombers with dropping in excess of 1,430 tonnes (1,408 tons) of HE and 9,500 canisters of incendiaries. These raids caused the deaths of 922 people and serious injuries to a further 1,927. In response the defence generated 486 night-sorties for the claimed destruction of three enemy aircraft, all to cat's-eyes fighters.[16] The recently introduced Fighter Night operations on the nights of 3/4 January over Bristol, the 10/11 January over Portsmouth and again over Bristol on the 16/17 January, were not regarded as a success, due in part to delays in the passing of instructions and the prolonged intervals between take-offs and early returns from patrol. Fighter Night operations led to a claim for one enemy aircraft destroyed and another damaged – a poor return for a seemingly complex operation.

The guns discharged 49,044 heavy AA rounds, and claimed twelve aircraft destroyed, which according to Brigadier Sayer in his *Army Radar*,[17] was about par for the course. German operational losses for January, of which most occurred at night, amounted to just twenty-eight in all theatres, with a further thirty-seven being lost in accidents.[18] One of the aircraft described as 'damaged' in *Luftwaffe* reports, concerned a Ju 88 from KG 51 commanded by its observer, *Oberleutnant* Kuechle, on the night of 11/12 January 1941, who described the combat in his report:

'We could see the fires in Portsmouth even before we reached Fécamp. We made our run from south-east to north-west. After releasing our bombs at 21.08 hours from an altitude of 15,500 feet [4,725 metres] we turned to port onto a compass heading of 140° for Fécamp. ... Over the target there was broken cloud cover at about 3,500 feet [1,065 metres], while the Isle of Wight was clear and could easily be seen in the moonlight. Between Portsmouth and the Isle of Wight, about 2–4 minutes after bomb release, the [radio] operator reported a night-fighter approaching from starboard at about 90° to our own course. Just before reaching us it pulled up, curved to port and swung onto our tail.

After the night-fighter's first run-in, during which it did not fire, our pilot started to descend at about 600 feet/minute [180 metres/minute], leaving the autopilot on. Fire was opened at about 100 yards (90 metres), almost simultaneously by our gunner and the night-fighter. At this the pilot increased the rate of descent and when hits were heard, we dived still more sharply. At 14,000 feet [4,270 metres] the airspeed indicator was showing 280 mph [450 km/hr]. While in this steep dive the pilot saw the night-fighter's tracer above his aircraft, on his course. He therefore swung off sharply to port. The night-fighter pulled to port and disappeared.

Immediately there was another attack, this time from below on the starboard side. It was probably a different night-fighter .... With the autopilot off and diving steeply, at an indicated airspeed of 310 mph [495 km/hr] ... we turned to starboard ... onto a southerly course. The flight engineer fired off half a drum of [machine-gun] ammunition and the fighter pulled away.

Using the autopilot and zigzagging constantly ... we flew towards Fécamp. We had dropped to an altitude of 6,500 feet [1,980 metres]. Since the starboard oil tank was showing empty and the starboard engine was running evenly, we gained height over the Channel .... As the hydraulic gauges were not reading, the pilot selected flaps and the undercarriage down to test the system; this showed that the main system was unserviceable. We radioed 'Undercarriage damaged, will land last'.

At 22.50 hours, after the undercarriage and flaps had been lowered on the emergency system, we came in to land. On touch-down the machine tried to swing to starboard, but this was checked by full port rudder and brakes .... The aircraft turned slightly to starboard as it drew to a halt. As soon as it stopped, the engines were cut and the electrics switched off, and the crew clambered out.[19]

On examination, the aircraft was found to have been hit thirty-eight times, eight or ten of which were cannon rounds. The starboard tyre was shredded, which would explain the aircraft's swing to starboard on landing. It is not known to whom the night-fighters belonged. However, hits by cannon shell indicates the attacker was a Beaufighter, and since the Isle of Wight was covered by both the Durrington and Sopley GCIs, it is probable the machine(s) came from 219 at Tangmere, or 604 at Middle Wallop.

The month saw a strengthening of the night-fighter defences, with the arrival at Squires Gate, near Blackpool, of the Defiant equipped No. 307 (Polish) Squadron, from their operational work-up at Jurby, on the Isle of Man. Along with No. 96 Squadron, recently moved to Cranage, to the south of Manchester, and flying a mixture of Defiants and Hurricanes, No. 307 provided the makings of a north-west air defence force to protect Liverpool and Manchester. No. 264 Squadron was moved for the second time in a few weeks, when it departed Gravesend for Biggin Hill on the 11 January, as Gravesend was not considered to be a good night-fighter station. The final Blenheim squadron, No. 68, was formed on 7 January, at Catterick, Yorkshire, with a mixture of Mk.Ifs and Mk.IVfs, under the command of the youthful Wing Commander, the Hon Max Aitken, DFC.[20]

Events on the other side of the English Channel brought some good news. Pressure on Germany's Italian Allies in the Mediterranean, necessitated the transfer of a number of *Kampfgruppen* and some 120 aircraft to X *Fliegerkorps* on Sicily. II./KG 1 and elements of KG 76 were also returned to Germany – moves that did not go unnoticed by Air Intelligence.[21]

Further good news for the defence occurred that month, with the deployment of the first six GCI stations. The successful completion of the trials with the first GCI station at Durrington put considerable pressure on the Air Ministry to produce an initial batch of mobile AMES Type 8 GCIs in the shortest possible time. Production of this batch was undertaken at a great pace by RAE and TRE, with additional staff and materials being drafted-in from outside these organsiations. By cancelling all leave and working day and night, with just a few hours off for sleep and refreshement, they were able to deliver the first set on Christmas Day, 1940, followed by the remaining five within the next fortnight. It is worth noting that this rate of production was never exceeded throughout the remainder of the war.

Whilst the sets were being manufactured, TRE staff were out searching the south-east of England for suitable GCI sites that had overlapping coverage one with the other and were

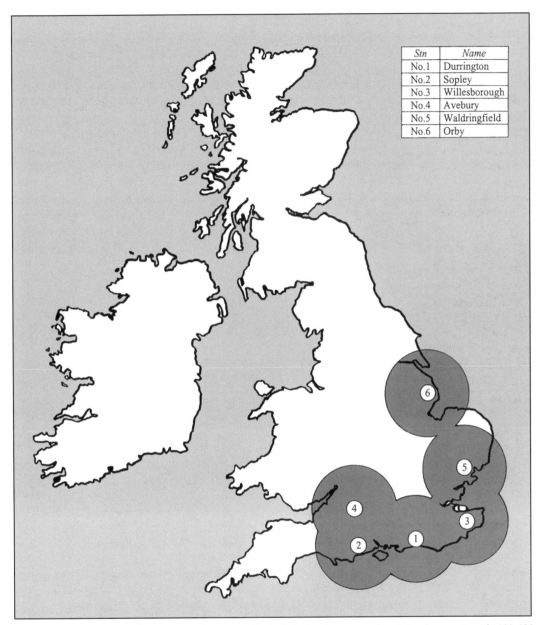

Stn	Name
No.1	Durrington
No.2	Sopley
No.3	Willesborough
No.4	Avebury
No.5	Waldringfield
No.6	Orby

Figure 10.   GCI stations and coverage at January 1941. Each GCI station covered a circle of approximately 100–120 miles (160–190 km) in diameter. These stations formed the core of a GCI network that would eventually envelope the whole country.

conveniently close to a sector station and a CH station. A number of these GCIs were connected to two sector stations to give some flexibility in their cover, which was affected by the weather conditions, both on the site and on the airfields. It also provided a degree of resilience in the event of bomb damage, or electrical failure. Even though these stations were regarded as 'mobile', some time was required to get them to an operational state, due to the need to calibrate the sets in accordance with the propagation characteristics of the surrounding ground – a period that could take several days.

The first site, Sopley, near Bournemouth and codenamed *Starlight*, was up and manned by 1 January 1941, with the next three completed by the middle of the month, and the sixth by 31 January. It was agreed that the experimental equipment at Durrington would be retained and brought up to an operational standard, leaving the sixth equipment as a spare at the disposal of Fighter Command.

The installation and operation of the first six mobile GCI stations, opened a new phase in the night air defence of Great Britain. After the trials and tribulations of the previous year, there was added confidence in the control system's ability to position a night-fighter behind an enemy bomber and bring the two close enough to gain a contact on the AI set. The technique evolved by AMRE and FIU required the Sector Controller to be in voice contact with the night-fighter, by means of a predetermined VHF R/T channel, whence, following take-off and climb, it would be directed towards a race-track holding pattern. Plots from the CH station(s) of prospective bombers were fed to the Sector Operations Room, who then fed course vectors, based on the CH station's plotting board, to the patrolling night-fighter that would place him in a suitable position.[22] At this point, the Sector Controller handed control of the interception to the GCI Controller, before tackling another hostile track.[23] On hand-over, the GCI Controller would check with the CH station to ensure there was no IFF response from the target and ascertain at what height he was flying. Using the PPI, the CGI Controller would guide the night-fighter by a series of precise steering instructions (vectors), until such time as the R/O detected the raider on his AI set. From this point on the R/O was responsible for directing the interception, until the pilot made a visual identification of the target.

The practicality and effectiveness of this technique was confirmed by daytime practises, before it was tried during the hours of darkness, and was to remain largely unchanged for the remainder of the war. The GCI controllers were drawn from the ranks of sector con-trollers, who already possessed the relevant experience in controlling aircraft by R/T. This practical means of selecting GCI controllers was devised during December 1940. By January, GCI controller instructors were teaching GCI techniques to sector controllers at Middle Wallop, Biggin Hill and Debden. The course of instruction required each trainee controller to act as observe in two or three night-interceptions, after which the trainee completed six daylight interceptions before 'going live'.[24]

The overall effectiveness of the GCI technique was not confined to its controllers. A great deal of expertise was also required on the part of the night-fighter crews, who had to acquire the necessary GCI experience, whilst perfecting their AI techniques with the AI Mk.IV equipment. This small, but growing band of controllers and aircrew, might reasonably be described as pioneers in the science of night interception, for much of what they learned was passed on to those who were training at the Radio Schools and the OTUs and shared within the night-fighter community.

February's night-fighter sortie-rate was down on the previous month, due to a reduction in the *Luftwaffe*'s operations brought about by the poor weather, with just 421 being flown by the cat's-eyes and single-seaters, against 147 by the AI equipped twins. These produced fifty-five visual contacts (thirty by the cat's-eyes and twenty-five by the AI twins) that led to thirteen combats and four enemy aircraft claimed as destroyed (two to the cat's-eyes and two to AI), plus eight claimed by the guns.[25]

During the month, the Hurricane equipped No. 85 Squadron at Debden, received its first Havoc Mk.I, twin-engined night-fighters. The Havoc I was based on Douglas DB-7 airframes that were originally intended for the French Air Force, but transferred to Britain following the French surrender in June 1940. Lacking the range to operate as bombers, the DB-7s were converted to night-fighters at the Burtonwood Aircraft Repair Depot, near Liverpool, and equipped with a solid nose designed by the Martin Baker Company, mounting eight 0.303 inch (7.69 mm) Browning machine-guns and their ammunition. With a crew of two, and fitted with AI Mk.IV, the Havoc had a maximum speed of 295 mph (470 km/hr) at 13,000 feet (3,960 metres), a service ceiling of 25,800 feet (7,865 metres) and a range of 966 miles (1,545 km).[26]

The combination of the increased numbers of AI equipped fighters and their close co-operation with the GCI stations, brought about a rapid improvement in the number of interceptions and enemy aircraft destroyed during March. Despite these improvements, the AOC-in-C strove to improve the situation still further, by requesting that the Air Ministry make available sufficient numbers of Beaufighters and/or Havocs to re-equip two of his Defiant squadrons. Unfortunately, calls on these aircraft by Coastal and Bomber Command prevented this request from being granted. Operations by No. 23 Squadron's Blenheims were also improving in their effectiveness, as their experience in intruder operations increased. During the first three months of the 1941 No. 23 claimed three enemy aircraft destroyed, four probably destroyed and three damaged, for the loss of two Blenheims.

The German's tactic of bombing an individual target on consecutive nights was emphasised during March, when Cardiff was raided on three nights during the first week, causing serious damage to the city and its docks. Fair to good weather during the first part of the month, combined with a full moon period, provided the *Kampfgruppen* with some of the best bombing conditions of the year so far, which they exploited to the full. Heavy raids were carried out against Liverpool, Clydebank, Hull, Glasgow and Southampton, to be followed later in the month by further attacks on London, Bristol and Plymouth. The weight of these raids was on a scale hitherto unequalled since the war began, with that on the night of 19/20 March, when 470 aircraft bombed London, being the worst.[27]

This raid caused considerable damage to public and commercial buildings in the City, as well as the docks areas of the eastern boroughs, mainly as the result of incendiaries. The public utilities, along with roads and railways were also severely affected. Casualties were correspondingly heavy, with 631 persons being killed.[28] Further attacks later in the month wrought similar levels of damage in Portsmouth (five raids in one week), Glasgow and Plymouth.

During March, the *Luftwaffe* flew 4,372 night-bomber sorties against Great Britain, with an additional forty-six being flown by the night intruders of *Nachtjagdgeschwader* 2 (NJG 2). A new feature of the night war was the introduction of light-bombers, Bf 110s and Ju 87 Stukas, which flew a further forty-eight sorties. Civilian casualties amounted to 4,259 killed and 5,557 seriously injured, bringing the total numbers since the Blitz began in September 1940, to 28,859 killed, 40,166 injured and 550 servicemen also killed.[29]

Fighter Command responded by flying 1,005 sorties and claimed twenty-two enemy aircraft destroyed; fifteen by AI fighters and seven by cat's-eyes.[30] Two nights in particular merit some consideration; the 12/13 and 13/14 March when Liverpool and Birkenhead were raided. On the first occasion, the *Luftwaffe* directed its efforts against the city and docks, during which the night-fighter defences destroyed at least five enemy aircraft: an He 111H-4 of 4./KG 27, an He 111P-4 of 5./KG 55, another He 111P-4 of 6./KG 55 and two Ju 88A-5s of

6./KG 76. Interestingly, only one of the five, a Ju 88A-5 from 6./KG 76, was shot down by a Beaufighter, the others all falling to cat's-eyes Defiants (three) and Hurricanes (one) – for further details see Appendix 4.

One of the KG 55 machines destroyed that night, was shot down by a Defiant, N1801, from No. 264 Squadron, flown by Flying Officer Frederick 'Desmond' Hughes,[31] with Sergeant Frederick Gash[32] as gunner. The pilot of the Heinkel, *Stabfeldwebel*[33] Brüning, describes the beginning of the flight. Having recently returned from leave with the rest of the crew; observer, *Feldwebel* Karl Düssel, radio operator, *Feldwebel* Konrad Steiger and flight mechanic, *Oberfeldwebel* Willi Weisse, Brüning's Heinkel, coded G1+GN, took-off from its airfield at Avord at 2100 hours:

> We crossed the English coast at 5,000 metres [16,400 feet]. The weather was good, with few clouds, excellent visibility and moonlight. I suppose it was perfect for night-fighter operations. There was no flak and no searchlights. On account of my experiences, I felt we could expect fighters, so I put my crew on the alert.[34]

The Defiant, took-off from its Biggin Hill base at 1945 hours and steered towards Beachy Head. Before reaching the coast, Hughes was vectored onto an enemy aircraft, but was unable to make visual contact. Returning to an orbiting pattern over Beachy Head, he was once again directed towards a second raider, which Sergeant Gash sighted some 800 yards (730 metres) ahead and 700 feet (640 metres) above the Defiant. The aircraft was positively identified as an He 111. Hughes closed to within 50 yards (45 metres) before Gash opened fire with a series of one second bursts with the turret's four Browning machine-guns.

> We were surprised with many direct hits and the aeroplane caught fire. Both engines were hit and stopped at once. The oil temperature shot up and the speed fell. The aircraft lost height and I was completely preoccupied with trying to control the machine …. My left hand on the throttle and my left ankle were injured, being outside the protection of the 8 mm armoured plate behind me, but I felt only light blows and no pain. Düssel seemed unhurt, so I told him to go back and see to Steiger and Weisse [who were both seriously wounded] and throw them out after preparing their parachutes …. There was no way I could bring the aeroplane down as the situation was desperate. Hopeless! Düssel came back and said he could not reach them as the gangway was ablaze.[35]

On Brüning's command to 'jump', Düssel left the aircraft through the side hatch, followed by Brüning, once he had trimmed the aircraft to fly straight and level. As he jumped he saw Hughes' Defiant some 10–20 metres away. The Heinkel was seen to spin, before crashing near the village of Ockley, Surrey. Brüning escaped by parachute and became a PoW, but his three comrades were killed. Hughes and Gash landed back at Biggin Hill at 2125 hours, with Gash having expended 1,000 rounds of 0.303 inch (7.69 mm) ammunition. Karl Brüning was repatriated to Germany in July 1947 and joined the post-war *Luftwaffe*, retiring in 1976.

Hughes and Gash's victory that night, showed, that given the right weather, adequate moonlight and the availability of GCI control, cat's-eyes night-fighters were able to provide a valuable contribution to night defence. Whilst a number of Defiant crews ran up good scores during the Winter Blitz, Hughes and Gash claimed five, and Flight Lieutenant Christopher Deanesly[36] and his gunner, New Zealander, Sergeant Jack Scott,[37] of No. 256 Squadron, four, the highest attainer was the legendary, Pilot Officer Richard Playne Stevens. Flying a Hurricane alongside his CO in 151 Squadron, Squadron Leader Jack Adams,[38]

Stevens specialised in solo, cat's-eyes night operations, where he used his considerable night flying experience (some 3,000 hours) tracking enemy aircraft by means of the shell bursts and searchlight beams that surrounded them. Steven's joined 151 in November 1940, and claimed his, and the Squadron's, first victory on the night of 15/16 January 1941, when he destroyed a Do 17Z-2 of 4.KG 3 and an He 111P-5 of 2./KG 53. By the end of May, he had claimed a further eight, making him Fighter Command's highest scorer of the Winter Blitz. Stevens was an exceptionally gifted and aggressive night-fighter pilot, who closed to an extremely close range before opening fire to ensure success, a fact that was acknowledged by the award of the DFC in February 1941, a Bar the following May and the DSO in December 1941. After service with 151, Stevens was posted to No. 253 Squadron at Hibaldstow, Lincs, for night intruder operations over Europe. He was killed on the night of 15/16 December 1941, when his Hurricane crashed near Gilze, in Holland, by which time he had added another three and a half enemy aircraft to his tally.[39]

Heavy raiding on the second night, that of 13/14 March, was once again directed against Liverpool and Birkenhead, with Glasow and Clydeside being added to the target list. Whilst very heavy damage was wrought in Scotland, that on Merseyside was less so and only rated as light. This night was, however, to record the greatest number of enemy aircraft destroyed during the Winter Blitz (see also Appendix 4), when a Do 17Z from the *Stab* KG 2, an He 111H-5 of 7./KG 26, an He 111P-2 from 7./KG 55, an He 111H-3x pathfinder of 1./KGr 100, a Ju 88C-4 night intruder of 4./NJG 2 and a Ju 88A-5 of 3./KfGr 106[40] fell to the fighter defences. Of these, five were claimed by Beaufighters and one by a cat's-eyes Spitfire of No. 72 Squadron.

The guns of Anti-Aircraft Command claimed seventeen of the enemy destroyed that month, making a combined total for the guns and night-fighters of thirty-nine, however, these only represented a loss-rate of 0.8 per cent,[41] a figure the *Luftwaffe* could easily sustain. Nevertheless, the figures served to show that Fighter Command's results were improving, as was the morale in the night-fighter squadrons and amongst the gun crews.

The long-range bombers concentration on ports and dock installations, particularly those on Merseyside and Clydeside, through which most of the vital food and war materiel from the US flowed, were by March beginning to cause the Air Ministry some concern. They accordingly ordered Douglas to redouble his efforts in defending those targets. In consultation with General Pile, Douglas ordered the redeployment of fifty-eight heavy AA guns from the industrial Midlands and the North, to the ports. He even persuaded the Royal Navy to part with eight guns from Scapa Flow! These were used to stiffen the Clydeside, Merseyside, Bristol/Avonmouth, Swansea, Cardiff and Newport defences. This deployment had its risks, for many of the industrial areas were forced to muddle through with fewer heavy guns; however, given the hammering the ports were taking, Douglas considered the risk justifiable.

By March 1941, Fighter Command had five GCI stations in full operation at Sopley; Waldringfield, Suffolk; Orby, Lincs; Durrington, Sussex and Willesborough, Kent, covering the eastern and south-eastern approaches to England. The station originally located at Avebury, Wilts, was dismantled and moved to Exeter to cover the approaches to the Welsh ports, with a seventh being commissioned at Sturminster, Dorset, by the end of the month and an eighth at Langtoft, Lincs, to reinforce the eastern approaches. The installation of IFF Mk.IIG in the AI and cat's-eyes night-fighters, which would respond to the GCI's interrogation pulses, was commissioned during the month and greatly eased the controller's task in identifying 'friendly' fighters. Limited GCI assistance could also be given to cat's-eyes fighters, by directing them towards the main bomber streams, where they could search visually for their targets. As has already been stated, this method of operation was to prove very effective on clear, moonlit nights.

The AOC-in-C shuffled the pack slightly during the month by moving 307 Squadron's Defiants to Colerne, No. 600's Beaufighters from Catterick to Drem, where it already maintained a detachment, and the semi-operational 256 Squadron from Colerne to Squires Gate, where it was declared operational in the defence of the North-West. No. 68 Squadron was also declared operational on the Blenheim If/IVf, whilst No. 23 Squadron took delivery of its first Havoc Mk.I (Intruders) at Ford, Hants.

The military situation in the Balkans necessitated the transfer of four *Kampfgruppen* to Austria, whilst a further three were stood down for retraining in the anti-shipping role at Brest, where they came under the command of the recently created *Fliegerführer Atlantik* (Flying Leader, Atlantic). The loss of these units was, however, to a degree, compensated by the introduction of a new bomber, the Dornier Do 217E-1, to I./KG 2. Although bearing a superficial resemblance to the Do 17, the '217 was an entirely new design, both structurally and aerodynamically. Powered by two 1,580 hp BMW radial engines, the Do 217 was designated a heavy bomber, with a crew of four and a defensive armament, for a German aircraft, that was regarded as quite formidable. With a maximum bomb-load of 5,550 lb (2,515 kg), carried internally, the '217 was capable of a spirited 320 mph (510 km/hr) at 17,060 feet (6,200 metres) and a service ceiling of 24,600 feet (7,500 metres). This medium bomber, for that is what is was, was capable of hauling a reasonable bomb-load to all targets in the British Isles, but carried a defensive armament that was inadequate, since little fire-power was concentrated in any direction. Nevertheless, it was a significant improvement on the Dornier 17 and a welcome addition to the *Luftflotten*.

Despite a poor start weather wise, April was set to be the heaviest yet for bombing, comprising sixteen major raids and five medium raids. On only three nights were there no raids. During the month, the *Luftwaffe* executed its first 1,000 tonne raid on the night of 19/20 April, when 712 bombers raided London, killing 1,200 people. This raid was in addition to one three nights earlier, when 685 aircraft delivered 890 tonnes, claiming 1,179 lives. Further major and heavy raids were also conducted against the ports; Bristol/ Avonmouth, Belfast, Glasgow/Clydeside, Liverpool/Birkenhead, Plymouth/Devonport, Tyneside, Sunderland and Portsmouth.

The bomber sorties statistics credited the *Luftwaffe* with flying 5,451 long-range bomber sorties, 205 intruder sorties and a further sixty-eight by light-bombers,[42] for the (claimed) loss of forty-eight enemy aircraft by night-fighters and thirty-nine by the guns – making April the most costly month for the *Luftwaffe* of the entire war, in terms of casualties over Great Britain (see also Appendix 4). Overall, twenty-seven were credited to the AI night-fighters and twenty to the cat's-eyes, plus one to a LAM Harrow of No. 93 Squadron.[43] Although the statistics gathered by Fighter Command continued to show a predominance in favour of the cat's-eyes fighters in terms of the numbers of sorties flown (842 cat's-eyes, against 342 AI),[44] the AOC-in-C was certain that AI in conjunction with GCI, offered the best prospect of overcoming the night menace. In this respect, he expressed his concern to the Air Ministry with the supply of AI-equipped, twin-engined night-fighters to his command, which threatened the performance of the defence. In fact, the delivery rate was so slow, that the actual numbers of Beaufighters available for operations decreased with every night of the month. Once again, he strongly appealed for a larger proportion of Beaufighter

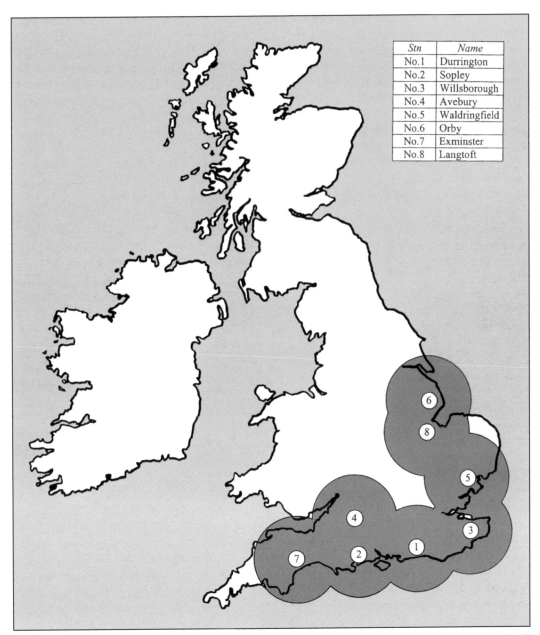

Stn	Name
No.1	Durrington
No.2	Sopley
No.3	Willsborough
No.4	Avebury
No.5	Waldringfield
No.6	Orby
No.7	Exminster
No.8	Langtoft

Figure 11.   GCI stations and coverage at May 1941. The GCI cover by May 1941 presented a barrier to *Luftwaffe* aircraft attempting to penetrate the east, south-east and southern coasts of Britain, requiring them to transit at least 100 miles (160 km) of overlapping radar protected airspace. As before, each GCI station had a coverage of approximately 120 miles (190 km).

deliveries to be diverted from Coastal to Fighter Command, or, alternatively, an increase in overall production.

With the coming spring and a reduction in the hours of darkness, Douglas perceived that *Luftwaffe* bombing operations would be concentrated in the south-east of the country. In anticipation of this change in German tactics, he sanctioned the move of No. 307 Squadron's Defiants from Colerne to Exeter, with a detachment to Pembrey, near Llanelli, in order to create space at Colerne for 600 Squadron's Beaufighters (from Drem and Prestwick). No. 29 was moved south from Digby to West Malling, Kent, whilst No. 141 travelled in the opposite direction to Ayr, to exchange its Defiants for Beaufighter Ifs. When at Ayr, No. 141 was to commence operations with the recently commissioned GCI station at St Quivox, near Prestwick. Two other GCI stations were also established that month; Hack Green, near Shrewsbury and Dirleton on the outskirts of Edinburgh.

Douglas was also concerned at the rising accident rate within his command, particularly at the OTUs. Although the overall accident rate was considered to be within limits, that at the OTUs was too high. Statistics for the first quarter of 1941 showed that Fighter Command's losses amounted to eighty-nine fatal accidents, which killed 106 aircrew. There were sixteen mid-air collisions and eleven pilots were killed when flying into high ground in bad weather, with most of the latter being attributed to the inexperience of the OTU trainees. Despite Douglas' concern, the accident figures showed little improvement, due in the greater part to the dangerous nature of night flying in bad weather from airfields surrounded by high ground.[45]

Whilst No. 23 Squadron was working-up with its new Havocs, conducting endurance flights to ascertain their range and performance characteristics, a number of these aircraft had been allocated to No. 93 Squadron to operate in the LAM role. This considerable expenditure of useless effort and the diversion of valuable aircraft on such a worthless scheme, was reinforced by Douglas asking the Air Ministry for permission to cease LAM operations. With common sense at last prevailing, the Ministry's permission was granted, and No. 93 Squadron was stood down to convert to the Wellington and a return to bomber operations. However, just as one hairbrained scheme was dropped, another rose to take its place and consume more Havoc airframes. This was the Havoc Turbinlite, conceived by Wing Commander W. Helmore, where the Havoc's nose gun-pack was replaced by a searchlight (and half a ton of batteries). Fitted with AI Mk.IV, the Turbinlite hunted conventionally for its target, accompanied by a pair of Hurricane fighters, one on each wing. When the enemy was within range of the searchlight it was switched on to illuminate the luckless target, which was then attacked by the Hurricanes. Like the LAM before it, the Turbinlite concept promised a great deal, consumed many hours of valuable research and development time, but delivered very little in the way of operational success. The development of the Turbinlight will be dealt with in Chapter 8.

Whilst the fortunes of the defence were on the rise, and the prospect for future success was high, the *Luftwaffe* returned in devastating form during May. The bombers opened their account on the night of 1/2 May, with three consecutive raids on Merseyside, which devastated the city and its docks, killing 1,900 people and injuring a further 1,450. These raids were followed by similar ones against Glasgow/Clydeside on consecutive nights, with London suffering the heaviest raid of the entire Blitz on the night of 10/11 May, when some 500 bombers caused extensive damage to the already battered capital and killed in excess of 1,000 people. The last raid of the Blitz, and the last one on Great Britain of any real material significance, was mounted against Birmingham on the night of 16/17 May. For this raid, the

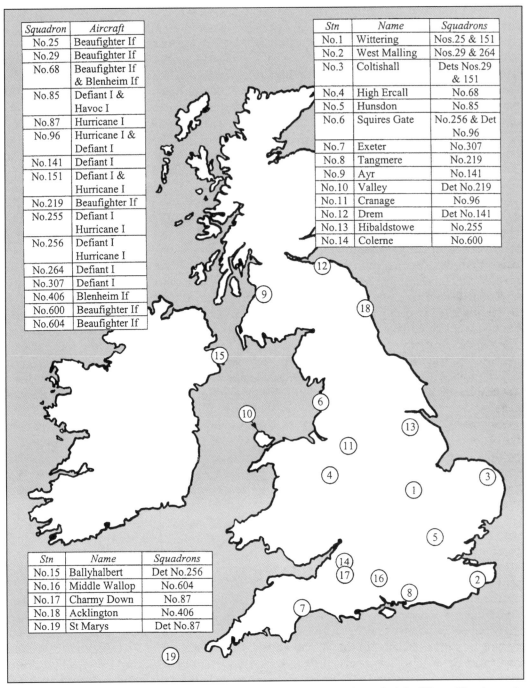

Squadron	Aircraft
No.25	Beaufighter If
No.29	Beaufighter If
No.68	Beaufighter If & Blenheim If
No.85	Defiant I & Havoc I
No.87	Hurricane I
No.96	Hurricane I & Defiant I
No.141	Defiant I
No.151	Defiant I & Hurricane I
No.219	Beaufighter If
No.255	Defiant I Hurricane I
No.256	Defiant I Hurricane I
No.264	Defiant I
No.307	Defiant I
No.406	Blenheim If
No.600	Beaufighter If
No.604	Beaufighter If

Stn	Name	Squadrons
No.1	Wittering	Nos.25 & 151
No.2	West Malling	Nos.29 & 264
No.3	Coltishall	Dets Nos.29 & 151
No.4	High Ercall	No.68
No.5	Hunsdon	No.85
No.6	Squires Gate	No.256 & Det No.96
No.7	Exeter	No.307
No.8	Tangmere	No.219
No.9	Ayr	No.141
No.10	Valley	Det No.219
No.11	Cranage	No.96
No.12	Drem	Det No.141
No.13	Hibaldstowe	No.255
No.14	Colerne	No.600

Stn	Name	Squadrons
No.15	Ballyhalbert	Det No.256
No.16	Middle Wallop	No.604
No.17	Charmy Down	No.87
No.18	Acklington	No.406
No.19	St Marys	Det No.87

Figure 12. Night-fighter Squadrons at May 1941. The order of battle at the end of the Winter Blitz shows the conversion of the AI fighter force to the Beaufighter, with the exception of 406 (Canadian) Squadron which was in the process of forming up, whilst 68 Squadron was destined to be the last squadron to use the Blenheim operationally. At this point the fighter force is equally balanced between AI and cat's eyes fighters.

Luftwaffe assembled 111 aircraft from *Luftflotte* 2 and fifty-three from *Luftflotte* 3, who directed the attack against the city's Nuneaton suburbs and its industrial centre, killing 113 people and damaging the works of ICI, the Dunlop Rubber Company and Wolseley Motors.[46]

Up to and including the 25/26 May, the *Luftwaffe* flew 4,835 night bomber sorties and a further 215 by the night intruders. In response, the defence flew 1,345 cat's-eyes and 634 AI night-fighter sorties, claiming ninety-six enemy aircraft destroyed, with the guns claiming another thirty-one. From the defence's point of view, these figures were, once again, a significant improvement on the previous month and the best for the entire war. The highlight of the month was a record breaking twelve enemy aircraft claimed destroyed on the night of 7/8 May. That night the raiders executed a heavy raid on Liverpool/Birkenhead, the east coast port of Hull and smaller raids on Manchester, Bristol, Plymouth, Great Yarmouth and Cleethorpes. In spite of their targeting the docks in Birkenhead and Hull, the majority of the bomb loads fell in the residential districts, killing thousands of people and making tens of thousands homeless.

In response to the bombers of *Luftflotte* 3 crossing the south coast and heading up through western England, on their long flight to Liverpool, and *Luftflotte* 2's crossing the North Sea to attack Hull, Fighter Command fielded 339 night-fighters to carry out night and dawn patrols across the bomber's paths. During the night, they claimed twenty-one aircraft destroyed (later reduced to twelve in contemporary accounts[47]), four probably destroyed and thirteen damaged, of which twelve fell on UK soil. A Fighter Night patrol was established and maintained over Liverpool throughout the night and claimed three enemy aircraft. The night's victories are shown at Appendix 4, but may be summarised as: ten He 111s, one Ju 88A and a single Do 17Z, with the Heinkels being shared fairly equally between the Beaufighters and Defiants. The ever reliable Pilot Officer Stevens, representing the Hurricane element, destroyed one Heinkel.

Air Marshal Douglas and General Pile had good reason to be pleased that theirs (and Dowding's) labours had finally been rewarded, tinged perhaps with some regrets that the *Luftwaffe*'s withdrawal from operations over Great Britain, had, to some extent, robbed them of final victory in the night battle. The civil population on the other hand, had they but known it, would have been relieved to know the worst was over and would undoubtly have been pleased to forgo further bombing to prove a military point!

CONCLUSIONS

It had always been Hitler's intention to invade the Soviet Union to rid the world of Bolshevism and create living space for Germany in the western provinces of Russia. By July 1940 the planned date for the invasion was set for the spring of 1941, but was later postponed to June. Codenamed *Barbarossa*, the OKW assembled 3,350 tanks and 7,000 artillery pieces into three Army Groups and deployed them to their marshalling points along the Russian border in Poland and eastern Germany. To support this force of some 120 divisions, the *Luftwaffe* transferred the bulk of *Luftflotte* 2 to the west to support Army Group Centre, whilst *Luftflotte* 3 was stripped of its aircraft and crews to bolster *Luftflotten* 1 and 4 in Army Groups North and South. On 24 June 1941, three days after the opening of the invasion, the *Luftwaffe* had 3,428 serviceable aircraft[48] operating along the Eastern Front. By 16 August, the western bomber force had been reduced to 203 aircraft under *Luftflotte* 3, of which only eighty-four were serviceable, distributed amongst seven *Kampfgruppen*.[49] The removal of so many bombers to the East, was, therefore, the principal reason for the ending of the Winter Blitz.

Although the bombing failed to bring about the total collapse of Britain's will to continue the war, it did prevent this country diverting its fighter resources to the Mediterranean, in particular to the defence of Malta and Greece. Its most significant effect was to be found in the level of war production, where aircraft deliveries were impeded and did not recover until February 1941, and heavy industrial output, ship building, iron, steel and armaments, was reduced. Communications and the stockpiles of oil and food supplies were also affected. As in the First World War, large numbers of servicemen and civilians, some 600,000 in all, were tied-up in the military and civil defence of the country. Damage to property amounted to the destruction, or damage, of more than 1,000,000 houses, whilst civilian casualties totalled 45,000 killed and some 50,000 injured.

Bad as these figures undoubtly were, they were insufficient to ensure a strategic victory for Germany. At no point during the Blitz was the damage imparted by *Luftwaffe* bombing sufficient to inflict permanent, or long term damage to industry, nor impair the country's ability to fight back. During the period of the Winter Blitz the *Luftwaffe* lost approximately 600 bomber aircraft to all causes, which represented an overall loss rate of 1.5 per cent – a level it could easily sustain. From the German viewpoint, whilst the campaign had failed to meet its objective, it had been conducted at relatively little cost to the *Luftwaffe*, and as in the previous war, it had forced the British to retain large numbers of troops and equipment on the Home Front.[50]

Having discussed the effects of night bombing, what then were the requirements of a night air defence system that evolved as a consequence of the Winter Blitz:

1. The defence required a long-range, early warning radar system to provide information and warning of the approaching bomber streams. This the defence possessed in quality and quantity in the CH system of coastal radar stations and its supporting communications and plotting infrastructure.

2. A shorter range, precision system was required to take over the responsibility for tracking the enemy once he had passed inland (through and behind) the CH screen. This requirement was met with the introduction of the first GCI station in December 1941, but until that time, the lack of such a device enabled the *Luftwaffe* to roam the country and bomb at will. Not only did GCI prove critical in guiding night-fighters to the locality of an enemy raider, it also proved sufficiently accurate on moonlit nights, to place cat's-eyes night-fighters into a smaller area of the sky and thus improve their chances of affecting a visual interception.

3. A good, reliable AI set, having the longest maximum range and, most importantly, the shortest minimum range, in order that a visual identification could be made whilst the target remained in AI contact. This requirement was not fully met until the introduction of AI Mk.IV in the Beaufighter, in September 1940.

4. If the aforementioned were not to be wasted, the defence needed to be equipped with a high performance, heavily armed fighter, capable of carrying the AI equipment and its power supplies. This requirement was met from September 1940 onwards by the superb Bristol Beaufighter. Although not delivered in any numbers until late in 1940, the Beaufighter was to make a significant contribution to the night defence, and remained effective to the spring of 1944.

5. The defence needed properly structured training facilities for air and ground crews. These requirements were not fully met until early in 1941, before which an attitude of 'make do and mend' prevailed within Fighter Command. However, when the first crews passed through the OTU system in late December 1940 and joined their squadrons, they were better equipped than ever before to operate as an efficient fighting team that was capable of destroying the enemy. In a similar vein, the impact of modest numbers of properly trained radar mechanics from the Radio Schools, together with the establishment of Special Signals Sections and their access to technical documentation and test equipment, greatly improved the operation and reliability of AI equipment.

6. Above all, it was realised that effective night-fighting depended on teamwork and efficient communications. Whilst the night-fighter crew may be said to represent the pinnacle of the night air defence triangle, they were themselves part of a team which depended on everyone fulfilling their part to the very best of their ability, and doing it consistently night after night.

When these objectives were met, as they were from March 1941 onwards, the numbers of enemy aircraft destroyed, rose dramatically.

Having established the requirements of the night air defence system, which of its two arms, the cat's-eyes fighter, or the AI fighters, were the most effective? A closer examination of the statistics (see Table 1), indicates that during the months, March to May 1941, when the AI and cat's-eyes fighters were operating with six or more GCI stations that whilst the cat's-eyes fighters destroyed the greatest number of enemy aircraft, they flew very nearly twice as many sorties as the AI fighters, but had fewer combats per sortie flown and destroyed less enemy per sortie flown. The AI fighters, therefore, provided the most effective antidote to the night bomber. Their results are all the more impressive when the weather conditions are taken into account. Whereas the AI fighters obtained contacts on most nights, irrespective of the weather conditions, the cat's-eyes fighters only achieved success on those nights combining bright moonlight and high concentrations of the enemy. As an aside, the cat's-eyes fighters also provided a useful role for the under-occupied day-fighters, who would otherwise have been sitting on their airfields 'twiddling their thumbs'.

It will be recalled that Douglas stated the need for twenty twin-engined, AI equipped night-fighter squadrons to defend Britain. By the end of May 1941 only six were Beaufighter

TABLE 1.   ANALYSIS OF NIGHT-FIGHTER SORTIES AND ENEMY AIRCRAFT CLAIMED AS DESTROYED JANUARY–MAY 1941[51]

Month	AI Squadrons			Cat's-eyes Squadrons		
	Sorties	Combats	Destroyed	Sorties	Combats	Destroyed
March	270	21	15	735	25	27
April	542	50	27	842	39	20½
May	634	74	37	1,345	116	59
Totals	1,446	145	79	2,922	180	106½

1 combat per 10 sorties
or
1 e/a destroyed per 18.3 sorties

1 combat per 16.2 sorties
or
1 e/a destroyed per 28.8 sorties

TABLE 2. *LUFTWAFFE* NIGHT BOMBER LOSSES OVER GREAT BRITAIN,
NOVEMBER 1940–MAY 1941[52]

Month	Estimated night sorties	Enemy aircraft claimed destroyed	Loss rate (%)
1940			
November	5,495	2	0.036
December	3,585	4	0.112
1941			
January	1,965	3	0.153
February	1,225	4	0.327
March	3,510	22	0.627
April	4,835	48½ [53]	1.003
May	4,055	96	2.367

equipped (see Figure 12), of which three were shortly to begin the process of converting to the Beaufighter (Nos 68, 141 and 456), one was on Havocs (No. 85) and the remaining eight still operated cat's-eyes Hurricanes or Defiants. These figures, therefore, add little credence to the AOC-in-C's bold statement made shortly after the end of the Blitz, and quoted in Basil Collier's *The Defence of the United Kingdom*, that 'he was confident that if the enemy had not chosen to pull out at the middle of May, Fighter Command would have inflicted such casualties on the *Luftwaffe*'s night bombers that continuance of the night offensive would have been impossible'.[54] In reality Fighter Command would never meet Douglas' requirement for twenty-eight operational, twin-engined, night-fighter squadrons. However, the statistics show a different story. Table 2 shows that up to February 1941, the *Luftwaffe*'s night bomber losses were negligible and taking into account the defence's best month, May 1941, the bomber loss rate could, even then, be sustained. Indeed, it was not until all the single-seat squadrons were re-equipped with the Beaufighter, or later with the superlative Mosquito in 1942, that the night-fighter crews of Fighter Command were in any way capable of carrying out their AOC-in-C's boast.

Notes

1. Winston Ramsey [2], *op cit*, p. 309.
2. A post-war analysis conducted by the Air Ministry's Air Historic Branch and quoted as Appendix 13 to Chapter 8 in *The Signals History, Volume 5*.
3. Conceived as the aerial equivalent of the naval minefield, to provide a curtain of slowly descending parachute supported mines, the LAM consumed a great deal of time and effort for a scheme that was at best, problematical as an anti-aircraft measure. Throughout the Winter Blitz, only two enemy aircraft were brought down by this means.
4. Nos 2 and 3 Signals Schools were retitled Nos 2 and 3 Radio Schools in January 1943.
5. John Rawlings, *Fighter Squadrons of the RAF* (Crecy Books, 1993), p. 517.
6. Later Air Chief Marshal Sir Keith Park, KCB, KBE, MC, AFC, and AOC-in-C Air Command, South East Asia, but perhaps best known for his defence of Malta during the siege.
7. Later Air Chief Marshal Sir Trafford Leigh-Mallory, KCB, DSO.
8. *The Signals History, Volume 5*, p. 132.
9. *The Signals History, Volume 5*, p. 132.
10. Sir Archibald Sinclair, KT, CMG, MP, was the Secretary of State for Air from May 1940 to May 1945.
11. *The Signals History, Volume 5*, p. 133.

12. Special Signals Officers were drawn from the ranks of civilian engineers and radio amateurs, who were given a rapid introduction to radar before being commissioned as Pilot Officers in the RAFVR.
13. No. 151 Squadron also operated a pair of Hurricanes flown by the CO, Squadron Leader Adams, DFC, and Pilot Officer R.P. Stevens.
14. Winston Ramsey [2], *op cit*, p. 311.
15. *Ibid*, p. 312.
16. Winston Ramsey [2], *op cit*, p. 376.
17. A.P. Sayer, *Army Radar* (London: The War Office, 1950), p. 50.
18. Winston Ramsey [2], *op cit*, p. 376.
19. Combat Report of *Oberleutnant* Kuechle of the *Stab*, KG 51, quoted in Wolfgang Dierich's *Kampfgeschwader 'Edelweiss'* (London: Ian Allan Ltd, 1975), pp. 42 and 43.
20. Max Aitken was Lord Beaverbrook's son and would later command the Banff Mosquito Strike Wing and No. 601 Squadron, AAF, post-war. He inherited his father's newspaper empire in 1964, but disclaimed his baronetcy.
21. X *Fliegerkorps* was transferred from *Luftflotte* 5 to Sicily and expanded by the acquisition of 2./KG 4, III./*Lehrgeschwader* (LG – an operational testing *Geschwader*), II and III./LG 1 and I.KG 26 – Winston Ramsey [2], *op cit*, p. 379 refers.
22. This was generally from behind and on the same course as the bomber, or with the bomber crossing from left to right, in order that the fighter could acquire it on the AI and turn onto its tail.
23. *The Signals History, Volume 5*, p. 193.
24. *The Signals History, Volume 5*, pp. 193 and 194.
25. Winston Ramsey [2], *op cit*, p. 420.
26. Reñe Francillon, *McDonnell Douglas Aircraft since 1920, Volume 1* (London: Putnam, 1988), pp. 273 and 294.
27. Winston Ramsey [2], *op cit*, p. 446.
28. *Ibid*, pp. 446 and 447.
29. *Ibid*, p. 446.
30. *The Signals History, Volume 5*, Appendix 13.
31. Later Air Vice Marshal F.D. Hughes, CB, CBE, DSO, DFC**, AFC, who commanded 604 Squadron in 1944/45.
32. Later Flight Lieutenant F. Gash, DFM.
33. A *Luftwaffe* rank broadly equivalent to an RAF Warrant Officer.
34. The written statement of Karl Brüning made in August 1978 and quoted in Winston Ramsey [2], *op cit*, pp. 468 and 469.
35. Combat report for Defiant N1801, dated 12 March 1941 and quoted in Winston Ramsey [2], *op cit*, p. 469.
36. Later Wing Commander E.C. Deansely, DFC, who commanded No. 256 Squadron, then No. 114 Wing in the Middle East and finally No. 575 Squadron.
37. Later Flying Officer W.J. Scott, DFM.
38. Later Group Captain J.S. Adams, DFC*, who later commanded Nos 256 and 226 Squadrons.
39. Kenneth Wynn, *Men of the Battle of Britain* (Croydon: CCB Associates, 1999), p. 482.
40. A coastal reconnaissance aircraft, operated by the *Luftwaffe* on behalf of the *Kriegsmarine*.
41. This figure is based on the numbers of aircraft raiding the country, quoted in Winston Ramsey [2] and those claimed shot down during March, quoted in *The Signals History, Volume 5*, Appendix 13.
42. Winston Ramsey [2], *op cit*, p. 502.
43. *The Signals History, Volume 5*, Appendix 13.
44. *Ibid*, Appendix 13.
45. Winston Ramsey [2], *op cit*, p. 503.
46. *Ibid*, pp. 624 and 625.
47. *The Signals History, Volume 5*, Appendix 13.
48. Alfred Price [1], *op cit*, p. 42.
49. Winston Ramsey [3], *The Blitz, Then and Now, Volume 3* (London: Battle of Britain Prints, 1990), p. 22.
50. Winston Ramsey [2], *op cit*, p. 627.
51. Taken from, *The Signals History, Volume 5*, pp. 134 and Appendix 13.
52. *The Signals History, Volume 5*, Appendix 13 and Winston Ramsey [2], *op cit*, pp. 154–627.
53. It should be noted that one enemy aircraft was claimed by a LAM Havoc of No. 93 Squadron, making the total for April 1941, 27 by AI means, 20½ by the cat's-eyes fighters and one by LAM=48½.
54. Quoted in Basil Collier's, *The Defence of the United Kingdom'* (London: HMSO, 1957).

CHAPTER SEVEN

The End of the Metric Line, AI Mk.V and Mk.VI

Autumn 1940–July 1942

Whilst the night-fighter squadrons were battling with the *Luftwaffe* during the dark nights of 1940/41, AMRE continued with the development of AI to meet Fighter Command's requirements for an improved system. For this reason, it is necessary to step back to the autumn of 1940 to understand a problem that was effecting the success of the night-fighter crews. Experience gained by the squadrons and at FIU in the operation of AI Mk.IV indicated that the success of an interception depended almost entirely on the skill with which the R/O guided his pilot into the final approach to the target. Without the closest co-operation between the two, there was a marked tendency for the fighter to weave from side to side as it closed with the target. It was thought this phenomenon might be attributed to the time delay in the passing of steering information from the R/O to his pilot.

The problem was not new; whilst assisting No. 25 Squadron at Northolt with their conversion to AI during the late summer and autumn of 1939, Robert Hanbury Brown realised it was very difficult to observe the bearing and range of a target and instruct the pilot in which direction to fly to effect an interception. To prove his point he carried out a simple experiment at Northolt. By following another aircraft visually and instructing the pilot, who acted blind throughout the pursuit, he found that when flying at 250 mph (400 km/hr), it was surprisingly difficult to keep on the tail of the target aircraft at distances less than 2,000 feet (610 metres). Any closer than that induced a weaving motion, whose swings increased with so much violence that forward vision was impaired to such an extent that the target aircraft was lost to sight.[1]

Hanbury Brown returned to the problem in 1940, following concerns being expressed at the failure of night-fighter crews to convert their AI contacts into visuals. It was suggested that the failures might be attributed to a German counter-measure that detected the AI transmissions and warned the crew to take evasive action. On the basis of his earlier findings, Hanbury Brown felt sure the cause had nothing to do with the enemy, but was instead related to the fighter weaving during the final part of the interception. In order that everyone appreciated the factors involved, Hanbury Brown prepared a paper entitled 'On Obtaining Visuals from AI Contacts'. In this document he clearly illustrated the factors involved and proved mathematically that when the fighter closed on its target, its track became unstable. He showed that the amount of 'weave' was dependent upon three factors:[2]

1. The actual and relative speeds of the two aircraft.
2. The minimum angular displacement between the two, as shown on the AI screens.
3. The time delay between the R/O calling out a course correction and the pilot carrying it out.

The paper concluded by advising fighter crews to resist the temptation to rush in to minimum range, but instead, to make their course corrections some 2,500 feet (760 metres) behind the target. When the pilot was satisfied he was on the same course as the target and slightly below his altitude, he should gradually close the range, whilst maintaining that course. In the closing stages of an interception, Hanbury Brown advised that the AI set should not be used to correct the course, but rather it should be employed to tell the pilot where the bomber was in relation to himself. If, when the minimum range was reached and the bomber was not visible, the pilot was advised to throttle-back and let the range open to 2,500 feet (760 metres) where he would most probably re-establish contact. He could then realign his course with that of the bomber and try again. By coincidence, or perhaps because it was found by practical trial and error, this approach method was already in use by FIU. Whatever the reason, the Hanbury Brown/FIU technique became the standard method of interception during the first half of the war.[3]

Having analysed the problem and proved it mathematically, Hanbury Brown set about the task of persuading AMRE's senior officers to support the development of a display that would more easily guide the pilot in the latter stages of an interception. This device, he argued:

> should improve all the factors which caused weaving. It would reduce the delay in acting on the radar information, improve the minimum angular displacement of the target which could be observed and introduce a greater element of prediction into the whole operation.[4]

The strength of Hanbury Brown's case was recognised by AMRE, who authorised the go-ahead for the development of a 'pilots indicator' as an adjunct to AI Mk.IV. As in the standard Mk.IV system, the R/O was provided with two screens; one showing range only information and the other, a D/F display, or 'spot indicator', which showed the position of the target relative to the fighter, as shown in Figure 13.

At the beginning of an interception, following the initial contact, the R/O selected the target return on the range tube by adjusting a 'strobe control', which moved a bright 'strobe' spot along the horizontal timebase until it coincided with the target return. This operation, termed 'strobing', was continued by the R/O as the fighter closed with the target. The quartered spot indicator provided the necessary elevation information. The information on the spot indicator tube was in turn repeated on a small 'pilot's indicator' CRT-based display (sometimes referred to as a G-scope), that was located in front of the pilot above the flight instruments, on the port side of the cockpit. By observing the pilot's indicator, the pilot sought to steer the aircraft such that the spot remained in the centre of his display.

The existence of the pilot's indicator was first brought to the attention of the Air Interception Committee at its ninth meeting on 18 July 1940 by the DCD, who explained that the new device was only suited to two-seat fighters, as its operation was dependent on the R/O. Not surprisingly, there was immediate opposition within the RAF to the introduction of yet another radar set, with AI Mk.IV still in its infancy and soon to be committed to production. They did not feel the need for a different type of R/O's display, and in particular, the

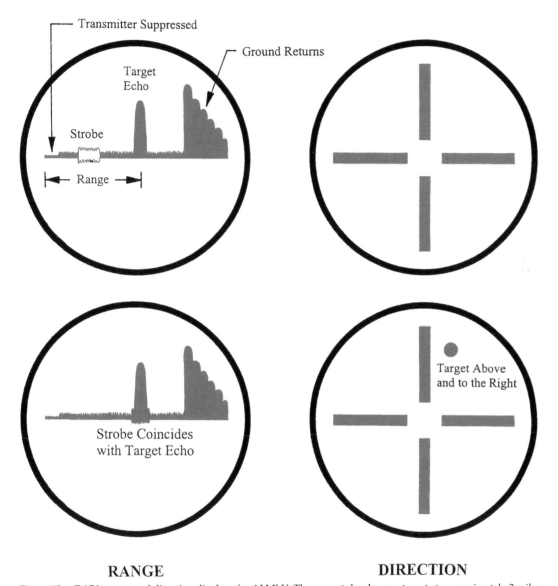

RANGE   **DIRECTION**

Figure 13.   R/O's range and direction displays for AI MkV. The range tube shows a target at approximately 3 miles (4.8 km), whilst the direction tube indicates it to be above the fighter and off the starboard nose. The strobe can also be seen to the left of the range echo, which the R/O will drive along the timebase until it is superimposed over the echo. At this point the radar information will be displayed on the pilot's indicator and remains thus for so long as the R/O tracks the target echo with the strobe.

introduction of the pilot's indicator in the cockpit. AMRE nevertheless continued to press their case for the adoption of the new set and finally persuaded the Air Ministry to sanction a series of trials by the FIU.

Blenheim IV, P4846, fitted with the new displays, was delivered to FIU at Shoreham by a pilot from the SD Flight on 7 October 1940. The first flight tests were undertaken personally by Wing Commander Chamberlain two nights later. A number of preliminary flights then

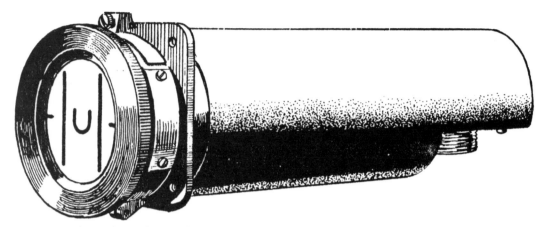

Figure 14.   Pilot's indicator for AI MkV.

followed, after which three satisfactory interceptions were accomplished. On the fourth flight the spot indicator display failed. A second series of tests were attempted the following day to rectify the fault, which was eventually traced to the grounding of the upper elevation aerial. With the fault cleared, Wing Commander Chamberlain, accompanied by Mr J.P.W. Houchin and Mr F. Blythen of EMI, and later by Messrs White and Taylor of AMRE, then completed the tests to everyone's satisfaction.

FIU's preliminary report[5] based on the flights undertaken on 8 October, concluded the indications in azimuth were very much more sensitive than those in elevation, and in the absence of range indications the pilot needed constantly to refer to his R/O. These short-comings apart, the pilot's indicator was considered to be a useful device which required further development to show range. FIU's recommendation was accepted by AMRE, who instructed Hanbury Brown to incorporate the necessary modifications to the pilot's indicator display.[6] The new version, which was ready for air testing by December 1940, retained the spot on the pilot's CRT; however, as the range reduced the spot grew 'wings' that reached their maximum span at the AI's minimum range, as shown in Figures 14 and 15.

The display also incorporated a 'U'-shaped mark in the middle and two vertical lines, or 'goalposts', on either side. At a range of 5,000 feet (1,525 metres) the wings just touched the sides of the 'U' and at 4,000 feet (1,220 metres) they exceeded it. As the range decreased further the wings steadily grew until at minimum range, normally 300–400 feet (90–120 metres), they were just touching the goalposts. Target elevation was indicated by the spot and its wings being positioned above, or below the 'U' mark.

The interception technique which evolved from the FIU trials followed that devised by Hanbury Brown and FIU for AI Mk.IV, whereby the R/O would talk his pilot to within 2,500 feet (760 metres) and dead astern of the target. At this point the R/O would continuously strobe the target and transfer the echo to the pilot's indicator, for him to complete the interception.

The first ranged example of the pilot's indicator was fitted to a Beaufighter that Wing Commander Chamberlain and Pilot Officer H.C. Randall collected from Middle Wallop on 8 December. By 13 December, the pilot's indicator had been transferred to a Boston DB-7 flown by Pilot Officer 'Dickie' Ryalls.[7] Testing continued at FIU until the end of the year, principally in the hands of Wing Commander Chamberlain, Flight Lieutenant Ashfield, Pilot

Officer Randall, Flying Officer Colin Clark,[8] Robert Hanbury Brown and Mr R.W. Taylor of AMRE/TRE.[9]

The subsequent report from FIU persuaded the members of the Night Interception Committee at their nineteenth meeting on the 12 December, to proceed with the manufacture, by TRE, of twelve hand-made pilot's indicators. These would then be installed in Havoc night-fighters and would be followed by a further thirty-six for use in a small number of Beaufighters, pending the delivery of production standard models. The interim models, adapted from AI Mk.IV sets, were designated AI Mk.IVA, with the production versions designated AI Mk.V. The hand built AI Mk.IVA sets were built by the Army's Air Defence Experimental Establishment (ADEE) and a contract for the follow-on batch of thirty-six was awarded to the Dynatron Company.

At a meeting at TRE on 30 December 1940, chaired by Dr Lewis, it was agreed, in line with the Night Interception Committee's requirements, that the new set be based as closely as possible to the design of AI Mk.IV, with the pilot's indicator being an 'add-on' unit. Pye Radio were tasked with the manufacture of the operator's unit, the receiver and pilot's indicators for the twelve ADEE sets. By 30 January 1941, Pye Radio were behind schedule with their delivery of the hand built Mk.IVA sets to ADEE. They had partly completed the manufacture of the operator's unit and completed the receiver, apart from the beacon facility, but no progress had been made on the construction of the pilot's indicator units. These problems, and the shortage of resistors, valves, CRTs, a suitable receiver motor-switch and circuit design shortcomings, served to delay the delivery and official type approval by several weeks. After making due allowances for these and the need for air testing, TRE estimated that at the very earliest, the first batch would not be ready for service use until the 1 May 1941.

The first two AI Mk.IVA sets were delivered to TRE from ADEE in late January 1941. After bench testing, one was sent to St Athan for No. 32 MU to install in a Beaufighter, whilst the second was probably sent to the SD Flight at Christchurch for installation in a Beaufighter. Delivery of further sets from ADEE and others from the thirty-six batch order from Dynatron, were scheduled at five each week, beginning from 10 February. The sets from the Dynatron line were also sent to No. 32 MU for installation in Beaufighters of Nos 219 and 604 Squadrons.

On 1 February, Flight Lieutenant Ashfield and Pilot Officer Ryalls delivered Beaufighter R2159 to Christchurch for the installation of the second AI Mk.IVA set. A second Beaufighter was similarly modified for FIU and collected by Flying Officer Clark and Pilot Officer Randall on 17 February. On its arrival at Shoreham, the Beaufighter underwent daylight trials in the hands of Wing Commander Chamberlain, Flying Officer Clark and Sergeant Ian McRae and was flown later that evening to ascertain its night performance. Testing continued at FIU for the remainder of February, with Hanbury Brown accompanying Wing Commander Chamberlain on a number of flights. By the end of the month, a number of AI Mk.IVA Havocs were also undergoing testing.[10]

The February trials of AI Mk.IVA were destined to be the last for Robert Hanbury Brown. A few nights later, when acting as Wing Commander Chamberlain's R/O in a Beaufighter, his oxygen supply became disconnected whilst they were flying at 20,000 feet (6,100 metres). Realising that something was wrong with his observer, Chamberlain quickly lost height and landed the Beaufighter. His swift action undoubtly saved Hanbury Brown's life. In hospital, Hanbury Brown's problem was diagnosed as mastoid related and surgery was recommended. Following a stay in hospital, during which he suffered periods of deafness,

Hanbury Brown was banned from high altitude flying, which effectively brought his AI career to an end. In June 1941 he returned to TRE, where he was transferred to a group led by John Pringle, who provided radar-related aids for the Army.[11,12] His place at FIU was taken by Mr Parry, who continued the trials in co-operation with Flight Lieutenant Ashfield, Flying Officer Clark and Wing Commander Douglas Morris, DFC,[13] who was on attachment to FIU before taking command of No. 406 Squadron, Royal Canadian Air Force (RCAF).

The manufacture of the full production standard AI Mk.V fared little better than that of AI Mk.IVA. Sharing a slightly modified version of EMI's AI Mk.IV modulator, with a completely re-engineered receiver and operator's indicating unit designed by Pye Radio and RAE, AI Mk.V entered production some time before February 1941. However, problems with the new receiver and the modulator, compounded by component availability, conspired to delay the delivery of the first two Mk.V sets to TRE until late February. When they were examined at Worth Matravers, it was discovered that both sets suffered from a series of shortcomings that would require rectification before they could be declared 'fit for purpose'. It was not, therefore, possible to forward the sets[14] to FIU for operational trials as was originally intended. Both units were subsequently condemned as being unfit for flying trials and returned to Pye Radio.

A third receiver was despatched to TRE on 1 March and like the earlier ones was also rejected, as it, too, was in need of considerable modification. The switch motor would not start properly, and it ran irregularly and generated interference, in addition to which a number of plugs and connectors were in the wrong place on the chassis, and the wiring was deemed insufficient for operation at high altitudes. Problems were also experienced with the growth of the wings on the CRT display, which appeared at ranges approaching 15,000 feet (4,570 metres). TRE once again returned the receiver to Pye Radio.

The shortcomings of the Mk.V system were aired at a meeting held at MAP on 12 March, where it was explained that the new AI was not yet ready for production because it failed to meet the TRE specification. The main culprit appeared to be the receiver motor switch, which was found to be deficient in terms of its overall speed (revolutions per minute – rpm) and tolerances in the manufacture of the switch contacts. Both were subsequently blamed on TRE's inexperience in specifying manufacturing requirements. MAP took a very serious view of the situation, as they had been led to believe the system required no further development and was ready for series production. Matters came to a head on the 29 March, when the Ministry proposed that since a great deal of work was required to bring the system to a production standard, the development of AI Mk.V should be terminated. It should be noted that at this point, orders had been placed with Pye Radio, EMI, Dynatron and E.K. Cole (transmitter) for in excess of 1,000 Mk.V sets. However, with the planned equipping of the Mk.II Defiant and the Mosquito NF.II with the fully automatic AI Mk.VI (of which more later) which was then not sufficiently advanced to take over from AI Mk.V, the DCD, Mr Watson Watt, supported by Mr Stewart of RAE, agreed the contracts for AI Mk.V be allowed to continue. Both men also agreed that some redesign of TRE's circuitry was necessary, which would defer production slightly (this was somewhat understating the situation).

With Mr Taylor of TRE assisting Pye Radio with a redesign of the receiver, where it was found necessary to replace two valves and rectify faults in the strobe circuitry, they hoped to have the prototype available for flight trials by 5 May. By 6 May bench testing was sufficiently advanced for the set to be promised to TRE for installation in a Havoc at Christchurch on 8 May. Installed in Havoc, BJ468, the first hand built Mk.V system was delivered to FIU on the evening of 19 May for formal flight trials. Testing began the

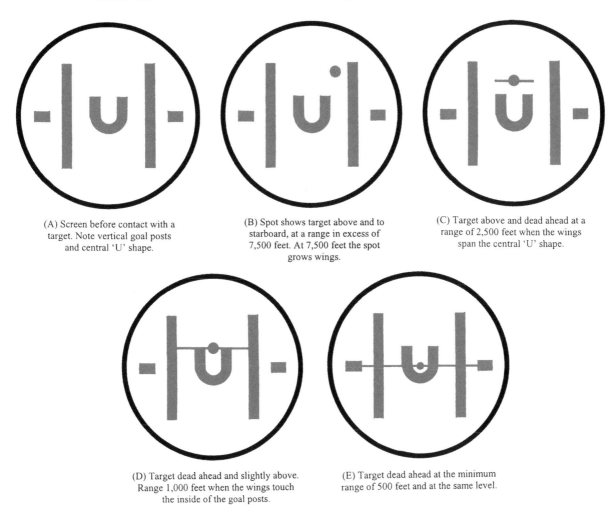

(A) Screen before contact with a target. Note vertical goal posts and central 'U' shape.

(B) Spot shows target above and to starboard, at a range in excess of 7,500 feet. At 7,500 feet the spot grows wings.

(C) Target above and dead ahead at a range of 2,500 feet when the wings span the central 'U' shape.

(D) Target dead ahead and slightly above. Range 1,000 feet when the wings touch the inside of the goal posts.

(E) Target dead ahead at the minimum range of 500 feet and at the same level.

Figure 15. Pilot's indicator displays for the production standard AI Mk.V.

following day under the personal command of the Unit's new CO,[15] Wing Commander Evans, with support being provided by Messrs Taylor and White of TRE, Dr Touch of RAE, and Mr Cope of Pye Radio, to coax the equipment through its trial period.

FIU's report[16] written at the conclusion of the trial on 22 May, established the maximum range as being 17,000 feet (5,180 metres), or 3.2 miles (5.18 km), and the minimum as 300 feet (90 metres). The wings began to grow at 7,000 feet (2,135 metres), passed the goal posts at 3,000 feet (915 metres), cleared the outside of the goal post by ⅛ inch (3 mm) at 2,000 feet (610 metres), and increased to ½ inch (12.5 mm) at minimum range. Azimuth indications were judged to be 'good' when the target was level, or above the Boston, but when below, a dip in the response was experienced to both port and starboard. The elevation response gave steady and constant indications below and above the Boston.

FIU considered the general construction of the Mk.V set to be superior to that of AI Mk.IVA, where its increased accessibility would improve maintenance. With respect to the pilot's indicator, they considered it a 'great improvement' on that used with AI Mk.IVA, with the spot indicator being steadier and free from 'agitation'. On the down side, the spot response was regarded as a little sluggish and its response 'slow'. There was also a lack of brilliance in both the R/O's and pilot's displays, which they considered to be 'adequate' for night operations. In conclusion, FIU recommended the equipment be fitted to a Beaufighter in order that a more complete series of trials be undertaken, following which the set should be used operationally at the earliest opportunity.

A similar report written by RAE, confirmed FIU's assessment that, on the whole, the new equipment worked well and gave no trouble during its twenty hours of testing. Only a few changes were thought necessary before the set was committed to production. They did, however, highlight a problem with 'dipping' of the azimuth response, which was eventually traced to the ingress of moisture in the lower elevation aerial elements. When these were dismantled and cleaned, the performance was restored. Greater attention would, therefore, have to be paid to the weatherproofing and matching of the Mk.V's aerial system. They also recommended the repositioning of the receiver to a more convenient position, since its present location behind the operator made it very difficult for him to manipulate the strobe control.

FIU's recommendation that the equipment be transferred to a Beaufighter, was accepted by Fighter Command. On the 24 May the set installed in BJ468 was fitted to Beaufighter, R2159, which was flown to Christchurch where Messrs Taylor and White cleared a number of faults that had arisen during the Boston tests; a lack of brilliance in the displays, the sluggish movement of the spot indicator and incorrect wing growth. R2159 was returned to Ford[17] on 27 May and commenced testing the following day. The tests were completed in three and a half hours.

The subsequent report[18] stated that whilst there was insufficient time to carry out a full series of tests, good returns were received from the target at ranges out to 14,000 feet (4,270 metres), when the Beaufighter was flying at 15,000 feet (4,570 metres). The minimum range was recorded as being somewhere between 300 and 500 feet (90–150 metres), whilst the responses from the pilot's indicator were broadly in line with those recorded in Boston, BJ468. These results, which were based on the standard aerial system employed in the Beaufighter, showed the indications in azimuth to be 'good', but 'poor' in elevation, due possibly to a known problem with the latter. The sluggishness of the spot indicator was reported as having been overcome since the trials in BJ468. Another problem leftover from the previous trials, that of CRT brilliance, had also been overcome and was now 'up to standard'. FIU did, however, comment adversely on the location of the gain control on the operator's display, which they considered 'useless' and in need of removal.

In conclusion, the report recommended that the set be retained by FIU to confirm the day's results, conduct proper maximum range tests and fly the equipment operationally at night. They clarified their report by saying:

> No policy recommendation should be implied as a result of today's test, but should this be confirmed, then an operational squadron should be equipped with the Mk.V as soon as possible. This Unit considers the equipment to be a great advance operationally.[19]

Testing continued for the remainder of May in the hands of Wing Commander Evans, Flight Lieutenant Stubbs, Flying Officer Clark and Mr Taylor. In their report[20] FIU recommend the R/O's spot display be provided with some form of calibration marks that would enable the operator to see at a glance the target's bearing in azimuth and elevation. This recommendation was eventually adopted and incorporated into production Mk.V sets.

The first confirmed kill for AI Mk.V was credited to the FIU on the night of 26/27 June 1941, when a Beaufighter If flown by Flight Lieutenant Ashfield, with Flying Officer Randle operating and Pilot Officer Morris observing, destroyed an He 111 off Beachy Head, Sussex.

With full type approval granted by RAE on 5/6 August, and the FIU's endorsement of its capabilities, there was little apparent haste on the part of Fighter Command or TRE to introduce the new set to operational service. This reluctance was due in part to the improved success and reliability of AI Mk.IV and the development of yet another 1½ metre AI. This set, designated AI Mk.VI, was the result of a collaborative venture by the TRE, RAE and EMI, to develop a fully automatic, or 'wandering strobe' AI for use in single-seat fighters. The development of this set was seen to offer a potential solution to the shortage of twin-engined night-fighters at the end of 1941 – a shortage that would be overcome if the set could be installed in the Defiant and Hurricane Mk.II.[21]

The initial order for 2,000 AI Mk.VI sets was placed with industry by December 1940 (well before AI Mk.V's development was completed), with the first examples being due for delivery in June 1941. This order was awarded in spite of the set not being cleared for production. With the production and development programmes for AI Mks IV, V and VI running concurrently, there existed a lack of co-ordination between MAP and the DSigs(Air) in their interpretation of Air Staff policy concerning the production of AI radars. The radio programme for December 1940 specified the production of:

- 600 AI Mk.IV sets for delivery during January 1941.
- Forty-eight pilot's indicators with which to modify the same number of Mk.IV sets to the Mk.IVA standard.
- 600 semi-automatic pilot's indicators from Pye Radio to be used in conjunction with AI Mk.IVA.
- 2,000 AI Mk.VI sets from EMI, with 200 being modified with beacon and homing facilities, under the designation AI Mk.VIA.[22]

The above rates were patently unrealistic and made no allowance for the production of AI Mk.V.

Nevertheless, the official policy[23] concluded, that when AI Mk.VI had passed its qualification trials it should replace AI Mk.IV as the standard airborne set. However, due to uncertainties surrounding the development of AI Mk.VI (of which more anon), it was considered prudent to limit the production of these sets to 1,000 units, in order to have adequate supplies of the Mk.IV equipment during the transition period.

At some point at or around May 1941 MAP placed an order for the manufacture of fifty AI Mk.V sets for operational trials with Fighter Command, with deliveries commencing the following October. Work on the conversion of No. 219 Squadron's Beaufighters to AI Mk.V, began immediately, but faults with the initial equipments delayed the completion of the programme for two months. By December, the Squadron had six Beaufighters flying night defence patrols with AI Mk.V. No. 219 Squadron's results were somewhat at odds with those previously described by FIU. They complained that the spot on the pilot's indicator was prone to 'swinging' and was sluggish and, overall, the interception angle was assessed as 'limited'. The fitting of further Beaufighters was interrupted whilst these faults were investigated and corrected. It is believed that No. 219 Squadron was the only Beaufighter unit to be completely converted to AI Mk.V. The predominant user of the Mk.V set was the Mosquito NF Mk.II, which entered service in the spring of 1942. In an effort to ensure there were sufficient sets available for the Mosquito force, production of AI Mk.V was stepped up.

TABLE 3.   ANALYSIS OF ENEMY AIRCRAFT DESTROYED
BY AI Mk.IV AND Mk.V, APRIL–JULY 1942[24]

AI Type	Destroyed	Probably destroyed	Damaged
Mk.IV	50½	15	32
Mk.V	13	3	7

By working through the Christmas period, Pye Radio succeeded in delivering 100 sets by the end of 1941.

The development and use of the Mosquito is dealt with later, however, it is interesting to compare the success of AI Mk.IV with that of the Mk.V. The direct comparison of the two sets (see Table 3), is not exactly fair. There were four times as many Mk.IV sets in service during the period, as there were Mk.Vs, which distorts the view. AI Mk.V was a more complicated set that demanded correspondingly higher standards of maintenance – a requirement that was not always achieved by some squadrons. This in turn led to a higher degree of unserviceability, which occurred during the early months of operations with each new squadron. The situation was further aggravated by prejudice on the part of the night-fighter crews, who preferred the 'patter' and teamwork developed between pilot and R/O with AI Mk.IV. See Appendix 2 for an outline description of AI Mk.V.

The pressing need to introduce an AI set suitable for installation in single-seat fighters, provided the incentive to adapt AI Mk.V's manual strobe to automatic operation. In pursuit of this objective, the Air Ministry established the AI Mk.VI Design Committee during the summer of 1940. This Committee, which comprised representatives from MAP, RAE, TRE, EMI, E.K. Cole and Pye Radio, was responsible for drafting the design specification for the new set. The requirement stipulated that whilst the new AI was intended for use in single-seat fighters, it should also be capable of operation in two-seat fighters, and would, therefore, be regarded as a direct replacement for AI Mk.IV. Air Ministry policy relating to the development of AI fluctuated considerably during the latter part of 1940 and the winter of 1941, due partly to the rapid progress being made in the development of centimetric technology and changes in the enemy's tactics. Taking these considerations into account, the Committee decided to abandon its original idea of basing the design on AI Mk.V, in preference to a completely new design.

During October 1940 EMI were awarded a contract to manufacture twelve prototype Mk.VI sets and to have these cleared for quantity production by 1 January 1941 – an impossible task. Investigations by RAE indicated that the equipment for the new variant could be accommodated in just four 9 × 8 × 18 inch (230 × 200 × 460 mm) units, that could be installed in Spitfires, Hurricanes and the Defiant. Mr Stewart of RAE was put in overall charge of the project, assisted by Messrs Bond (aircraft installations), Hunt (valves), Howard (receiver) and Grenfell (indicating unit), whilst the responsibility for liaison between RAE and industry was placed in the hands of Mr Padden. The group was asked to keep the number of valves to the absolute minimum and not to employ any of the Acorn types. It was also agreed that EMI and RAE would build prototypes of the receiver chassis, with the best features of both being incorporated into the final design.[25]

EMI began work on the design immediately and had breadboard models of the auto-strobe and receiver ready for testing by late October. These were quickly followed by prototypes of the transmitter and modulator chassis. A prototype pilot's indicator, operating in the manual mode, was successfully tested by Wing Commander Chamberlain and Sergeant McRae in an FIU Blenheim on the night of 8/9 October. The test results showed the elevation sensitivity to be poor, when compared to that of the azimuth response. Notwithstanding this criticism, FIU expressed themselves as being impressed with the system and considered the apparatus to be 'quite workable'.

The success of the FIU flight trials prompted the RAE team to press hard for the bread-board auto-strobe to be incorporated into the system as soon as possible, in order that they might evaluate its operation. These units were then tested with a receiver taken from an ASV set (possibly ASV Mk.II), while another was taken from AI Mk.IV.

Whilst EMI and RAE were developing the hardware, work was proceeding at TRE with the fitting of an aerial system in a Defiant and designing its equipment installation. In order that things might move along at a quicker pace, EMI were asked to build a second breadboard auto-strobe unit for airborne testing. This somewhat rudimentary unit was duly completed by EMI and made ready for RAE's inspection by 12 October. It was later modified by the parent company to incorporate a 'prevention' (also known as an 'auto-flyback') circuit, that would return the auto-strobe to zero (its starting point along the timebase) whenever it accidentally latched on to the ground return.

A whole raft of design modifications,[26] submitted to MAP during October, threatened to seriously delay the Mk.VI programme. To counter these proposals and establish an agreed specification, MAP called a meeting under the chairmanship of one of the DCD's deputies (DDCD), Group Captain Chandler. Following a series of discussions it was agreed that further delays in the development programme could not be upheld, and bearing in mind the urgency to get AI Mk.VI installed in Defiants, only minor modifications would be approved. However, a second meeting held at EMI's Hayes Factory on 1 November, succeeded in having this policy reversed, when the Company confirmed it would be possible to incorporate anti-jamming circuits in the receiver and a series of modifications in the rest of the system.[27] These changes would later conspire to strangulate the development of AI Mk.VI and prolong its delivery to Fighter Command.

The design of the auto-strobe circuit was devolved to TRE, who guided EMI through the design of a breadboard prototype and had it ready for testing by 4 November. Unfortunately it failed to work with the lower performance, non-strategic valves, defined in the specification, but was later made to function correctly by the incorporation of a number of feedback circuits. By 22 November the unit was ready for flight testing, along with examples of the modulator, transmitter and Pye's receiver. Wooden space models of the units were also made available to aid RAE's Defiant installation design team.

With development of the production standard auto-strobe circuit also well advanced (it had been built alongside the prototype), EMI were confident the production design would be completed on time (January 1941). Progress overall was behind schedule, due to TRE's request for the inclusion of pre-set tuning controls on the transmitter's front panel, which required the first design to be scrapped and started again. It was hoped that if the bread-board transmitter worked well during the flight trials, the lost time might be recovered and yet meet the design date.

Further delays were added on the 26 November, when a meeting of the AI Mk.VI Design Committee confirmed the need for the blacking-out of the pilot's indicator during strobe search and added another; namely, the strobe should be released from a distant target when a closer one presented itself. A control for this facility was later added to the pilot's control unit, which drove the strobe through the distant echo until it hit the ground return, whence it continued to search and latch-on to the nearest target.

With the system slowly evolving and the development schedule sliding ever backwards, MAP turned its attention towards series production. To this end they drafted a tentative AI Mk.VI production requirement in December; this called for the manufacture of 1,500 complete sets by EMI, plus 500 transmitters and modulators from E.K. Cole and 500 receivers, strobe units and pilot's indicators from Pye Radio.

The first complete set of AI Mk.VI equipment was delivered to TRE during the early days of December for installation in a Blenheim test aircraft at Christchurch. Following the completion of its ground tests, the aircraft underwent flight testing between 7 and 23 December, which highlighted a number of shortcomings. First, the wings appeared to grow at too great a rate; second, the minimum range was a disappointing 700 feet (215 metres); third, the maximum range was barely 10,000 feet (3,050 metres); fourth, the spot tracking on the pilot's indicator was too slow; and finally, the elevation indications were 'bad'.

EMI agreed the elevation problem was due to the poor aerial system and could be overcome. They also considered the transmitter and receiver units were by no means the best that could be provided, and modifications to them would ease the minimum range performance. Improvements in the suppression circuits in the receiver were likely to improve the minimum range, but trials with a modified chassis failed to overcome the problem. Modified once more with two additional controls, the receiver chassis finally met its design criteria. The problem with wing growth was also traced to the receiver. Work on the receiver, transmitter and modulator units to incorporate the necessary changes was estimated to take another two to four weeks to complete, for by now the programme was running some four to six weeks behind schedule.

Flight trials of a new modulator, which enabled the strobe circuits to hold a target down to 150 feet (45 metres), were undertaken by the SD Flight at Christchurch in January 1941. These showed that whilst the minimum range of 400 feet (120 metres) was now met, the auto-strobe still failed to hold the target automatically. Success did, however, accompany the trials of new versions of the transmitter and receiver, both of which performed well.

Whilst the development of AI Mk.VI continued at TRE and with EMI, the Air Staff were turning their attention to the introduction of the new set. Their priority was to get the system fitted as quickly as possible into Fighter Command's aircraft, to which end discussions had taken place in the previous November and December with Boulton Paul, concerning the Defiant. The Air Staff's second priority was the Mosquito, followed by the Havoc, Beaufighter, Hurricane and Spitfire in that order. Authority was also given to evaluate the system in the Hawker Typhoon and Tornado. However, with the increase in availability of two-seat night-fighters, the Air Staff's policy changed again in April 1941, when the VCAS ruled that with the exception of the Defiant II, AI was not to be fitted to single-seat fighters. Priority was therefore given to the fitting out of Defiant IIs, Mosquitoes, Havocs and Beaufighters in that order. In reality, AI Mk.VI would only be installed in the Defiant II, the Hurricane II, and a handful of dual-controlled Mosquito Mk.IIIs. With 2,000 sets already on order, a quantity of just 200 would have been adequate to fulfil Fighter Command's requirements. However, for whatever reason, the production order was not reduced.[28]

In January 1941, TRE established a second team to expand AI Mk.VI's capabilities to include beacon homing (AIH) and IFF facilities. An experimental version of the strobe circuit, modified for beacon working, underwent testing towards the end of the month, with which RAE hoped it would be possible to home onto a beacon at ranges up to 100 miles

(160 km). Two schemes were proposed; the first, Scheme A, envisaged the system working with the existing beacon network, whilst the second, Scheme B, proposed the AI-band (188–198 MHz) be spilt into two equal parts (188–193 MHz and 193–198 MHz), with one being reserved for beacon working and the other for AI. Both schemes were also required to work with AI Mk.IV and Mk.V. At a meeting at MAP on 13 February it was decided to adopt Scheme B, with the proviso that it should be capable of changing from one half of the band to the other, in the event of enemy jamming.

In respect of IFF, the identification of a friendly aircraft (that is one showing the correct IFF code) in single-seat fighters, would cause the auto-strobe to reject the target. In the case of two-seaters, a lamp was caused to flash in sympathy with the code, or the target was to be automatically rejected. The adoption of these two facilities (AIH and IFF) in effect produced a second version of the set, which was correspondingly allocated the designation AI Mk.VIA. Priorities called for the production of the original AI Mk.VI to take precedence over AI Mk.VIA, with the first 300 being produced as Mk.VIs by April 1941 before changing to the Mk.VIA the following September. These timescales were later amended by the Air Ministry to read 'fifty Mk.VI sets by 15 May, a further 150 by 1 July, and the remainder following as production built up'.

Testing of AI Mk.VI by the SD Flight continued throughout February and on into March, during which continuous wave (CW) interference plagued the programme. It was not by any means certain that the interference emanated from a local source, or was being encountered from outside the country.[29] The problem was eventually overcome during March, by the simple expedient of adjusting the receiver's automatic voltage control. TRE's reported on the 24 March, that their trials showed the system was consistently capable of achieving a minimum range between 250 and 300 feet (75–90 metres), but the maximum range was still a disappointing 11,000 feet (3,830 metres). The equipment was then removed from the Blenheim test aircraft and returned to EMI for further examination and a thorough overhaul.

Back at Hayes, EMI's engineers traced the range fault to flash-over (arcing) in the HT supply in the cable that ran from the modulator to the transmitter. This was replaced by one made up from 'Unimet' ignition cable, before the set was returned to Christchurch for further testing. These were successfully accomplished by 1 April, whence the transmitter and modulator were cleared to operate at heights up to 35,000 feet (10,670 metres).

During the SD Flight/TRE testing period, the opportunity was also taken to test various elements of the Mk.VI system at FIU. By 30 March, Wing Commander Chamberlain, ably assisted by Flying Officer Clark, Flying Officer Randle, Flight Lieutenant Ashfield, Pilot Officer Ryalls and Robert Hanbury Brown, had amassed some forty hours flying with the pilot's indicator. Based on their experiences, Wing Commander Chamberlain proposed that the wings should grow from 7,500 feet (2,285 metres) and not from 5,000 feet, as stated in the RAE specification. This recommendation was accepted by Fighter Command and the set was corresponding modified to sprout wings at 7,500 feet.

With the equipment installed in Blenheim IV, L4839, further trials were conducted by Wing Commander Chamberlain and Flying Officer Randle during the first days of April. On these flights the RAF crew were accompanied by Mr Blumlein and Mr Houchin from EMI's development team, who respectively sat alongside the pilot and observer. The trials confirmed the set worked satisfactorily and auto-strobing of the target operated correctly from 12,000 down to 150 feet (3,660 down to 45 metres). A final version of the equipment was then installed in a Battle to better represent the aerodynamic shape of the Defiant. These trials conducted between 13 and 20 April proved the final version of the strobe circuit, with no more trouble being experienced with its locking to ground returns or arcing. A further set of tests during the week ending 18 April proved the final versions of the transmitter, modulator and receiver. With the Battle flying at 10,000 feet (3,050 metres), the minimum range was recorded as 'better than 250 feet (76 metres)', with the target echoes being captured as they emerged from the ground returns at maximum range. Pressure testing of all the Mk.VI's modules during the early part of May cleared the system to operations at 35,000 feet (10,670 metres), following which RAE approved the system for quantity production.

With Mk.II Defiants rolling off Boulton Paul's Wolverhampton production line in May 1941, the race was on to deliver AI Mk.VI sets at the earliest opportunity. MAP's concerns in this regard were fully supported by the AOC-in-C Fighter Command, who commented 'I trust that all these aircraft may be issued with Mk.VI and that every possible effort will be made to that end'.[30] Unfortunately, the AOC-in-C's aspirations were not to be met, as the delayed development programme inevitably ate into the production schedules, which continued to slip further behind.

During June, reports were received of the high rejection rate amongst the VT90 valves supplied by GEC, with 400 transmitters already built and awaiting an investigation to identify the cause. This problem also impacted on the delivery of AI Mk.V, ASV Mk.II and various naval sets. Flight trials continued to uncover further problems in the rate of the wing's growth, that limited the set's range to 14,000 feet (4,270 metres), due to failures in the auto-strobe circuits, whilst the aerial motor-switch continued its poor performance. These and other problems were discussed at a progress meeting at MAP on the 27 June, where RAE expressed the view that the principal delay could be attributed to the unusually high number of technical faults in the original design – a design which they had already signed-off! On the positive side, TRE confirmed the delivery of the first set of equipment for the prototype installation in the Defiant.

These and other factors continued to slow the production rate and delay the equipment deliveries. EMI's production target had been to deliver fifteen hand-built pre-production models of AI Mk.VI by 28 June. A revised schedule promised the first three models to TRE by early July and the remaining twelve by the 14 July. However, when these were delivered to TRE it was discovered the problem with the wings growth had not been corrected, nor had that with the auto-strobe circuits. Needless to say, the deliveries of the pre-production sets slipped further behind. Problems were also being experienced in designing the Defiant's aerial system at Christchurch, which added yet more delays to the programme.

It was not until August that the equipment fit on Defiant AA357 was completed and the aircraft delivered to FIU for testing between 9 and 12 August. As the aircraft had yet to be fitted with flame damping exhausts, the testing was done during daylight hours. Following completion of ground testing the Defiant was flown with EMI's Mr Houchin acting as observer.[31] During these tests it was discovered that external interference caused by other AI equipped aircraft in the vicinity and AI navigation beacons, prevented the Defiant 'seeing through' the electronic mist and locking onto its target. This required yet another circuit modification in the modulator, which delayed the trials until early September, but cured the problem at the expense of further delays to the production schedule. Further testing of AI Mk.VI ceased on 26 September, when AA357 went unserviceable (u/s) with engine trouble.

It was to be 5 December 1941 before the first, fully cleared, AI Mk.VI pre-production sets were ready for delivery to MAP. It had been intended to convert the Defiants of Nos 96 and

264 Squadrons to AI Mk.VI and later a third unit, No.151 Squadron. However, before this work was completed the Air Staff took a decision on 2 May 1942, to discontinue the use of the Defiant in the night-fighter role, in the light of sufficient quantities of Beaufighters and Mosquitoes being delivered to Fighter Command.[32] Squadrons 151 and 264 converted to the Mosquito NF.II in April and May 1942 and 96 Squadron to the Beaufighter Mk.II. By October 1942 the Defiant had been removed from Fighter Command's front line inventory and with it passed the need for a single-seat fighter AI. It is not known how many AI Mk.VI sets were delivered to MAP, however, its known that the original production order for 2,000 was later reduced to 1,125, whilst AI Mk.VIA was never built in quantity and did not work properly in the twelve sets that were produced by EMI. For some inexplicable reason, the 'break clause' in MAP's Mk.VI contract was not invoked and the full 1,125 were subsequently produced. After modifications they were delivered to Bomber Command as tail-warning radars under the codename *Monica*,[33] for installation in the Command's four-engined 'heavies'.[34]

Although it had an operational life that lasted barely four months, the Defiant/AI Mk.VI combination is reputed to have claimed the destruction of at least one enemy aircraft; an He 111 destroyed by a 264 Squadron Defiant on the night of 17/18 March 1942 – a claim that is not confirmed in contemporary records.

In addition to the Defiant, AI Mk.VI was originally scheduled for testing in the Hurricane and Typhoon/Tornado day-fighters. To this end, Hurricane II, Z2905, was equipped with a mock-up of AI Mk.VI units at Hawker's Kingston Factory by 3 March 1942 (see Appendix 5) and delivered to FIU on 3 May for flight testing. The initial evaluation was carried out by Wing Commander Evans and Flight Lieutenant Park, before handing over to Squadron Leader Hayes, Flying Officer Ryalls and Pilot Officer McCulloch to complete the testing. A second Mk.VI equipped Hurricane Mk.II, MB288, was flown from Farnborough to FIU by Flying Officer Davidson and tested against a Beaufighter target aircraft throughout the remainder of the month. These and other trials by FIU proved successful and enabled the fitting-out of a batch of Hurricane IIcs to proceed in July 1942. In the event, the low priority allocated to this project only enabled a small number, sufficient to equip a single flight, to be delivered in the spring of 1943. Several of these aircraft[35] were dispatched to India by sea and issued to No.176 Squadron, where a flight was formed to provide the night defence of Calcutta from May 1943. The Hurricanes served alongside the Squadron's Beaufighters and claimed a small number of Japanese bombers destroyed.[36]

A final attempt to reintroduce the single-seat fighter to night operations was made during the early months of 1943, following a report on the suitability of the Typhoon for night flying by Flight Lieutenant Roland Beamont.[37] Beamont's report,[38] which was ostensibly aimed at proving the case for the Typhoon as a night offensive weapon, was based on two flights he undertook from Duxford on the nights of 24 and 26 July 1942. From these he concluded the Typhoon could be operated under GCI conditions on moonlit nights and for Fighter Night and searchlight box operations. Encouraged by Beamont's report, the Air Ministry ordered a series of feasibility studies to assess the Typhoon as a night-fighter. To this end a standard Typhoon, R7881, was set aside by Hawkers and equipped with an aerial system similar to that used on the Defiant and Hurricane. Designated Typhoon NF.Mk.Ib, R7881 was first flown from Hawker's Langley airfield by the Company's chief test pilot, Philip Lucas, on 23 March 1943. The requirement for the new night-fighter stated that the AI equipment was to be capable of installation in a container the same size and shape as a standard 44 gallon (200 litres) under-wing fuel tank. Provided with two such 'tanks' ballasted to represent the operational weight of 11,900 lb (5,400 kg), R7881 recorded a maximum speed of 368 mph (590 km/hr). It was then flown to TRE for the installation of its 'packaged' AI Mk.VI, before undergoing a prolonged evaluation at FIU that lasted well into 1944, but which produced no operational requirement.[39]

CONCLUSIONS

The concept of the pilot's indicator, however well intentioned, served to introduce added complexity to radar production and testing at a time when Fighter Command needed all the AI radars it could lay its hands on.

AI Mk.V was designed at a time when the radio industry was committed to the production of a single AI radar, AI Mk.IV, and Fighter Command was trying to introduce maintenance and testing procedures in the squadrons to improve its reliability. Bearing in mind that Robert Hanbury Brown and FIU had already established a procedure to overcome weaving in the latter stages of an interception, was the introduction of another AI set into the Command's inventory, with all that entailed in respect of documentation, fault finding and the supply of spare parts, worth the investment? Although it did eventually mature into a useful device, its introduction to service was not greeted with outright enthusiasm within the night-fighter community. There were those who thought it reduced the R/O's job to little more than that of a 'strobe chaser', and threatened the loss of the vital link between the pilot and his operator. Finally, could Fighter Command have done without AI Mk.V, since it did not enter service until 1942, by which time the threat had passed? The answer to these questions is that Fighter Command could probably have got by without AI Mk.V.

Although a well engineered radar, AI Mk.VI was committed to production before its development and testing were satisfactorily completed. There were, however, mitigating circumstances that required the rapid development of this AI for use in single-seat fighters, at a time (late 1940) when the RAF was desperately short of night-fighters. Nevertheless, although based on the earlier Mk.V, the development of such a sophisticated AI was so protracted that its production schedules stretched into 1942 and rendered it unnecessary when the fighter for which it was designed, the Defiant, was withdrawn from service. The blame for the delay may be apportioned equally to the main contractor, EMI, and to TRE, as the design authority, for their over optimistic development timescales and the consequent need to modify the equipment on the production lines to accommodate changes found in testing. The need to add AIH and IFF facilities to the set required the establishment of a second design team, a separate designation, AI Mk.VIA, and further resources in production. The Air Staff and MAP are not above criticism. Why did they permit the simultaneous development and production of five AI sets; AI Mks IV, IVA, V, VI and VIA, within an industry that lacked the capacity and a component supply chain that struggled to cope with the demands placed upon it. Also, when the Air Staff announced the withdrawal of the Defiant in April/May 1941, why were the AI Mk.VI contracts not cancelled and the components thus saved diverted to a more useful purpose?

In summary, AI Mk.VI's technical shortcomings may be attributed to:

1. Its maximum range being poorer than that of AI Mk.IV, due to the auto-strobe's inability to lock on to weak (long range) target returns.
2. The spot on the pilot's indicator was insufficiently agile to cope with targets taking evasive action.
3. The strobe circuits were complex and difficult to fault-find and maintain.

4. The pilot's workload was high, due to the need simultaneously to monitor and interpret the pilot's indicator and look outside the cockpit into the darkness to catch sight of the enemy.

However, the principal reason for its demise and the termination of metric radar development, was the introduction of centimetric techniques during the latter months of 1940, which rendered all previous airborne radars, obsolete, or at best, obsolescent.

Notes

1. R. Hanbury Brown, *op cit*, p. 64.
2. *Ibid*, p. 65.
3. *Ibid*, pp. 64 and 65.
4. *Ibid*, p. 65.
5. *FIU Report No. 40*, dated 10 October 1940.
6. R. Hanbury Brown, *op cit*, pp. 66 and 67.
7. Later Squadron Leader D.L. Ryalls, by then a flight commander with No. 219 Squadron, was killed on the night of 26 December 1944, when, it is believed, his Mosquito was shot down by a Messerschmitt Me 262B jet night-fighter.
8. Later Flight Lieutenant C.A.G. Clark and a flight commander with No. 137 Squadron, was lost at sea on 30 October 1941.
9. *FIU ORB*, pp. 12–16.
10. The background to the design and testing of AI Mk.IVA and AI Mk.V are taken from AVIA13/1050 as shown in the bibliography.
11. R. Handbury Brown, *op cit*, p. 67.
12. Robert Hanbury Brown remained a member of the TRE staff until the war's end, where he worked on radar techniques for airborne forces and served in the USA. Post-war he left Government service and joined Professor Bernard Lovell at Jodrell Bank to study radio astronomy. In 1962 he moved to Australia to join the Narrabi Observatory and study quantum physics with Dr R.Q. Twiss. He was elected as a Fellow of the Royal Society and made a Companion of the Order of Australia.
13. Later Air Marshall Sir Douglas Morris, KCB, CBE, DSO, DFC, AOC-in-C Fighter Command from May 1962 to March 1966.
14. Each of the two models was fitted with a different CRT for competitive evaluation; Mullard's A41/G4 and the VCR138, but since neither unit met the original specification, no trials were undertaken.
15. Wing Commander Chamberlain was replaced as FIU's CO on 10 May 1941, on posting to HQ Coastal Command and was awarded the OBE the following September.
16. *FIU Report No. 63*, dated 23 May 1941.
17. FIU transferred its base to Ford on 24 January 1941.
18. *FIU Report No. 65*, dated 28 May 1941.
19. *FIU Report No. 65*, dated 28 May 1941.
20. *FIU Report No,72*, dated 25 June 1941.
21. *The Signals History, Volume 5*, p. 136.
22. *The Signals History, Volume 5*, pp. 136 and 137.
23. *Ibid*, p. 136.
24. *Ibid*, p. 138.
25. The background to the design and testing of AI Mk.VI are taken from AVIA13/1047 and AVIA13/1048 as shown in the bibliography.
26. The provision of an IFF Mk.II or III transponder, AI beacon homing, anti-jamming circuitry, pre-set tuning of the transmitter and various training and maintenance requirements.
27. The incorporation of anti-jamming circuits, a 10 MHz tuning range in the receiver, auto-fly-back of the strobe, the provision of an extra indicator for the multi-seat variant and the blacking-out of the pilot's indicator whilst the strobe was searching for a target.
28. *The Signals History, Volume 5*, p. 138.
29. It should be noted that the SD Flight's base at Christchurch on the south coast, was a relatively short distance from occupied France, and that enemy jamming on the 1½ metre waveband could not be discounted.

30. *The Signals History, Volume 5*, p. 138.

31. The *FIU ORB* fails to identify the pilots who flew AA357 on the first air tests, however, it does identify Flight Lieutenant Ashfield, Flying Officer Clark and Flying Officer Ricketts as being involved.

32. *AI* in *The Signals History, Volume 5*, p. 26.

33. *Monica* was to have an undistinguished career in Bomber Command before being forcibly removed by the order of the AOC-in-C in the summer of 1944.

34. *The Signals History, Volume 5*, p. 139.

35. Hurricanes HV709, HW435, KX359 and KX754 are known to have been amongst the batch thought to have been modified to the AI Mk.VI standard.

36. Three Army Type 97 bombers on 15 May 1943 and two more on the 19 May.

37. Later Wing Commander R.P. Beamont, CBE, DSO*, DFC* and CO of No. 609 Squadron and No. 150 (Typhoon) Wing and English Electric test pilot from 1947–1979.

38. Report by Flight Lieutenant Beamont to Squadron Leader G.L. Sinclair, DFC, HQ No. 11 (Fighter) Group, dated 25 July 1942.

39. Francis Mason, *The Hawker Typhoon & Tempest* (Bourne End: Aston Publications, 1988), p. 49.

CHAPTER EIGHT

Consolidation and Baedeker

June 1941–October 1942

The decline in *Luftwaffe* night operations from June 1941 was matched by a steady increase in the number of night-fighter squadrons, and an expansion of the GCI network to cover the whole of England by the year's end. Although a much reduced component of the night bomber force, the *Kampfgruppen* retained in western Europe (Belgium, Holland and Occupied France), under the command of *Luftflotte* 3, continued to press home small and medium level attacks on the British Isles. These raids, which typically involved up to eighty aircraft, were carried out against Manchester, Birmingham and the Midlands, Southampton, Hull, Tyneside, Dover and across the eastern counties of England, between June and October. Although small by comparison to the raids of the Winter Blitz, these attacks continued to cause a steady stream of casualties amongst the civilian population. However, by October, raids against inland targets in England more or less ceased whilst the enemy concentrated its efforts against the English coastal counties and their ports.[1]

Throughout this period the original five AI night-fighter squadrons, Nos 25, 29, 219, 600 and 604 Squadrons, flying the Beaufighter If, were strengthened by new units and fresh crews from the OTUs. The first of these, No. 68, began its conversion from the Blenheim IVf to the Beaufighter If at High Ercall in May 1941, under its CO, Wing Commander the Hon Max Aitken. On the night of 16/17 June, the Squadron claimed its first enemy aircraft when Flight Lieutenant D.S. Pain, and Flying Officer Davies destroyed an He 111H-3x of 3./KGr 100 that crashed near Bratton, Wilts. Although operating the Beaufighter If through-out the Winter Blitz, 600 Squadron received the first of a new variant of the Beaufighter, the Mk.IIf, during April, with which to begin its conversion. Based on the Mk.I airframe, but with R-R Merlin XX in-line engines in place of the Hercules radials, the IIf differed little in terms of its performance and capability (see Appendix 3), but suffered a more pronounced swing on take-off that needed even greater care on the part of the pilot. AI Mk.IV was fitted as standard, with the aerial system being identical to that of the Mk.If. The Squadron was operational on the IIf by June, when the last Mk.I was flown out from Colerne for disposal. No. 406 (Canadian) Squadron received Blenheim I and IVfs in May for twin-engined practise, and Beaufighter IIs later in the month. No. 406 was unique in two ways; it was the first Commonwealth night-fighter squadron in Fighter Command, and second, it was established as a twin-engined night-fighter squadron without any previous experience in night operations. Under its CO, Wing Commander D.G. Morris, the Squadron formed at Acklington on 10 May and destroyed its first enemy aircraft on the night of 1/2 September, when Flying Officer R.C. 'Moose' Fumerton[2] and his R/O, Sergeant L.P.S. Bing, brought down a Ju 88A-4 of *Stab* III./KG 30 at Bedlington, Northumberland.

Nos 600 and 406 Squadrons were followed by No. 255 Squadron, a mixed Defiant and Hurricane unit, which began its conversion to the Beaufighter II at Hibaldstow, Lincs, in July 1941. The next unit, No. 409 (Canadian), became the second Commonwealth squadron to be committed to night-operations, when it exchanged its Defiants for Beaufighter IIs at Coleby Grange during August. The first of the allied squadrons from Occupied Europe, No. 307 (Polish), received Beaufighter IIs at Exeter during August and was declared operational the following October in the defence of the south-west. The final Beaufighter II unit to convert in 1941, No. 456 (Australian), completed its work-up at Valley before being declared fully operational on 5 September and placed on the readiness roster.

One more squadron began its conversion to the AI role in the spring of 1941. No. 85, based at Debden, Essex, which had previously operated a mix of Defiants and Hurricanes, received its first Havoc night-fighters during February 1941. Commanded by Squadron Leader Peter Townsend,[3] the Squadron had participated in the day-battle during August and September 1940, before transferring to cat's-eyes night-fighting with the Hurricane in the autumn. Developed from the Douglas DB-7 bomber, the Havoc Mk.I was intended as a stopgap fighter pending the availability of suitable quantities of Beaufighters (see Appendix 3). Equipped with AI Mk.IV and eight 0.303 inch (7.69 mm) Browning machine-guns in a nose pack designed by the Martin Baker Company, the Havoc was issued to No. 85 Squadron and to no other squadron in Fighter Command. Another version designated Havoc Mk.II, equipped with AI Mk.IV and a twelve-gun nose pack was also used by 85 Squadron. In performance terms, the Havoc realised its maximum speed at 13,000 feet (4,265 metres), some 2,000 feet (610 metres) lower than the Beaufighter and consequently performed less well (see Appendix 3). Nevertheless, it had a comparable service ceiling and carried a useful armament.

The Squadron claimed its first victim with the Havoc on the night of 13/14 June, when a Havoc I flown by a Canadian, Flight Lieutenant Gordon Raphael,[4] with AC1 Nat Addison[5] operating the radar, despatched an He 111H-4 of 3./KG 28 over the Isle of Grain.

By October 1941 Fighter Command's order-of-battle had strengthened considerably, with a total of twenty squadrons dedicated to night defence. Of these, thirteen were equipped with AI-fighters and seven with cat's-eyes Defiants, with three mounted on the more powerful Mk.II version (see Appendix 3). With respect to May 1941, this represented an overall 25 per cent increase in strength, but more importantly, a 54 per cent increase in AI-fighters with better trained crews. The GCI defences during this period had not been neglected and continued to expand across the country. By November there were thirty-two stations operational, covering an area that stretched from Cornwall in the south-west, along the south coast and north over the border into southern Scotland and across to Northern Island (see Figure 17). Developments were also in hand at TRE to improve the original AMRE Type 7 GCI to handle multiple targets.

Having established the radar equipped night-fighter as the best means of destroying the enemy's night-bombers, the Air Ministry and the Air Staff were side-tracked in 1941 to the development of a bizarre concept – the searchlight-carrying Turbinlite Havoc. Devised at a time when AI-fighters were in short supply, the Turbinlite hunter-killer concept called for an

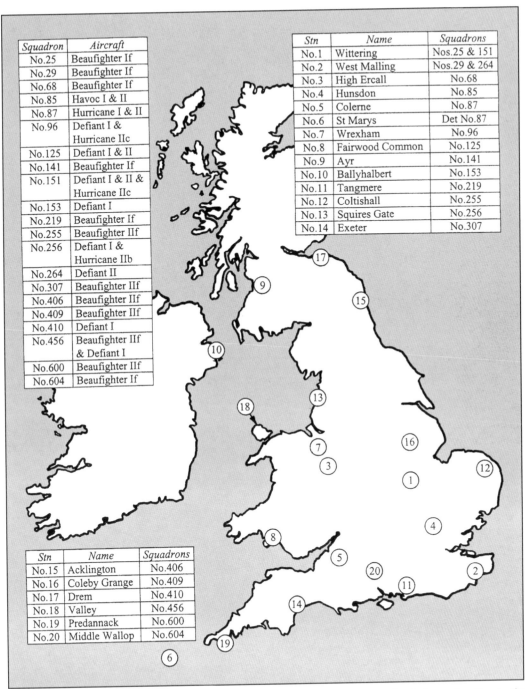

Squadron	Aircraft
No.25	Beaufighter If
No.29	Beaufighter If
No.68	Beaufighter If
No.85	Havoc I & II
No.87	Hurricane I & II
No.96	Defiant I & Hurricane IIc
No.125	Defiant I & II
No.141	Beaufighter If
No.151	Defiant I & II & Hurricane IIc
No.153	Defiant I
No.219	Beaufighter If
No.255	Beaufighter IIf
No.256	Defiant I & Hurricane IIb
No.264	Defiant II
No.307	Beaufighter IIf
No.406	Beaufighter IIf
No.409	Beaufighter IIf
No.410	Defiant I
No.456	Beaufighter IIf & Defiant I
No.600	Beaufighter IIf
No.604	Beaufighter If

Stn	Name	Squadrons
No.1	Wittering	Nos.25 & 151
No.2	West Malling	Nos.29 & 264
No.3	High Ercall	No.68
No.4	Hunsdon	No.85
No.5	Colerne	No.87
No.6	St Marys	Det No.87
No.7	Wrexham	No.96
No.8	Fairwood Common	No.125
No.9	Ayr	No.141
No.10	Ballyhalbert	No.153
No.11	Tangmere	No.219
No.12	Coltishall	No.255
No.13	Squires Gate	No.256
No.14	Exeter	No.307

Stn	Name	Squadrons
No.15	Acklington	No.406
No.16	Coleby Grange	No.409
No.17	Drem	No.410
No.18	Valley	No.456
No.19	Predannack	No.600
No.20	Middle Wallop	No.604

Figure 16. Night-fighter Squadrons at October 1941. With the *Luftwaffe*'s attention firmly focused on Russia, the night air defence situation improved rapidly, with many of the cat's-eyes squadrons beginning their conversion to AI-equipped fighters and Commonwealth units being raised on the Beaufighter II. From hereon the rump of the *Luftwaffe*'s bomber force in the west would encounter greater difficulty in penetrating UK airspace.

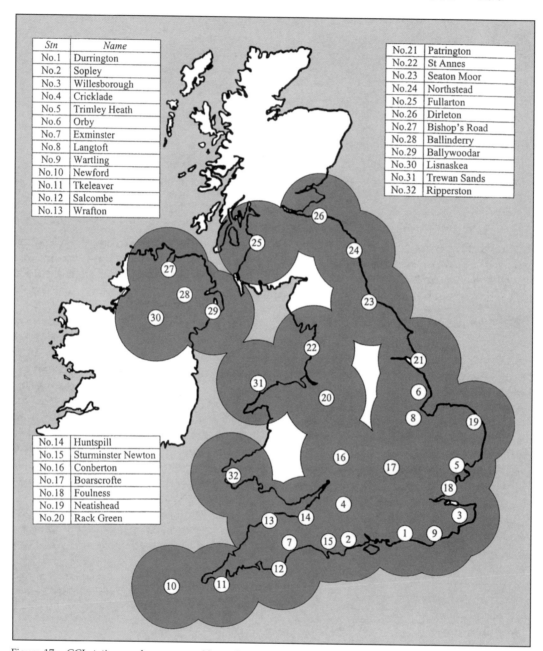

Stn	Name
No.1	Durrington
No.2	Sopley
No.3	Willesborough
No.4	Cricklade
No.5	Trimley Heath
No.6	Orby
No.7	Exminster
No.8	Langtoft
No.9	Wartling
No.10	Newford
No.11	Tkeleaver
No.12	Salcombe
No.13	Wrafton

No.21	Patrington
No.22	St Annes
No.23	Seaton Moor
No.24	Northstead
No.25	Fullarton
No.26	Dirleton
No.27	Bishop's Road
No.28	Ballinderry
No.29	Ballywoodar
No.30	Lisnaskea
No.31	Trewan Sands
No.32	Ripperston

No.14	Huntspill
No.15	Sturminster Newton
No.16	Conberton
No.17	Boarscrofte
No.18	Foulness
No.19	Neatishead
No.20	Rack Green

Figure 17.   GCI stations and coverage at November 1941.

aircraft to be fitted with a powerful 2,700 million candlepower Helmore[6] searchlight and AI radar, but no armament, that would fly in concert with a pair of single-engined fighters, usually Hurricanes. When a target was detected on its radar the Turbinlite was required to close in the normal fashion until it was judged the enemy was within range of the search-light (900 feet/275 metres), whereupon the light would be turned on and the formating

Hurricanes would attack in the classic day-fighter manner. The concept, which sounded fine in theory, failed to take into account the bomber pilot's natural reaction to roll and climb/ dive out of the beam, thus robbing the Hurricanes of their target.

To prove the feasibility of the concept, Fighter Command established No. 1422 (Night Fighter) Flight at Heston on 12 May 1941 and equipped it with Havocs and a band of specially trained electricians to maintain the lights. The choice of the Havoc as the carrier aircraft may be attributed to:[7]

- Its availability in relatively large numbers.
- Its ability to carry the searchlight, batteries and power supplies that weighed some 2,000–3,000 lb (910–1,360 kg).
- It was capable of having its nose modified to take the flat glass plate that fronted the searchlight.
- It possessed the necessary performance to intercept the enemy's bombers.

The workshop facilities at Heston were also responsible for the conversion of the first sixteen Turbinlites, which were completed by the end of July. The remaining 'production' was the responsibility of the Butonwood Repair Depot, on Merseyside, with the total number of Turbinlites delivered being of the order of 100 aircraft. Boston IIIs were also used for Turbinlite conversion (see Appendix 3).[8]

No. 1422 Flight's Havocs were fitted with AI Mk.IV or V and an aerial system similar to that used in the Havoc fighter, with the exception of the arrow-head transmitting aerial, which was duplicated either side of the flat nose panel to ensure adequate coverage on either beam. White span-wise strips were painted above and below each wing and formation-keeping lights were installed to assist the Hurricanes with station keeping during the hunting phase of the interception. Keeping station on the Havoc was easy for the Hurricanes, since their level speed was superior to that of the Douglas twin. Flight trials showed that when a target was detected by radar, the Hurricanes were instructed by R/T to move towards the enemy and make visual contact before the light was exposed. This method enabled the fighters to get to a firing position, whilst the Turbinlite maintained contact with the enemy and radioed any course corrections. Only when the fighters were properly positioned was the light exposed and the target illuminated.

The CO of No. 1422 Flight, Wing Commander A.E. Clouston,[9] regarded the Turbinlite's handling as 'virtually unaltered' from that of the Havoc, whilst 'the extra drag and speed reduction due to the flat nose were very small due to the Townsend Ring' fitted around the nose.[10]

It was originally planned that the Hurricanes would be drawn from day-fighter squadrons, with those involved in Turbinlite operations having an attached flight of Havocs. To this end, Fighter Command created ten Turbinlite Havoc Flights, Nos 1451 to 1460 (see Table 4).

In practice it proved difficult for the Hurricane squadrons to carry out their day-fighter duties and provide sufficient aircraft and crews to train in the night role. Consequently, the ten Havoc flights were expanded in September 1942 to full squadron status, with each having separate Havoc and Hurricane flights as shown in Table 5.

These squadrons were destined to be short lived. A reduced *Luftwaffe* presence, coupled with a poor interception record, an even poorer kill ratio and the availability of large numbers of Beaufighters and Mosquitoes, signalled the end of the Turbinlite experiment. By the late autumn of 1942, with few operational patrols being flown and many of the crews

TABLE 4.   TURBINLITE FLIGHTS AND ASSOCIATED HURRICANE SQUADRONS,
MAY 1941–SEPTEMBER 1942[11]

Flight	Established	Location	Associated Units
No. 1451	22 May 1941	Hunsdon	No. 3 Squadron
No. 1452	7 July 1941	West Malling	No. 32 Squadron
No. 1453	10 July 1941	Wittering	No. 151 Squadron
No. 1454	4 July 1941	Colerne	No. 87 Squadron
No. 1455	7 July 1941	Tangmere	Nos 1 and 3 Squadrons
No. 1456	24 Nov 1941	Honiley	No. 257 Squadron
No. 1457	15 Sept 1941	Colerne	No. 247 Squadron
No. 1458	6 Dec 1941	Middle Wallop	No. 245 Squadron
No. 1459	20 Sept 1941	Hunsdon	No. 253 Squadron
No. 1460	15 Dec 1941	Acklington	none allocated

TABLE 5.   TURBINLITE SQUADRONS, SEPTEMBER 1942–FEBRUARY 1943[12]

Squadron	Flight	Location	Disbanded
No. 530	ex-No. 1451	Hunsdon	25 Jan 1943
No. 531	ex-No. 1452	West Malling	31 Jan 1943
No. 532	ex-No. 1453	Wittering/Hibaldstow	1 Feb 1943
No. 533	ex-No. 1454	Colerne/Charmy Down	25 Jan 1943
No. 534	ex-No. 1455	Tangmere	25 Jan 1943
No. 535	ex-No. 1456	Honiley/High Ercall	25 Jan 1943
No. 536	ex-No. 1457	Colerne/Fairwood Common	25 Jan 1943
No. 537	ex-No. 1458	Middle Wallop	25 Jan 1943
No. 538	ex-No. 1459	Hunsdon/Hibaldstow	25 Jan 1943
No. 539	ex-No. 1460	Acklington	25 Jan 1943

posted to more appropriate day and night-fighter units, the squadrons were run down and disbanded early in 1943.

As far as is known, only one enemy aircraft was confirmed destroyed by a Turbinlite team, as Alfred Price recounts:

Early on the morning of May 1st [1942], Flight Lieutenant C.Wynn in a Havoc of No. 1459 Flight and Flight Lieutenant D. Yapp of No. 253 Squadron took off from Hibaldstow near Scunthorpe. The two aircraft headed northwards under Kirton Sector Control, before being handed over to the GCI station at Patrington near Spurn Head. The ground station directed the pair on to a suspected enemy bomber, until it was within range of the Havoc's AI radar. The Turbinlite aircraft and its satellite closed in and, as [on a] previous occasion, the Hurricane pilot caught sight of the enemy before the light had been switched on. This time the light remained off and Yapp closed to about 100 yards [90 metres]; he was just about to open fire when he was seen and the bomber went into a violent diving turn to starboard. Yapp followed the spiralling bomber, which he recognised as a Heinkel 111[13], firing several bursts which he saw striking the engines and the fuselage. After a fairly long burst the Heinkel entered cloud but shortly after that a large glow lit up the cloud for about two minutes; almost certainly this was from the wreckage of the German bomber burning on the surface of the sea (the position was some 20 miles [32 km] to the north east of Flamborough Head) and it was credited as destroyed.[14]

Squadron Leader Winn claimed a 'probable' on a Do 217 the following July and Pilot Officer Gunn 'damaged' another '217 a few nights later. A crew from No. 534 Squadron are also known to have destroyed a Stirling on the night of 4/5 May 1942, thankfully with no casualties![15]

Overall, the Turbinlite scheme provided little in operational terms and diverted valuable resources in terms of air and ground crews, radar, airframes and money, at a time when it may have proved more fruitful to produce conventional radar equipped Havocs. Fortunately, the redundant Helmore lights were not wasted. They were passed to Coastal Command for installation in aircraft engaged in the Atlantic U-Boat battle.

The autumn of 1941 continued with limited raids on England's coastal towns and ports, which enabled Sholto Douglas to continue his work of strengthening the night-fighter defences and General Sir Frederick Pile to redeploy his guns and searchlights to better effect. Whilst the GCI network was undoubtly the most effective way of directing night-fighters, the AMES Type 7 stations were only capable of directing one night-fighter at a time and there existed the possibility that the system could become saturated. To counter this possibility and enable greater numbers of night-fighters to participate in the battle, Sholto Douglas and Pile established the *Smack* system of searchlight control, late 1941.[16]

The *Smack* system required the repositioning of the searchlights into a series of 'Fighter Boxes', that were deployed in front of the gun defended areas which covered the approaches to important targets (industrial towns and cities, ports and naval bases). The size of these boxes varied in accordance with the available area in front of the gun defended areas, with, ideally, a depth of 32 miles (51 km) and a width of 14 miles (22 km) being required. The centre of the fighter box was indicated by a vertically pointing searchlight around which a night-fighter would 'orbit'. To increase the area covered, many of the searchlights were located individually and not in batteries, with a good number being radar controlled.[17] The first 12 miles of each box was identified as the 'Indicating Zone', whose primary purpose was to indicate the general direction of the enemy's approach. The remaining 20 miles (32 km) of the box was designated the 'Killing Zone', whose purpose was continuously to illuminate the enemy for cat's-eyes attacks (see Figure 18).

When an enemy aircraft penetrated the Indicator Zone, the vertical orbit searchlight was depressed to an angle of 20° and pointed in the direction of the raider. Simultaneously with this action, the night-fighter pilot was ordered to intercept the target along the direction indicated by the depressed orbiting searchlight. As the target moved further inland and entered the Killing Zone, its was continuously illuminated by one or more lights to enable the fighter to begin a cat's-eyes interception. However, if the fighter was equipped with AI and was able to make radar contact, the pilot could request the dousing of the searchlight by broadcasting a code-word. It was not unusual for the AI-fighter to intercept its target before it reached the end of the Killer Zone, provided it had the necessary 20–25 per cent speed advantage, which aircraft such as the Beaufighter and particularly the Mosquito possessed. However, should the pilot not be successful by the time he reached the end of the Killer Zone, the borders of which frequently overlapped those of the gun defended area, the fighter was required to break-off the pursuit and leave the target to the guns.[18]

The *Smack* system was fully implemented early in 1942, and in concert with the GCI network, remained the principal means of night interception for the remainder of the war.

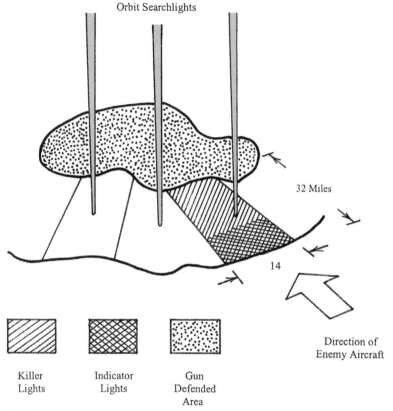

Figure 18.   The *Smack* system.

1942

The defence was further strengthened in the spring of 1942 with the introduction of the superlative Mosquito NF.II. Based on the design of the basic bomber/photographic reconnaissance model (see Appendix 3), the NF.II was capable of speeds up to 370 mph (590 km/hr) when carrying AI Mk.V radar, four 20 mm cannon, four 0.303 inch (7.69 mm) machine-guns and a requisite fuel load. The Mosquito entered Fighter Command service with No. 157 Squadron on 26 January, when the CO, Wing Commander Gordon Slade, delivered a dual-control Mk.II, W4073, to the unit's Castle Camps airfield in Cambridgeshire (Cambs). Deliveries were slow, with the first fighter Mk.IIs, W4087 and W4098, being delivered from No. 32 MU on 9 March, albeit without all of their operational equipment installed. By the end of that month, the MU had delivered fourteen aircraft and the first night training flights had begun.[19]

A second unit, the Wittering based No. 151 Squadron, received its first Mosquito II, DD608, on 6 April, following which its Defiant equipped 'A' Flight stood down from operational flying to begin its conversion to the new fighter. By the end of April the Squadron had a complement of sixteen Mosquitoes, including one dual-control aircraft, W4077, which enabled the station commander, Group Captain Basil Embry[20] to 'acquire' DD628 for his personal use. Despite both squadrons being keen to engage the enemy, there was little trade for them during April, when the time was put to good use in training the crews and ironing out a few of the aircraft's shortcomings.[21,22] This in retrospect, proved to be time well spent, for on the night of 23 April the *Luftwaffe* returned to the offensive over England.

A raid by 234 aircraft of Bomber Command on Lübeck on the night of 28/29 March, during which approximately 190 acres of the ancient parts of the town and 30 per cent of the built-up area were destroyed and some 320 of its citizens killed, brought forth an angry reaction from Hitler.[23] Outraged at the scale of the destruction, Hitler, in a signal to the *Luftwaffe* High Command (the *OberKommando der Luftwaffe* – OKL), ordered:

> that the air war against England is to be given a more aggressive stamp. Accordingly, when targets are being selected, preference is to be given to those where attacks are likely to have the greatest possible effect on civilian life. Besides raids on ports and industry, terror attacks of a retaliatory nature are to be carried out against towns other than London.[24]

In order to carry out the Füehrer's instruction, *Luftflotte* 3 was expanded by the incorporation of each *Kampfgeschwader's* fourth (IV) *Gruppe*,[25] the secondment of the minelaying and anti-shipping units and the transfer of two *Kampfgruppen* from Sicily and Russia. One of these units was the old KGr 100, now upgraded to a full *Kampfgeschwader* and returned to *Luftflotte* 3 for the occasion. To increase the effect even further, double night sorties were to be flown whenever possible.

The first of the so-called 'Baedeker Raids'[26] was flown against Exeter on the night of 23/24 April by forty-five aircraft drawn from KG 2, KfGr 106 and I./KG 100. Few people were killed on this night because a cloud layer masked the city and scattered the bombing. A Do 217E-2 from 5./KG 2, which crashed near Axminster, Devon, was destroyed by a 604 Squadron Beaufighter, crewed by Pilot Officer Tharp and Sergeant King. The following night, sixty aircraft returned to assault the city in two waves, killing eighty-eight people and injuring a further fifty-five, for no loss to the themselves. The situation could have been much worse, since the city had no balloon barrage and good moonlight enabled the crews to bomb from as low as 5,000 feet (1,525 metres).[27]

The offensive continued on the night of 25/26 April, when, using almost every available bomber in the west, 205 tons (210 tonnes) of bombs were dropped on Bath and, in error, on the south-east districts of Bristol, causing twenty-six significant fires and killing at least sixty-six people. The city was attacked the following night, where damage was reported as 'heavy' and a further fifty to sixty people were killed. This raid was intercepted and resulted in the destruction of an elderly 12./KG 2 Do 17Z by Squadron Leader John Topham[28] and Flight Lieutenant Strange of No. 219 Squadron and a Ju 88A-6 of 12./KG 3 by Flying Officer Wyrill and Sergeant Willins of No. 225 Squadron. Interestingly, both the aircraft destroyed were from their respective *Geschwader's* IVth *Gruppe*.

A change of target was enacted on the night of 27/28 April, when Norwich was intentionally raided for the first time. The raid, which lasted for thirty minutes, started thirty-five fires and killed fifty-three people. No enemy aircraft were lost to the defences. A second raid took place two nights later, with a reported casualty list of thirty killed and 'heavy damage' to the city. Once again, no enemy aircraft were reported as being shot down. Bath was raided again on 29/30 April, whilst York was hit for the first time. The only success that evening occurred over Elvington, Yorks, when a Hurricane IIc of No. 253 Squadron, flown by Warrant Officer Mahe, destroyed a Ju 88D-1 of 1./KfGr 106 in the early hours of 30 April.

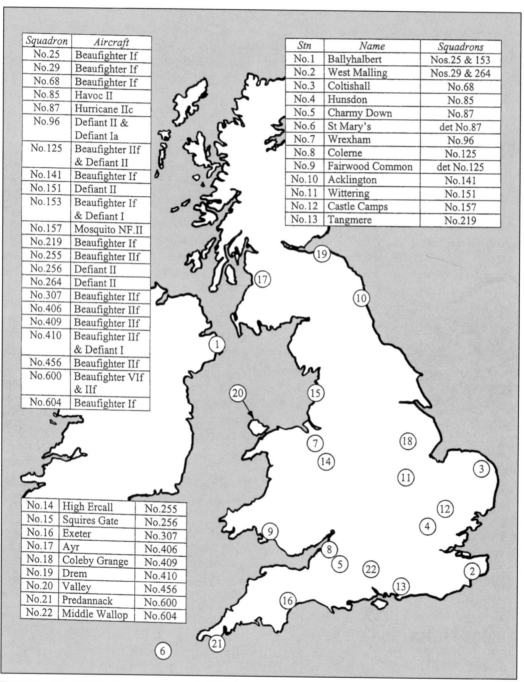

Squadron	Aircraft
No.25	Beaufighter If
No.29	Beaufighter If
No.68	Beaufighter If
No.85	Havoc II
No.87	Hurricane IIc
No.96	Defiant II & Defiant Ia
No.125	Beaufighter IIf & Defiant II
No.141	Beaufighter If
No.151	Defiant II
No.153	Beaufighter If & Defiant I
No.157	Mosquito NF.II
No.219	Beaufighter If
No.255	Beaufighter IIf
No.256	Defiant II
No.264	Defiant II
No.307	Beaufighter IIf
No.406	Beaufighter IIf
No.409	Beaufighter IIf
No.410	Beaufighter IIf & Defiant I
No.456	Beaufighter IIf
No.600	Beaufighter VIf & IIf
No.604	Beaufighter If

Stn	Name	Squadrons
No.1	Ballyhalbert	Nos.25 & 153
No.2	West Malling	Nos.29 & 264
No.3	Coltishall	No.68
No.4	Hunsdon	No.85
No.5	Charmy Down	No.87
No.6	St Mary's	det No.87
No.7	Wrexham	No.96
No.8	Colerne	No.125
No.9	Fairwood Common	det No.125
No.10	Acklington	No.141
No.11	Wittering	No.151
No.12	Castle Camps	No.157
No.13	Tangmere	No.219

No.14	High Ercall	No.255
No.15	Squires Gate	No.256
No.16	Exeter	No.307
No.17	Ayr	No.406
No.18	Coleby Grange	No.409
No.19	Drem	No.410
No.20	Valley	No.456
No.21	Predannack	No.600
No.22	Middle Wallop	No.604

Figure 19. Night-fighter Squadrons at April 1942. The night-fighter force at the beginning of the Baedeker raids in April 1942 was at its strongest since the war began and would remain at this level until the drawdown of fighter forces in preparation for the creation of the 2nd Tactical Air Force at the end of 1943. April 1942 also marks the beginning of the decline in the Defiant cat's-eyes squadrons.

An attack pattern for the Baedeker raids was clearly established by the end of April 1942, when the targets were identified as:

- Towns and/or small cities, many of which were of historical or cultural significance.
- Those that were poorly, or less well defended, where attacks could be made from low altitude.
- Being located close to, or within a short distance, of the coast.

Equally, the raids were of a short duration that rarely lasted more than one and a half hours and were undertaken during periods of good moonlight to concentrate the bombing.

With the bombers taking few casualties, raiding continued on into May, with raids on Exeter, Cowes, Isle of Wight, Norwich, Hull, Grimsby, Poole, Southampton and Canterbury. The raid on Exeter, on the night of 3/4 May, by thirty aircraft, inflicted heavy damage on the city, whilst that on Norwich on 8/9 May was reported as being 'not serious'. Serious damage was, however, caused in Hull on 19/20 May, when twenty-one people were killed and damage was caused to the docks and the Blackburn aircraft factory at Brough. Finally, the raid on 31 May/1 June on Canterbury, which lasted for some two hours, caused a number of large fires and killed approximately 100 people.

Throughout May the night-fighter defences began to take a steady toll of the raiders (see Appendix 4): four on 3/4 May, two on 7/8 May, one on 18 May, two on 23 May and one on 31 May/1 June. All were claimed by Beaufighter squadrons, whilst the Mosquitoes continued their operational training and work-up. The first operational Mosquito patrol occurred during the raid on Norwich on the night of 27/28 April, when Flying Officer Graham Little and his R/O, Flight Sergeant Walters, in DD603, from 157 Squadron, gained two radar contacts on friendly aircraft, but none on the enemy. No. 151 Squadron flew their first patrol on 30 April, when Flight Lieutenant Pennington flew DD613 in a search for the enemy. No. 157 Squadron came close to success on the night of 8/9 May, when, under the direction of Trimley GCI, Squadron Leader Ashfield, late of FIU, was vectored onto a 'hostile' but lost it. Although being one of the most experienced night-fighter pilots, Ashfield had difficulty in intercepting the Do 217, which crossed the coast at high speed and at low level (below 3,000 feet [915 metres]):

> ... for all its speed ... the Mosquito ... was none too fast, for the Do 217s were in the habit of going home at over 300 mph [480 km/hr], so that a speed advantage of even 60 mph [96 km/hr] meant a twenty-minute 100-mile [160 km] chase to catch a Dornier with only a 20-mile (32 km) start.[29]

Difficulties were also experienced by the GCI controllers in tracking these aircraft.

A third squadron, No. 264, flying the Defiant from Colerne, Wilts, received its first Mosquito, W4086, on 3 May and a Mk.III trainer, W4053,[30] a few days later to begin their conversion. The CO, Wing Commander Hamish Kerr, converted his crews to the Mosquito following two hours twin 'practice' on the Airspeed Oxford, or in some cases, directly from the Defiant. Kerr and Flight Lieutenant Lesk in DD642 flew the Squadron's first night patrol on 13 June.[31]

With three Mosquito squadrons added to Fighter Command's order of battle by June 1942 and their crews eager to claim the type's first confirmed victory, the *Luftwaffe* responded by reducing the night-sortie rate. However, towards the end of the month enemy activity increased and with it came the Mosquito's first success. On the night of 24/25 June, New Zealander, Wing Commander Irving Smith[32] the CO of No. 151 Squadron and his R/O,

Flight Lieutenant Sheppard, flying W4097, having already claimed an He 111 probably destroyed over the North Sea, were instructed:

> ... to search 4 miles [6½ km] due east, where he immediately picked up a Do 217E-4 ... of KG40. He closed unseen to 100 yards [90 metres], fired about four rounds from each cannon, and then the Dornier dived into the sea and exploded. Eight minutes later Smith contacted a Do 217 of I./KG 2 by radar and fired a long burst from 200 yards [180 metres], after which the bomber's wings were almost completely enveloped in flame. Even so the enemy fired a short burst, without hitting the fighter. Smith closed in again and finished off his foe.[33]

July proved to be another quiet month for the defences as the strength and serviceability of the Baedeker raiders gradually diminished. Nevertheless, the enemy managed to assemble sufficient aircraft and crews for three raids on Birmingham, three on Middlesborough and one on Hull, at a cost of sixteen aircraft destroyed. Whilst these raids represented the greatest effort expended by the *Kampfgruppen* since April, they delivered modest results and frittered away large numbers of untrained crews and their instructors from the IVth *Gruppen*. Amongst the losses were two Ju 88s and their crews from III./KG 26 that had recently completed their torpedo courses in Italy, prior to joining *Fliegerführer Atlantik* for attacks on allied shipping. The use of these specialists on bombing raids repressented an uneccessary waste of skilled manpower and deprived other *Luftwaffe* commands of their services – a fact the *Fliegerführer Atlantik* himself, Major Martin Harlinghausen, personally brought to the attention of *Reichsmarshall* Goering.[34]

Aware of the strength of the night defences, the *Luftwaffe*'s planners avoided routes that would take them over land, preferring instead to divert the bombers at low level out over the sea approaches, before turning them inland along the shortest possible line to the target. This was particularly true of the south coast, where the gun and fighter defences were amongst the strongest in the country. These tactics, which were first employed during raids on Birmingham in July, were enhanced by forward basing aircraft on the French coast. On the Birmingham raids the *Kampfgruppen* took off from their bases in France and turned west at low level to give Land's End a clear berth, before steering north up St George's Channel and along the Irish east coast. When abreast the Blackwater and Arklow light-ships, the aircraft turned east to intercept the Welsh coast at high speed and fly a direct course to their targets in Birmingham, which they bombed at a little over balloon barrage height. On completion of their task, the bombers, now lighter and therefore faster, reduced their altitude to near ground level and continued on an easterly heading to exit the English coast over Norfolk. Those aircraft that were short on fuel were able to land in Holland and return to their bases later in the night, whilst others returned directly to their French airfields.[35]

These tactics provided the defence with a headache, as the floodlighting beams of AI Mk.IV and Mk.V lost the low flying bombers in the ground returns – a situation that would not be remedied until the introduction of centimetric radar in 1943. Only when the bombers rose to a higher altitude over their targets were they vulnerable to interception, and even then, having released their ordnance they proved difficult to catch as they accelerated 'down hill' and out across East Anglia.

By August, however, the offensive had more or less collapsed, with just seven light raids being undertaken against Norwich, Swansea, Colchester, Ipswich and Portsmouth. This figure dropping even lower the following month, when just two towns, Sunderland and

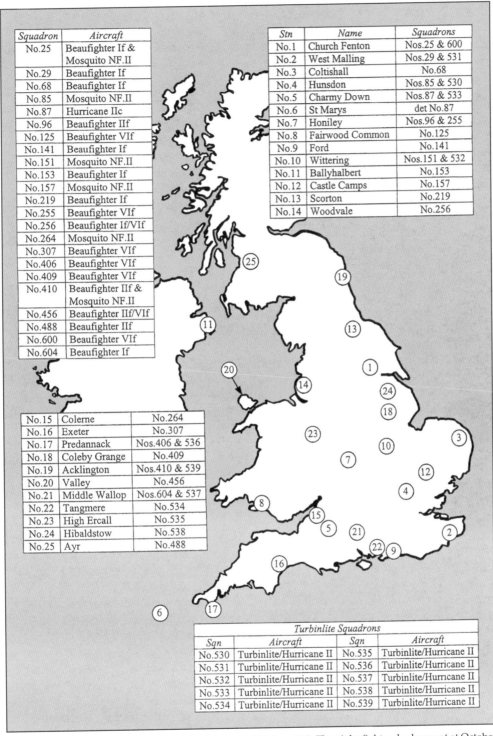

Squadron	Aircraft
No.25	Beaufighter If & Mosquito NF.II
No.29	Beaufighter If
No.68	Beaufighter If
No.85	Mosquito NF.II
No.87	Hurricane IIc
No.96	Beaufighter IIf
No.125	Beaufighter VIf
No.141	Beaufighter If
No.151	Mosquito NF.II
No.153	Beaufighter If
No.157	Mosquito NF.II
No.219	Beaufighter If
No.255	Beaufighter VIf
No.256	Beaufighter If/VIf
No.264	Mosquito NF.II
No.307	Beaufighter VIf
No.406	Beaufighter VIf
No.409	Beaufighter VIf
No.410	Beaufighter IIf & Mosquito NF.II
No.456	Beaufighter IIf/VIf
No.488	Beaufighter IIf
No.600	Beaufighter VIf
No.604	Beaufighter If

Stn	Name	Squadrons
No.1	Church Fenton	Nos.25 & 600
No.2	West Malling	Nos.29 & 531
No.3	Coltishall	No.68
No.4	Hunsdon	Nos.85 & 530
No.5	Charmy Down	Nos.87 & 533
No.6	St Marys	det No.87
No.7	Honiley	Nos.96 & 255
No.8	Fairwood Common	No.125
No.9	Ford	No.141
No.10	Wittering	Nos.151 & 532
No.11	Ballyhalbert	No.153
No.12	Castle Camps	No.157
No.13	Scorton	No.219
No.14	Woodvale	No.256

No.15	Colerne	No.264
No.16	Exeter	No.307
No.17	Predannack	Nos.406 & 536
No.18	Coleby Grange	No.409
No.19	Acklington	Nos.410 & 539
No.20	Valley	No.456
No.21	Middle Wallop	Nos.604 & 537
No.22	Tangmere	No.534
No.23	High Ercall	No.535
No.24	Hibaldstow	No.538
No.25	Ayr	No.488

Turbinlite Squadrons			
Sqn	Aircraft	Sqn	Aircraft
No.530	Turbinlite/Hurricane II	No.535	Turbinlite/Hurricane II
No.531	Turbinlite/Hurricane II	No.536	Turbinlite/Hurricane II
No.532	Turbinlite/Hurricane II	No.537	Turbinlite/Hurricane II
No.533	Turbinlite/Hurricane II	No.538	Turbinlite/Hurricane II
No.534	Turbinlite/Hurricane II	No.539	Turbinlite/Hurricane II

Figure 20. Night-fighter and Turbinlite Squadrons at October 1942. The night-fighter deployment at October 1942, whilst it appears significant, is exaggerated by the presence of the five Turbinlite squadrons, whose contribution to the country's defence was negligible and a drain on Fighter Command's resources. It should also be noted that the Defiant had disappeared from the inventory, bringing to an end the deployment of cat's-eyes night-fighters in Fighter Command.

Kings Lynn, were hit, before finally coming to an end in October with a single raid on Canterbury.[36]

CONCLUSIONS

Overall, the Baedeker Offensive may be regarded as a costly exercise in retaliatory bombing and one which failed to achieve its objective of terrorising the British public. It is estimated that in the fourteen significant raids between April and October 1942, some forty *Luftwaffe* bombers were lost, along with approximately 150 aircrew killed or taken prisoner. These losses, of which many were partially trained aircrews and, more importantly, their experienced instructors, could not easily be replaced and were therefore frittered away for no real return. Taken in the context of the losses being incurred by the *Luftwaffe* in the east, these were not large, but they did nevertheless represent a significant proportion of the bomber force available in western Europe.

From the RAF's viewpoint, the raids served to keep the defences on their toes and provide valuable operational experience to the new generation of night-fighter crews joining Fighter Command after the Winter Blitz. The Baedeker period saw the introduction of the first three squadrons of the Mosquito II, which, by the end of 1942, was beginning to make its presence felt within the night-fighter community and would eventually render the Beaufighter obsolete by the end of 1943. The tactics of the German bomber crews in flying to their targets at high speed and at a low level over the sea approaches, were difficult to counter, but did provide the impetus for the introduction of improved radars with narrower beams that were not vulnerable to the sea returns. Although the crews were not aware of it, such a radar was being developed and would be in service with Fighter Command during the early months of the new year.

Notes

1. Alfred Price [2], *op cit*, p. 127 and Winston Ramsey [3], *op cit*, pp. 36–77.
2. Later Wing Commander Fumerton, DFC*, who returned in August 1943 to command No. 406 Squadron.
3. Later Group Captain P.W. Townsend, CVO, DSO, DFC*, who commanded No. 605 Squadron at Ford in June 1942 and served as an Equerry to HM King George VI from February 1944 until the King's death in 1952.
4. Later Wing Commander G.L. Raphael, DFC*, who commanded No. 85 Squadron from May 1942–January 1943.
5. Later Warrant Officer N. Addison, DFC, DFM.
6. The Helmore light was developed by Wing Commander William Helmore and manufactured by the GEC Company.
7. L.M. King-Brewster, *Turbinlite Operations* (lost source), p. 25.
8 *Ibid*, p. 25.
9. Later Air Commodore A.E. Clouston, CB, DSO, DFC, AFC.
10. L.M. King-Brewster, *op cit*, p. 25.
11. Taken from John Rawlings, *Fighter Squadrons of the RAF*, pp. 463–6.
12. *Ibid*, pp. 463–6.
13. Contemporary accounts state this to have been a Do 217.
14. Alfred Price [2], *op cit*, pp. 134 and 135.
15. George Evans, *Bless 'Em All* (Privately published, 1998), pp. 133–7.
16. Alfred Price [2], *op cit*, p. 131.
17. *Ibid*, p. 131.
18. Alfred Price [2], *op cit*, pp. 131 and 132.
19. Martin Sharp and Michael Bowyer, *Mosquito* (London: Faber & Faber, 1967), p. 148.
20. Later Air Chief Marshal Sir Basil Embry, KBE, CB, DSO, DFC, AFC, and AOC-in-C Fighter Command from April 1949–April 1953.

21. The design and installation of flash-eliminators on the Browning guns, problems with the night-flying exhaust shrouds and engine cowlings and tail-wheel shimmy.
22. Sharp and Bowyer, *op cit*, pp. 148 and 149.
23. Martin Middlebrook and Chris Everitt, *The Bomber Command War Dairies* (Viking/Penguin Books), pp. 251 and 252.
24. Quoted in Alfred Price [2], *op cit*, p. 132.
25. Each *Kampfgeschwader* contained a IVth *Gruppe* that provided the operational training for new crews, which were not usually committed to combat until fully trained.
26. The term was invented by the German Press to describe the historic towns highlighted in the pre-war *Baedeker's Guide to Great Britain*.
27. The record of this and subsequent raids are taken from Winston Ramsey [3], *op cit*, pp. 101–57.
28. Later Air Commodore John Topham, DSO, DFC*, and the CO of No. 125 Squadron in October 1943.
29. Sharp and Bowyer, *op cit*, p. 150.
30. W4053 was the prototype Mosquito turret fighter that had been returned by RAE to de Havilland and converted to the T.Mk.III standard.
31. Sharp and Bowyer, *op cit*, p. 151.
32. Later Group Captain I.S. Smith, CBE, OBE, DFC*, who commanded No. 487 Squadron and led the raiding force during the famous attack on the Amiens Prison on 18 February 1944.
33. Sharp and Bowyer, *op cit*, p. 152.
34. Winston Ramsey [3], *op cit*, p. 103.
35. Winston Ramsey [3], *op cit*, p. 103.
36. *Ibid*, p. 103.

CHAPTER NINE

Centimetric Development
February 1940–August 1942

The limitations of 1½ metre radar that restricted its detection range to the height of the carrier-aircraft above the ground, caused profound problems for night-fighter crews searching for low flying minelaying aircraft operating in coastal waters. Tactics to overcome the problem were evolved within the night-fighter community, particulalry No. 29 Squadron, but these delivered few positive results from a select group of crews.[1] It was known that if the radar beam could be narrowed into a cone of radio energy, the range of the set would no longer be dependent on its height above the ground. Taking the analogy of a searchlight, or a torch beam, it is evident that when a reflective mirror is placed behind a light source, such as an electric bulb, the light is concentrated into a narrow beam. This analogy is also true of radio waves. Placing a reflecting dish behind the transmitting dipole will convert the beam from a 'floodlight' into a cone of radio energy. Decimetric radar development showed that an aerial's size was approximately half that of its wavelength, i.e. 75 cm. However, if this could be reduced to just 5 cm (a true wavelength of 10 cm) a reasonably narrow beam could be produced. Unfortunately, this in turn required an RF-source with an output in the kW range, which, in the mid-1930s, was beyond the capability of current technology.[2]

The possibility of building radar sets operating on centimetric wavelengths had been the subject of discussion at the 16th meeting of the Tizard Committee on 25 February 1936. At this meeting Watson Watt described the technical limitations involved in the generation of centimetric radio waves with sufficient power to drive a transmitter and aerial system. Nevertheless a number of theoretical studies undertaken prior to the outbreak of war had shown that it was indeed possible to design devices capable of generating the requisite shorter wavelengths. Similarly, the general nature of aerials and feeders capable of operating at these wavelengths was also well understood. These studies were, however, largely theoretical and poorly co-ordinated within the academic and industrial communities and certainly outside the knowledge of most radio engineers and designers.

The development of a national television service during the early 1930s stimulated research into valves that would operate efficiently at ever higher frequencies. The miniaturisation of valves to overcome the transient flow of electrons between the cathode and anode[3] was achieved in 1933, with the introduction of RCA's Acorn and Western Electric's 'Doorknob' valves. These offered the possibility in 1938, as Bowen recalls, of designing a 30 cm, pulsed, radar set:

using Samuel 2B250 valves from Western Electric in both the transmitter and receiver. As we were thinking only of the airborne application, the antenna were quite small and

124

consisted of cylindrical parabolas 2 foot [610 mm] square. With these, the range of detection on ground objects turned out to be only three-quarters of a mile [1.2 km]. This was too small compared with our existing 1.5 metre systems and the project was not continued.[4]

A similar requirement for shorter wavelengths existed within the Admiralty for the detection of U-Boats by the warships of a convoy's escort. From the very beginnings of airborne radar at Bawdsey the Director of Scientific Research (DSR) at the Admiralty, Charles Wright,[5] and his scientific staff, worked alongside Bowen and encouraged him in the development of centimetric wavelengths, which they considered were best suited to airborne and naval radar. From 1938 onwards it was the Admiralty who were the driving force behind the search for shorter wavelengths, whilst AMRE, which Bowen recalled, showed 'practically no interest' and 'anyone talking about centimetre waves was thought of as some kind of a crank'.[6,7]

The Admiralty were not discouraged by the apparent lack of interest in centimetric waves amongst AMRE's senior staff, as they had one significant administrative advantage over their colleagues at Bawdsey – they headed the Communications Valve Development (CVD) Committee that controlled the design and development of all valves on the behalf of the three service ministries. Following discussions between Bowen and Wright in the spring/summer of 1939, during one of the DSR's visits to Bawdsey, the pair agreed the basic requirement for a centimetric radar, as Bowen recalls:

> As we have seen, by far the most limiting factor in airborne radar at 200 Mc/s [MHz] was the enormous signal which came from the ground directly below the aircraft …. The only way of improving the system would be to project a narrow beam forward to get rid of the ground returns. I estimated that a beam width of 10 degrees was required to achieve this. Given an aperture of 30 inches [760 mm] – the maximum available in the nose of a fighter – this called for an operating wavelength of 10 centimetres. This agreed with Sir Charles's own assessment of what was required for Naval purposes.[8]

Following discussions with GEC during a visit by Tizard to the Hirst Laboratories at Wembley in November 1939 on the subject of centimetric wavelengths, and a further visit by Watson Watt in his capacity as DCD, GEC were persuaded to accept a contract on 29 December for the design of a short-wave AI set. Shortly after these meetings, the CVD Committee placed a development contract with the Mond Laboratory at Birmingham University[9] for the design of 10 cm transmitting and receiving valves.[10]

Impressed by the work done by EMI on the design of the modulator by Alan Blumlein and his team, Bowen, on behalf of the Air Ministry, co-ordinated a series of meetings in January 1940 between the Company and GEC, with the objective of encouraging an active collaboration on the development of the centimetre AI set. The meetings had a secondary objective on the part of the Air Ministry, who wished to see more companies involved in the development and manufacture of radar equipment. The first meetings were held during January 1940, with Drs C.C. Paterson, Gossling, E.C.S. Megaw and Jesty representing GEC and Dr Schoenberg, Alan Blumlein and Dr E.L.C. White, EMI. The alignment of these two companies in the field of radar research and development (R&D) brought together the foremost experts in the field of valve and circuit design and set the seal on British radar development for the remainder of the war.[11]

The Mond Laboratory group, led by Professor Mark Oliphant,[12] based their research on work carried out in 1936 at Stanford University by W.W. Hansen on the properties of resonating cavities. This device, termed a 'Rhumbatron', was made to project a stream of electrons across a resonant cavity, the direction of which could be reversed every half-cycle by external electro-magnets which forced it through the cavity before escaping. Although not taken up immediately by the radio industry, the rhumbatron principle was applied to the next valve in the chain, the Klystron.

This valve employed two rhumbatrons with the electric stream being applied first to one cavity and then the other. In the first cavity the stream was velocity modulated before passing into a second, where it was used to excite a resonator that was tuned to an appropriate frequency. Power was extracted from the second cavity by means of a loop, or similar device. The Klystron (also known as the 'Sutton Tube' after its inventor, Dr R.W. Sutton) could be made to operate as an HF amplifier, with the first cavity connected to an external source, such as that from an aerial. Alternatively, if the first cavity was driven by the power from the second cavity's catching loop, then the valve behaved as an HF oscillator. Further work during the spring of 1937 by Hansen and another American, Varian, produced Klystrons that were radiating radio energy in the 12 cm band. It was Oliphant's interest in this device, which he saw during a visit to the US in 1938, that was to have such far reaching consequences for British radar technology.[13]

Oliphant's Group first set about the task of trying to extend the power range of the Klystron. By the end of 1939 they had succeeded in running a Klystron-based oscillator with an output of 400 Watts, on a wavelength of 10 cm (3,000 MHz, or 3 GigaHertz [GHz]). However, their goal of producing a device that would give KiloWatts of power eluded them. Whilst the majority of Oliphant's scientists concentrated on developing the Klystron, two others, Dr John Randall and Dr Harry Boot,[14] spent their time exploring other avenues of interest. Observing their colleagues' lack of success, Randle and Boot set about examining the feasibility of using resonant cavities as a source of centimetric waves. From Hansen's work it was known that resonant cavities made from solid blocks of copper, demonstrated low losses at high frequencies, were stable in operation and were capable of dissipating large quantities of heat. Set against these advantages was the problem of getting sufficient power into the electron stream, or beam, in cavities of such small dimensions. Asking themselves the question, 'was there any means of using the desirable features of cavity resonators, whilst at the same time, avoiding the limitations of the electron beam?', the pair turned to work already undertaken in the US on the Magnetron valve.

Simple Magnetrons had been run in the US and Japan during the 1920s and these had produced small amounts of power on wavelengths as short as 10 cm. These devices comprised a cylindrical anode with a wire cathode running along its axis. The anode was then subjected to a powerful magnetic field, whose line of force ran along in the direction of the cathode. This arrangement made the electrons leaving the cathode pursue a curved path away from the anode, which they never reached. The oscillations produced by this technique were dependent upon the time it took for the electrons to describe their curved path. A further development, also established during the 1920s, split the anode cylinder into two half-cylinders. The resultant 'Split-Anode Magnetron', was capable of maintaining its oscillations when connected to a tuned circuit.[15]

Hansen's papers described resonators made from hollowed spheres and cubes which could not be matched to the cylindrical form of the Magnetron. Referring back to work carried out by Hertz[16] many years previously, Randall and Boot devised a version of his original loop-wire resonator circuit and a short-circuited three quarter wave line. Using these principles the pair designed a three dimensional version that comprised a cylinder 1.2 cm in diameter, with a 0.1 × 0.1 cm slot cut down one side. Six of these were then symmetrically spaced around a central cathode, to give an anode of approximately 10 cm. The overall length of the resonators was 4 cm. Air was extracted by an external pump and large electro-magnets were added to create the necessary magnetic field. With an applied potential of 16,000 volts, the resultant 'Resonant Cavity Magnetron' first ran on the bench at Birmingham University on 21 February 1940. It worked first time, demonstrating an output of 400 Watts on a wavelength of 9.8 cm (3.06 GHz). Although the valve obviously worked well, there existed some doubts as to whether it could be employed in a practical radar set, since Magnetrons were known to suffer from frequency instability. This problem was eventually overcome in August 1941, when one of Oliphant's scientists, J.Sayers, discovered that by providing a strap (a piece of copper wire) between alternate cavities, the oscillations were stabilised.[17]

During April 1940 Dr Megaw was approached in great secrecy with a proposal that GEC design a sealed-off version of the Cavity Magnetron, suitable for quantity production. Megaw's design transformed Randall and Boot's prototype by introducing two technical innovations. First, with the assistance of Mr S.M. Duke, he devised a method of sealing the Magnetron's cavities by means of compression-fused gold wire rings. Second, he replaced the original's tungsten filament cathode, by one that was oxide coated. Two of these 'Sealed Magnetrons' were delivered by GEC during June 1940, and when connected to a 10 kilovolts (kV)[18] supply, demonstrated a pulsed output of 10 kW. From this point onward, the possibility of building AI and other types of radars that operated on centimetric wavelengths had become a reality.[19]

Returning to 1939, the GEC team at Hirst began their investigations into centimetric radar by using the VT90 Micropup valve, which represented the best in conventional thermionic valve design, as a starting point. By reducing the spacing between the Micropup's electrodes and incorporating it as an integral part of an overall circuit design, two of the valves working in push-pull mode proved capable of generating some 10 kW of power on a wavelength of one metre (300 MHz). This was later 'squeezed' to give several kWs on a wavelength of 25 cm (1.2 GHz). By the summer of 1940, GEC had the prototype of a viable 25 cm radar mounted on the roof of their Wembley laboratory.

1940

Short-wave research, although significantly reduced by the need to support the installation work at St Athan, had not entirely disappeared within the Airborne Group. Whilst at St Athan, the Group was strengthened by the addition of two scientists who were later to play key roles in the development of centimetric radar. The first, Dr Bernard Lovell,[20] was recruited from Manchester University where he worked as an assistant lecturer in Professor P.M.S. Blackett's physics department, before joining AMRE in August 1939. In February 1940 Lovell was joined by a Cambridge marine biologist, Alan Hodgkin,[21] who was recruited by AMRE at Blackett's instigation to improve the prospects of them restarting airborne research at St Athan.[22]

With the help of A.H. Chapman, one of Blackett's technical assistants, Hodgkin and Lovell built and experimented with a horn-type aerial operating on a wavelength of 50 cm. The pair visited Wembley on 5 March, where they were shown the VT90 Micropup valve in operation by one of GEC's staff, Mr M.R. Gavin, that was then capable of producing several kWs of peak power. It was this device that convinced Lovell and Hodgkin of the practicality of creating narrower beams for AI and ASV radar sets.[23]

Returning to St Athan to further refine the horn design and continue their work as 'radar fitters', the two scientists were informed on the 29 March by an official from the DCD's office of a reorganisation of the Airborne Group and its members. The Group was to be dispersed with Gerald Touch going to RAE to continue his work in radar manufacture, where he was joined by other members involved in the maintenance of airborne radar, whilst the remainder were sent to RAF stations to support AI and ASV in the field. Lovell, Hodgkin and John Pringle, a zoologist from Cambridge, who had spent much of his time at Pembroke Dock supervising ASV installations on Sunderland flying boats, were to rejoin the main AMRE establishment before it was transferred to Worth Matravers in May 1940.[24] Bowen, it appeared would play no further part in the development of AI at AMRE.

Shortly after AMRE's arrival at Worth Matravers, Rowe established a small team under the leadership of Dr Herbert Skinner, to undertake an investigation into centimetric radar technology. However, this initiative to return AMRE to the radar research field was imperilled by the rapidly deteriorating military situation on the Continent. By the time Skinner's Group was assembled on the site at Worth Matravers the military situation in Europe had deteriorated. On 10 May, German troops invaded and secured The Netherlands and Belgium, and three days later crossed the Meuse and broke into France. With the French Army and the BEF under considerable pressure, Britain's pressing need was to withdraw the BEF and build up its armed forces and materiel in the south of England. With the need to build considerable quantities of radar equipment, Rowe, with the backing of the Air Ministry, ordered all leave in the Establishment cancelled, and directed that the greatest possible effort be put behind getting those radars then in development, into operational service as quickly as possible. It fell to Lewis, as the only AMRE officer aware of the developments at Birmingham, other than Rowe, to persuade his superintendent to retain Skinner's Group for the time being.

Skinner had been a lecturer in the Physics Department of Bristol University, before joining the Air Ministry during the winter of 1939/40. Following work on the installation of CH stations in the north of Scotland, Skinner was moved to the AMRE's main base at Dundee, where in February 1940, he was charged with the task of investigating centimetric radar techniques. Before leaving Dundee he was joined by another scientist, J.R. 'Jimmy' Atkinson. On their arrival at Worth Matravers the team's strength was increased by the addition of Lovell, Hodgkin, A.H. Chapman, Dr Philip Dee,[25] Dr W.E. 'Bill' Burcham,[26] a Canadian, A.G. Ward and Reg Batt. Dee, who had previously worked in the Cavendish Laboratory in Oxford, where he was senior to W.B. Lewis, was a 'distinguished physicist'[27] before joining the Air Defence Department of the RAE, in Exeter, where he ran a group that developed the PAC (Parachute and Cable) rocket for the defence of merchant ships. Burcham was a young scientist whom Dee had brought with him from Exeter, as were Ward and Atkinson, whilst Chapman and Batt were scientific assistants: Chapman from Manchester University, where he assisted Blackett in his cosmic ray experiments, and Batt from the Post Office Engineering Department.[28]

The Group established themselves in a hut on St Alban's Head alongside a CHL station, on the edge of the 200 feet (60 metre) cliffs that overlooked the English Channel. While the team's members sorted themselves out and got organised for their various tasks, Lovell erected his horn aerials and with a Split-Anode Magnetron as the source, had sufficient power to begin plotting their polar diagrams. He also began an examination as to the best radio-transparent materials with which to build a 'radome' to protect the horn aerials on a Blenheim IV trials aircraft. On the 22nd Dee travelled to Birmingham where he was shown Randle and Boot's work on the Cavity Magnetron, but such was the secrecy surrounding the device, he was forbidden from informing his colleagues of its existence. Indeed, on his return to Worth Matravers, he worked all day with Burcham using one of the Klystrons as the microwave source.

In his search for the right radome material, Lovell, who was now responsible for the aerial system, was advised of the suitability of parabolic 'mirrors' as potential replacements for the horns on the new system:

> Although the horns gave good polar diagrams they were more than a yard [0.9 metres] long and it seemed impossible that aerials of that kind could ever be used in a night-fighter. In the early days of June, Skinner had persuaded me to use a cylindrical (sectional) parabola and on the 11th I measured the polar diagram. The parabola was only 22 centimetres deep and, to my amazement, the polar diagram was the same as the horn of more than ten times the depth ... on 12th June, I moved the dipole across the mouth of the parabola and found that it shifted the beam direction by 8 degrees for 5 centimetre movement. This seemed an enormous step forward in producing a narrow beam from an aerial which one could imagine fitting in a night-fighter aircraft – and furthermore the possibility of moving the beam by displacing the dipole at the focus seemed to be such an exciting discovery that I wrote: 'This makes me regard the aerial problem as 75 per cent solved'.[29]

On 19 July, the first Cavity Magnetron (henceforth 'magnetron') designated Type E1188, was delivered to Worth Matravers to be installed in Dee's experimental set. The following day, by great good fortune, two spun parabola mirrors, that had been manufactured by the London Aluminium Company to Lovell's requirements arrived by train and were deposited at the nearby Swanage railway station. By 22 July Lovell had these erected and working to a klystron source to measure their resultant polar diagrams. He also confirmed his theory, that a moving dipole did indeed displace the angle of the beam by $\pm 25°$ without a serious signal loss.

The magnetron was quickly installed in the set, which Atkinson converted to pulsed operation by means of a modulator that he and Burcham designed, whilst Skinner and Ward collaborated on the receiver. By 8 August, with the equipment 'lashed' to a pair Lovell's 3 feet (0.9 metre) diameter parabola mirrors (one for transmission and the other for reception), echoes were obtained from a nearby coastguard's hut. On the evening of 12 August, with the parabolae mounted on a dual-swivel built in the Worth Matravers workshops, Burcham and Lovell witnessed the first detection of an aircraft on a wavelength of 10 cm, when, quite by accident at 1800 hours, an aircraft few along the coast a few miles away from the site:

> The next day (13 August) there were more echoes from aircraft. Dee had returned [he was absent from the site on the 12 August] and brought down Watson Watt and Rowe to see these historic echoes. In the afternoon we sent one of our junior assistants, Reg Batt, with a tin sheet and told him to cycle along the cliff in front of us. The ground rose slightly to the face of the cliff and where the young man cycled the ground was behind the tin sheet as viewed from our paraboloids. As we swivelled the paraboloids to follow

Batt and the tin sheet a strong echo appeared on the cathode-ray tube .... None of us had any inkling on that afternoon of the immense significance of that somewhat casual experiment.[30]

It was only later that the team appreciated the real significance of this experiment. It had been expected that with Batt cycling so low down on the radar's horizon the echoes from the tin sheet would have been lost in the ground returns. This characteristic of centimetric radar was later put to very good use by Lovell in his design of the H2S ground mapping radar for Bomber Command.

Sometime during late July or early August 1940, Rowe placed Dee in charge of the 10 cm project, which by then had been christened 'AI Sentimetric', or AIS for short. The leadership of the project caused Dee some concerns as he constantly strove with Rowe and Lewis to define his terms of reference. When Sir Frank Smith, the Controller of Telecommunications Equipment (CTE) at MAP, visited Worth Matravers on 8 August, Dee took the opportunity to complain to him about the lack of co-ordination within the AIS project. Whilst Dee's Group was hard at work developing their 10 cm AI system, a GEC team under the leadership of Mr G.C. Marris[31] was striving to build a 25 cm set under contract to the Air Ministry. Understandably, and unaware of Dee's work, the GEC team thought their system superior to that of AMRE and were equally reluctant to see the fruits of their endeavour wither away. On 22 August, Marris and one of his scientists, Mr D.C. Espley, were invited to Worth Matravers to see a demonstration of AIS during which Hodgkin, who was responsible for the receiver, tracked a Battle light-bomber tail-on, at a range of 2 miles (3.2 km). Of this event, Lovell commented in his diary that both men 'were very much sobered down by Hodgkin's apparatus'.[32] Dee's complaint and Hodgkin's demonstration had the desired effect. By the month's end, Rowe was instructed by the DCD to place the development of all centimetric radar in Dee's hands. This action effectively killed-off 25 cm radar research as far as AI was concerned and concentrated everyone's minds on 10 cm.[33]

To complicate matters further, Dee's Group was moved during late September 1940, from their home on the Worth Matravers site to another in a vacant girls school, Leeson House, in Langton Matravers, where they set up their laboratories in one of the old classrooms and a part of the stable block. Whilst the conditions at Leeson House were no worse than those at the main site the Group was separated from its workshops, requiring them to build one of their own using borrowed equipment. The site did, however, have one significant advantage, 'it looked down across the town of Swanage to the sea and in the distance to the Isle of Wight. This area, sometimes know as "centimetre alley", proved an excellent place for testing 9 or 3 centimetre radar on aeroplanes or ships and submarines'.[34]

By the late summer of 1940 only two problems remained to be overcome; beam scanning and a suitable display system and the duplexing of the transmitter and receiver to a common aerial. Concerning the scanning, one solution was to rotate the parabolic dish about its vertical axis whilst at the same time rocking it up and down to produce some form of helical scanning (see Appendix 1 for a description of helical scanning). A rotating helical scanning system suffered from two distinct disadvantages. First, because the dish was spun vertically half the scanning time was wasted when the dish was pointing backwards and away from the target. Second, the proposed system required the development of a rotating joint to carry the RF signal from the magnetron to the dish without any significant loss of power. The latter problem was further complicated by the need for the joint to work at high altitude without flash-over. At a meeting between Dee, Hodgkin and Espley's GEC group[35] at

Wembley on 25 October, GEC undertook to develop the helical scanner. They devised a method which neatly overcame the 'backwards pointing problem' by placing two dishes back-to-back, alongside a mechanism that switched the magnetron power to the forward facing dish. Dee and Hodgkin were hopeful the system might have been completed by Christmas 1940, but further discussions quickly established that it would take much longer. Unable to tolerate any delays in the test programme Hodgkin and Dee sought an alternative solution.[36]

The solution lay with a well known engineering company, Nash & Thompson, who built the hydraulically powered gun turrets in use in the Wellington and Lancaster bombers. One day, in the middle of July 1940, Hodgkin was introduced to Mr A.W. Whitaker of Nash & Thompson and appraised him of the work being undertaken at AMRE on the design of rotating scanners. Being, like Hodgkin, a Cambridge man,[37] the two got into conversation as to the best means of mechanically scanning the dish. By that time Hodgkin had gone so far as to persuade the Worth Matravers workshops to assemble a crude scanner in which the dipole was made to vibrate vertically at a rate of five times a second, whilst the dish was oscillating about an arc of 120° every two seconds. Whitaker immediately became interested in the project and promised to build an oscillating scanner to Hodgkin's specification, which now required the dipole to stand still, whilst the scanner moved in the horizontal and vertical planes.[38]

By the following November, Nash & Thompson had built a prototype model which worked well in the horizontal plane, but not quite so well in the vertical. The problem lay in a violent vibration that was set-up by the interaction of the vertical and horizontal motions, which prevented the scanner from running smoothly. The design was abandoned whilst Lovell and Hodgkin sought an alternative solution.

Discussing the matter further, Hodgkin had the idea of fixing the dipole and rotating the dish at high speed in a spiral fashion to produce a cone of radiation that lit up the sky ahead of the fighter. Hodgkin explained his idea to Whitaker, who got his design staff to build a scale model that was followed a few weeks later by a full size 28-inch (710 mm) dish which rotated at 1,000 rpm. Its eccentricity was made to deviate very rapidly over an angle of ±30°, giving an overall (conical) beam width of ±45°. By coupling a magslip to the scanner drive mechanism the resultant signal could be used to provide a timebase for the CRT that was in sympathy with the spiral motion. He assumed, correctly, that the target would appear as an arc on the CRT display whose radius from the centre of the tube would correspond to the range and its displacement from the centre would be its position relative to the fighter. As the fighter closed the radius of the arc would decrease as its circumference increased, as shown in Figure 21.[39]

During the late spring of 1940 AMRE were allocated Avro-built, Blenheim IV, N3522, which was delivered to Farnborough for the installation of a radio-transparent radome to protect the scanner in flight. Having already had a spell working at Farnborough Hodgkin was allocated the responsibility for the installation and ensuring the aircraft was ready for flight trials by the year's end. The task of manufacturing the radome had been sub-contracted by RAE to a company named 'Indestructo Glass' who operated from premises in west London. Once again Hodgkin was breaking new ground and delays were inevitable. No one had previously built transparent domes from materials that were light enough and yet strong enough to withstand the aerodynamic pressure created by the Blenheim as it flew along at over 200 mph (320 km/hr). Indestructo Glass proposed the use of 8 mm perspex, whilst RAE considered a composite material made from polystyrene fabric and Egyptian cotton, bound together with Formaldehyde resin, or one supplied by Messrs J. Burns Ltd made from a resin and paper composite. Tests at AMRE cleared the perspex and poly-styrene/Egyptian cotton composite as being 'satisfactory' from the radio absorption view-point, with RAE stating a preference for perspex.[40] This decision and a delay in manu-facturing the rear metal ring that provided the primary attachment point for the radome, delayed its delivery until December, and even then, additional reinforcement of the structure was required. The problem was finally solved in the spring of 1941 with the delivery by Indestructo of satisfactory models.[41]

The solution to the aerial duplexing problem was found by A.H. Cooke from the Clarendon Laboratory, who suggested that the electron-gun from a reflex klystron be removed and the tube filled with helium or hydrogen under a moderate pressure. The resultant 'Soft Sutton Tube', or 'Rhumbatron Switch', as it became known, was eventually produced as the CV43. Filled with water vapour the tube would flash-over during the transmitter pulse but would not conduct the weaker pulse from the receiver. One of the rhumbatron switches was built into a co-axial line system by Skinner, Ward and Mr A.T. Starr, such that the gas discharge action made the receiver present a high impedance path to the transmitter pulse. This arrangement worked extremely well and was to be employed successfully when built into co-axial line, or waveguide sections. From there on, rhumbatron switches were provided as standard features on very nearly all of the early types of centimetric radar systems.[42]

Dee's Centimetric Group was strengthed from the summer of 1940 by the addition of three new members, S.C. Curran, who had previously worked with Dee at Exeter, joined AMRE in November to work on the design of the modulator, whilst Dr A.C. Downing joined Hodgkin and another new member, J.V. Jelly, on the installation design in the Blenheim. At this juncture it should be noted that AMRE was retitled the Telecommunications Research Establishment, or TRE, in November 1940.

Burcham and Atkinson who were responsible for the transmitter were provided with a later version of the magnetron, the eight cavity Type E1189 (CV38), which worked on the slightly lower wavelength of 9.1 cm (3.3 GHz) and produced a peak output of 15 kW under test into a dummy load.[43] This was required to work to the modulator whose pulses triggered its operation. In light of their previous experience on 1½ metre AI, GEC were invited to collaborate on the design of the AIS modulator, where the main problem concerned the generation of the 15 kW peak pulse. This work had occupied Burcham and Atkinson throughout the summer with their ideas being centred around a thyratron and a pentode, as opposed to the GEC team who favoured the use of a single thyratron. TRE considered their design to be superior to that of GEC since it potentially offered higher PRFs and better pulse shapes. It was eventually agreed that the TRE proposal should form the basis of the modulator design. On test the modulator delivered a 1 μsec pulse with a peak power of 50 kW – more than enough the drive the magnetron. With the 1 μsec pulse and a PRF of 2,500 Hz, the minimum range was estimated at a little under 500 feet (150 metres).[44] It was desirable to use PRFs of a high value, but limitations on the rise and fall times of the pulse and the need to refresh the display at regular intervals precluded this. In the prototype installation the modulator was located towards the rear of the Blenheim's fuselage, with the magnetron pulse being conveyed to the transmitter in the nose by means of cables, plugs and sockets. Under test the transmitter delivered 10 kW.[45]

The AIS receiver contained an IF amplifier operating at 45 MHz, a value that had previously been chosen for use on 1½ metre AI. For the centimetric set a new mixer and local oscillator stage was designed by Herbert Skinner, who experienced some difficulty with the local oscillator which had to operate on a frequency of only 45 MHz different to that of the magnetron. After a series of trials with a split anode magnetron and a klystron Skinner and his co-designer, A.G. Ward, admitted defeat and called in Mr R.W. Sutton from the Admiralty Signal School in Portsmouth, who was the acknowledged authority on thermionic valves. By October 1940 Sutton had devised what later came to be known as the 'Reflex Klystron'. Unlike the ordinary klystron the reflex had only one resonant cavity and had its electron beam reversed, such that by careful choice of anode potential the beam was reflected back in the correct phase and with an increase in power. Outputs of 300 milliWatts[46] (mW) in the centimetre band were demonstrated. Many thousands of these valves were manufactured and later used successfully on all types of centimetre radar sets.[47]

1941

Blenheim N3522 was delivered to Christchurch in December 1940 for the installation of the AIS system and the Nash & Thompson scanner. This work was completed sometime during February 1941, with N3522 being temporarily fitted with a reinforced radome pending the delivery of the final version later in the year. Following system tests and calibration on the ground N3552 was made ready for flight trials in the first days of March. The Blenheim made its first AIS assessment flight on 10 March with Flight Sergeant Barrington at the controls and with George Edwards of GEC and Alan Hodgkin as observers. They had intended to carry out their task using a Battle light-bomber as their target, but when this failed to materialise the opportunity was taken to track an unknown aircraft which conveniently appeared on the scene. This was detected at a range of 7,000 feet (2,135 metres) when the Blenheim was flying at an altitude of 5,000 feet (1,525 metres). The display performed much as Hodgkin had expected but with one additional benefit:

> [The target] aeroplane appeared as an arc but we were surprised and pleased to find that the ground echo from the main beam showed up as an artificial horizon, which rose and fell as the pitch of the spiral scan increased and decreased. Later, when making interceptions, it proved helpful to know from this 'radar horizon' how much the pilot had banked the aircraft in response to the radar operator's instructions. We hadn't anticipated this useful property of the display, but as soon as we thought about it we realised that the result could have been predicted by an elementary exercise in solid geometry. In addition to the main ground return there was an altitude ring produced by radiation scattered directly downwards. In the early trials this was unpleasantly large and one of our first tasks was to reduce the altitude return to manageable proportions.[48]

Later trials against the Battle recorded ranges of 3 miles (4.8 km) when the Blenheim was flying at less than that height about the ground (in fact below 5,280 feet/1,610 metres). By changing the dipole's polarisation from horizontal to vertical and with improvements in the construction and stiffening of the perspex radome, and the incorporation of a 'rejecter circuit'[49] on the receiver's output, Edwards and Hodgkin reduced the altitude ring to manageable proportions over the following months.

The early trials with AIS adequately demonstrated the main advantage that had been claimed for centimetric radar; namely, the ability of the set to detect targets at ranges greater than the height of the carrier aircraft above the ground. However, earlier ground trials had

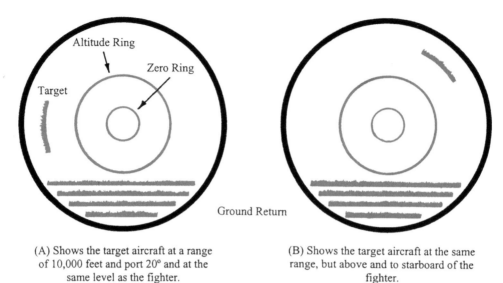

(A) Shows the target aircraft at a range of 10,000 feet and port 20° and at the same level as the fighter.

(B) Shows the target aircraft at the same range, but above and to starboard of the fighter.

Figure 21.   Indicating unit displays for AIS and AI Mk.VII and Mk.VIII.

led Dee's team to expect ranges approaching 10 miles (16 km), when in practice, only 3 miles (4.8 km), or less, was being achieved. The trouble was recognised as being due to heavy losses of transmitter power in the dummy loads of the hybrid transmit/receive system that was installed prior to the introduction of the rhumbatron switch, and the rapid deterioration of the tuning crystals in the mixer stage of the receiver. Both problems were overcome during June when Ward's receiver group installed the rhumbatron switch. By early June flight tests were demonstrating ranges close to those predicted.

The development testing of AIS occupied the six months from March to August 1941 during which Edwards, Downing and Hodgkin flew seventy trial flights in N3522. The majority of these lasted between one and two hours and were flown at heights varying from 500–15,000 feet (150–4,570 metres), over land and sea. Flying above 15,000 feet was not encouraged on account of the equipment arcing in the thin air found at these altitudes.

Towards the end of the trials period in July, at the behest of Sholto Douglas, the decision was taken to equip four Beaufighters with AIS for operational testing and service evaluation by FIU. Contracts were placed with GEC for the electronic units and with Nash & Thompson for the scanners.

A major reorganisation of the Air Ministry's trials units took place during the summer of 1941 and brought to an end the operational life of the SD Flight and its association with radar testing.[50] The first elements of this change took place on 1 March, when the Target Flight of ADEE and 'H' Flight of No. 1 Anti-Aircraft Co-operation Unit were brought together at Christchurch. On 1 August the SD Flight was amalgamated with Fighter Command's Fighter Experimental Establishment (FEE) based at Middle Wallop and the RAE's Blind Landing Detachment at Farnborough, to form the Telecommunications Flying Unit (TFU) at Hurn, near Bournemouth. The new Unit and its fifty-eight aircraft and crews were placed under the administrative control of No. 10 Group, Fighter Command, in whose operational area Hurn was located. The transfer of aircraft, personnel and their technical facilities to Hurn began immediately and was completed by 10 November 1941. On the departure of the

SD Flight Christchurch airfield reverted to the status of a satellite for Hurn under the command of Squadron Leader Theobald.

Burcham was placed in overall charge of the AIS installation in Beaufighter If, X7579,[51] which was delivered to Christchurch on 4 September. At the same time Hodgkin was despatched to the Bristol Aeroplane Company's Filton factory to assist the Company's engineers with the design of a perspex nose for the Beaufighter. Following its initial flight testing at Christchurch, X7579 was ferried to FIU's base at Ford for service testing on 27 November along with a second machine.[52] Burcham and a small party from TRE accompanied the Beaufighter to supervise the trials, service the equipment and instruct FIU's personnel.

FIU flew its first familiarity flights with X7579 on 30 November, in the hands of Wing Commander Evans, Flight Lieutenant Ashfield, Flight Sergeant Philips and Flying Officer Park. Testing proper began on 1 December and lasted until the 14 December, after which FIU were permitted to undertake operational flights against enemy minelayers operating in the Thames Estuary. During the first of these on the night of 12 December an FIU crew detected and chased a Junkers Ju 88, as the ORB records:

> By night Wing Commander Evans with Pilot Officer Mitchell mounted an anti-minelayer patrol over the Thames Estuary. Located, intercepted and damaged a Ju 88. The aircraft's windscreen was marked by oil from the enemy aircraft. This was FIU's first anti-minelaying patrol and the first recorded operational success of AIS.[53]

The experimental AIs in the two Beaufighters provided valuable training for the RAF's air and ground crews in centimetric radar techniques. The R/Os 'rapidly adapted themselves to the radial timebase presentation and ranges of 3½ miles (5.6 km) were obtained'. In FIU's experience, based on their assessment of the previous marks of AI radar, the new equipment 'gave less trouble than any prototype AI set' and 'warranted early introduction into two night fighter squadrons for use against low flying enemy aircraft'.

Whilst the experimental equipment in the two Beaufighters proved very useful Rowe recognised the need for some of TRE's staff to become involved with the testing of AIS systems under service conditions. In this respect Dee proposed to Rowe that Mr J.A. Radcliffe, who had previously served under Rowe at Dundee, be recalled from his post at the Army's Radar School at Petersfield to set-up up what would become the Post Design Services Group, or PDS. Radcliffe accepted the invitation and rejoined TRE on 27 July at Christchurch were he worked with Burcham. When his new group was formed it moved to FIU to work alongside the RAF on the trials of centimetric AI.

When Hodgkin was sure the equipment was working successfully, TRE conducted a series of demonstration flight in N3522 for visiting senior officers and fellow scientists. Amongst them was Air Chief Marshal Sir Philip Joubert de la Ferté, the ACAS, who had once famously commented during a centimetric radar demonstration that '1½ metre AI seemed satisfactory' for the RAF's needs and that AMRE 'should give priority to devising a 10 cm gun-laying equipment'. He fortunately changed his mind the following day after talking with Dee and Skinner. These visits which included the CO of No. 604 Squadron, Wing Commander John Cunningham, Squadron Leader Derek Jackson, who was then working on Fighter Command's behalf assessing the feasibility and consequences of radar jamming, and Group Captain Raymond Hart, who had set up the CH training school at Bawdsey.

However, the most important aspects of these visits was the comments and suggestions as to the future application of airborne centimetric radars. During April and August 1941,

Edwards and Downing discovered the ease with which AIS detected small objects on the sea and the way the coast stood out against the sea returns. The former was of considerable interest to the Admiralty and Coastal Command and on Test Flight No. 24 on 30 April 1941, HM Submarine *Sealion* was employed as a target for ASV trials, followed by HMS *Sokol* between 10 and 12 August. These flights and Bomber Command's requirements for a ground mapping radar, first identified on 26 October 1941, would lead Hodgkin and Lovell into the fields of alternative centimetric systems, for which another Blenheim, V6000, was made available to TRE in November 1941.[56]

CONCLUSIONS

The summer of 1940 marked a significant milestone in the development of airborne radar. The effective disbandment of Bowen's Airborne Group and the dispersal of all but two of its members to other parts of the Air Ministry's empire, or into the RAF, was to some extent countered by the restoration of radar research with the establishment of Skinner's Centimetric Group at Worth Matravers in May 1940. The Air Ministry's decision to encourage academia and the British electronics industry to become actively involved in the development of components for centimetric radars and the design of equipment, set the trend for the remainder of the war and the post-war period. In these areas, the pioneering work of Oliphant, Randle and Boot and GEC in the development and manufacture of the resonant cavity magnetron would prove to be not only one of the most significant technological inventions of the war, but also the key to the future of British, and, as we shall see in the next chapter, American radar.

With regard to airborne radar, Skinner, and later Dee's work in leading the Centimetric Group that took Randle and Boot's magnetron and installed it in an experimental set that would, in a relatively short period, be turned into a reliable prototype, requires our respect and admiration. It is perhaps one of Britain's better characteristics that usually surfaces in times of adversity, that a cobbled-together group of former academics, engineers, industrial specialists, technicians and administrators could produce the century's most important breakthrough in high frequency technology.

However, this was achieved at the cost of the man who had done more than anyone to establish Britain at the forefront of airborne radar development, and who, after May 1940, would find himself effectively side-lined from leadership in the field. With no prospect of work in AMRE, Edward Bowen, at the behest of his mentor, Sir Henry Tizard, was to travel to the US, where he would use his knowledge and experience in establishing radar research in America, and, ironically, ensure that country's pre-eminence in radar before the war was ended.

Notes

1. The technique, which required the night-fighter to fly at 6,500 feet (1,980 metres) and a search range of barely 1,000 feet (305 metres), was related to the Author by Mr Richard 'Jimmy' James, DFM, who flew with No. 29 Squadron at Digby and West Malling.
2. Jeremy Howard-Williams [2], *The Development of Air Intercept Radar 1939–45*, The American Airpower Heritage Museum & Midland College, 1993.
3. The operating frequency of a valve was dependent on the inter-electrode capacitance of the valve and the circuitry that surrounded it. By 'bolting together' two triodes in a 'push-pull' configuration, it was possible to raise the operating frequency to 200 MHz, but not much further.

4. E.G. Bowen, *op cit*, p. 142.
5. Later Sir Charles Wright, DSR from 1934–1946.
6. E.G. Bowen, *op cit*, p. 143.
7. Howse, Derek, *Radar at Sea* (London: Macmillan Press, 1993), p. 67.
8. E.G. Bowen, *op cit*, p. 143.
9. Similar contract were also placed with Professor Bleaney at Oxford and a team at Bristol University.
10. E.G. Bowen, *op cit*, p. 144.
11. E.G. Bowen, *op cit*, p. 144.
12. Later Sir Mark Oliphant, KBE, FRS.
13. Sir Bernard Lovell, *Echoes of War* (Bristol: Adam Hilger, 1991), p. 33.
14. Later Professor Sir John Randall and Professor H.A.H. Boot.
15. Sir Bernard Lovell, *op cit*, p. 33.
16. Professor Heinrich Hertz (1857–1894) proved that electricity could be transmitted as a series of electro-magnetic waves. The unit of frequency, the Hertz, is named in his memory.
17. Sir Bernard Lovell, *op cit*, p. 34 and 35.
18. 1 kV=1,000 volts.
19. Sir Bernard Lovell, *op cit*, pp. 35 and 36.
20. Later Professor Sir Bernard Lovell, KBE, OBE, FRS, is best known for his post-war work in the field of radio astronomy at Manchester University and Jodrell Bank.
21. Later Professor Sir Alan Hodgkin, KBE, FRS, who was elected to the Presidency of the Royal Society to succeed Blackett in 1970, was a Nobel Prize winner in 1963 and the Master of Trinity College, Cambridge, from 1978–1984.
22. Sir Bernard Lovell, *op cit*, pp. 1–24.
23. *Ibid*, p. 25.
24. *Ibid*, p. 25.
25. Later Professor Philip Dee, CBE, FRS, and the Head of Natural Philosophy at Glasgow University.
26. Later Professor W.E.Burcham, CBE, FRS, and the Oliver Lodge Professor of Physics at Birmingham University.
27. Sir Bernard Lovell, *op cit*, p. 29.
28. *Ibid*, p. 29 and Reg Batt, *The Radar Army* (London: Robert Hale, 1991), p. 14.
29. Sir Bernard Lovell, *op cit*, p. 39.
30. *Ibid*, pp. 41 and 42.
31. G.C. Marris was one of GEC's leading scientists at the Hirst Laboratory, who reported to the Laboratory's director, Paterson, for all work relating to communications and broadcasting, whilst D.C. Espley reported to Marris on matters relating to television.
32. Sir Bernard Lovell, *op cit*, p. 45.
33. *Ibid*, pp. 45 and 46.
34. Alan Hodgkin, *Chance & Design* (Cambridge: Cambridge University Press, 1992), p. 169.
35. C. Cherry, D.O. Hawes, R.J. (later Sir Robert) Clayton and Bernard O'Kane.
36. Alan Hodgkin, *op cit*, pp. 173 and 174.
37. Whitaker had studied Physics at Cambridge in the 1920s before taking up a career in engineering.
38. Alan Hodgkin, *op cit*, pp. 161 and 162.
39. Sir Bernard Lovell, *op cit*, p. 47 and Alan Hodgkin, *op cit*, p. 176.
40. Letter from AMRE to Chief Superintendent RAE, dated 11 November 1940, and filed in AVIA7/100.
41. Alan Hodgkin, *op cit*, pp. 165 and 166.
42. Sir Bernard Lovell, *op cit*, p. 63.
43. The date the E1189 magnetron was provided to Burcham and Atkinson is not stated by Lovell or Hodgkin, but it was installed in the prototype AIS system for the flight trials in March 1941.
44. A 1 μsec pulse represents an equivalent distance of 980 feet (300 metres) out and back, making the minimum range half this figure, or just under 500 feet (150 metres).
45. Sir Bernard Lovell, *op cit*, pp. 59 and 60.
46. 1 mW=1,000th of a Watt.
47. Sir Bernard Lovell, *op cit*, p. 61.
48. Alan Hodgkin, *op cit*, p. 181.
49. This unit, comprising three valves and a delay circuit, was connected to the receiver output to dampen the ground returns – AVIA26/6 provides a detailed circuit description.

50. The RAF's SD Flight was re-established in July 1951 at Sculthorpe, Norfolk, to undertake clandestine high altitude PR flights over the Soviet Union using specially adapted US RB-45C Tornado reconnaissance aircraft. The Flight undertook two incursions into Soviet airspace in April 1952 and April 1954, after which it was disbanded.
51. This aircraft was built at Bristol's Weston-super-Mare factory from a batch of 500 Beaufighters.
52. The serial number of this Beaufighter is not known, but it almost certainly was a Mk.If.
53. *FIU ORB*, pp. 28 and 29.
54. Sir Bernard Lovell, *op cit*, p. 45.
55. Later Professor Derek Jackson, DFC, AFC, FRS.
56. Alan Hodgkin, *op cit*, pp. 186 and 187 and Sir Bernard Lovell, *op cit*, p. 91.

CHAPTER TEN

Centimetric Crosses The Pond

September 1940–October 1941

At this juncture it is necessary to leave the story of AIS in Britain and look across the Atlantic towards the development of radar in the US and the impact it was to have on the production of AI in the years to come.

Although by June 1940 Sir Henry Tizard had resigned from active participation in the development of radar, his influence nevertheless lingered on. Throughout the first months of the war and especially during the early part of 1940, Tizard had been aware of Britain's limited industrial capacity and its ability to meet the needs of the ever expanding armed forces. This lack of capacity, particularly in the supply of electronic components, led him to conclude that without some form of outside assistance the country's manufacturing base would soon reach the point of saturation. In reality there was only one country that had both the capacity and the technological expertise to supplement British production – the USA. Whilst American industry was already committed to supplying Britain with arms, the possibility that it could be persuaded to increase its industrial output to maintain Britain in the war would require a bold gesture on the part of the Government. One solution, as Tizard saw it, was to offer the US government the technical details of Britain's 'war inventions'[1] in exchange for their co-operation in the production of military equipment.

Whilst this proposal, which was first suggested by Tizard early in 1940, was not well received in some quarters of Government, where Watson Watt and Sir Frank Smith were certainly against it. In order to 'test the waters' Tizard arranged for his friend and colleague, Professor A.V. Hill, to be posted to the British Embassy in Washington as a 'supernumerary air attaché'. Once there Hill's discreet contacts with the US government proved fruitful and elicited a response to the effect that 'the President would welcome a proposal from Britain to share all scientific knowledge of weapons and equipment'.[2] With the support of the Ambassador, Lord Lothian, the British government was encouraged to send a scientific mission to the US.[3]

With much of what Tizard had said becoming a fact by the summer of 1940, and with Churchill's and President Roosevelt's support, it was agreed towards the end of July that Britain would send a technical mission to the US in the very near future. Tizard was subsequently chosen to lead a mixed party of scientists and service officers with experience in the relevant weaponry fields and invited Bowen to lead on radar related matters. Whilst

Tizard was flown to America ahead of the main party to make the preliminary arrangements for what became known as the 'Tizard Mission', Bowen and the other members[4] prepared themselves for an Atlantic sea crossing during August. Whilst the other members made their preparations Bowen went to Birmingham University on 7 August to be given a thorough briefing on progress with the magnetron, followed by a visit to Wembley on 12 August to collect a single example, the last of twelve produced in the first batch of magnetrons, and a set of drawings.

Bowen and the rest of the mission set out from Liverpool on board the *Duchess of Richmond* on the evening of 29 August and reached Halifax, Nova Scotia, on the morning of 6 September. After a brief stay in Ottawa to brief the Canadian National Research Council (NRC), Bowen and Cockcroft arrived in Washington on the evening of the 11 September to join Tizard and the rest of the Mission.

During the early period of their discussions, Bowen and Tizard briefed officers[5] from the US Army and Navy on British developments in the fields of AI, ASV and CH, with the Americans reciprocating by describing their progress and conducting their guests around the US Navy's (USN) Naval Research Laboratory (NRL) at Anacostia and the US Signal Corp's Laboratory at Fort Monmouth. Nevertheless, Bowen, Tizard and Cockcroft were agreeably surprised at the Americans' level of progress in the radar field.

During this period no mention was made of the magnetron, but Bowen and Tizard did hint that Britain had developed a means of generating 'substantial pulse power at centimetric wavelengths'.[6] At this point the Americans agreed that the US Signal Corps and the USN would assume responsibility for the development of metric radar, whilst the US Government's National Defense Research Council's (NDRC) newly established Microwave Committee would direct centimetric research. Before he left to begin his preparations to depart the US, Tizard, in company with the NDRC's Dr Vannevar Bush and Dr Karl Compton, formalised this agreement on behalf of Britain, leaving Bowen and Cockcroft to continue the discussions with the Microwave Committee's Chairman, Dr Alfred Loomis. On the evening of 19 September at a meeting attended by Dr Loomis, Compton, Carrol Wilson, Admiral Bowen, Cockcroft and Bowen, in Loomis' apartment in the Wardman Park Hotel, Bowen:

> quietly produced the magnetron and those present at the meeting were shaken to learn that it could produce a full 10 kilowatts of pulsed power at wavelengths of 10 centimetres.[7]

Whilst the Americans were fully aware of the advantages of radars operating on centimetric wavelengths from their research at the Bell Telephone Laboratory, the General Electric (GE) Company and at the Massachusetts Institute of Technology (MIT), they had yet to come anywhere near devising a component that would generate the necessary frequencies with sufficient power. Over a weekend gathering at the home of Dr Loomis[8] in Tuxedo Park, New York, on 28/29 September, which amongst others comprised Wilson, Hugh Willis a member of the Microwave Committee and the Director of Research at Sperry Gyroscope Company, it was agreed that Loomis on behalf of the Microwave Committee would sponsor the production of thirty magnetrons by the Bell Telephone Company (henceforth – Bell), whilst others pressed Sperry and the Radio Corporation of America (RCA), GE and Westinghouse to become involved in the programme. Under a subsequent agreement Bell was allocated the task of producing magnetrons, Sperry the parabolic reflectors (scanners), RCA the CRTs, power supplies and IF amplifiers and GE the magnetron's magnets. RCA and Westinghouse later joined Sperry in the production of the reflectors.[9]

Bowen's 'Magnetron No. 12' was run for the first time in Bell's Whippany Laboratory, New Jersey, on 6 October, where, surrounded by an *ad hoc* set of components it generated approximately 15 kW on 9.8 cm. Much pleased with the result, Bowen returned to Washington. The following day he was urged to return to New York as something serious had transpired. Arriving at Bell's headquarters in West Street, Bowen was ushered to a conference room on the top floor of the building to be told that Magnetron No. 12 did not comply with the GEC drawings supplied to Bowen when he left Britain. X-ray examination of the component by the Americans clearly showed No. 12 to have eight cavities and not the six described in GEC's drawings. A trans-Atlantic telephone call to Megaw quickly resolved the American's concerns. Megaw had placed the order for the first twelve magnetrons with the Wembley Laboratory's experimental shop in the latter part of July and had instructed the foreman to make ten with six cavities, one with seven and the twelfth with eight! In his haste to collect a magnetron before departing Britain, Bowen had been given No. 12, or selected No. 12 himself, and hence the confusion.[10]

Faced with a decision to proceed with the eight cavity magnetron, or switch to the original six cavity device, those present in the conference room, Bell director Mervin Kelly, Wilson, Bell's lawyers and Bowen, agreed that Bell should proceed with the former as it was already proven to work. For this reason the magnetrons supplied by GEC had six cavities and those by Bell had eight. Bell had their first magnetron working in the laboratory within a few weeks and the full thirty completed shortly thereafter.[11]

With the magnetron's capability duly proven Bowen and Cockcroft were invited to place Britain's needs, with respect to radar, before their American hosts. With the Battle of Britain drawing to a close and the Night Blitz getting underway the pair expressed the country's requirement for a centimetric AI set to counter the *Luftwaffe*'s night raids, followed by a long-range navigation system for use in Britain's bombers and, finally, a 10 cm gun-laying radar for AA use. With a little prodding Bowen and Cockcroft were persuaded to draft outline specifications for all these equipments. Being at the same stage as Britain was before the war in terms of radar research the Americans were interested to hear their guests describe their experiences in establishing the facilities at AMRE/TRE. The British approach to the recruitment of civilians, drawn from academia to staff its laboratories and experimental establishments, struck a chord with the Microwave Committee members, and with Bowen's encouragement the Committee approved the creation of a national radar laboratory.[12]

The plan was approved by NDRC on 25 October with a budget of $455,000 being voted for the first year. A site for the new facility centred around MIT, at Cambridge, Massachusetts, since it already had a track record in microwave research and the location of a new group there would attract little attention from the outside world. MIT made 10,000 square feet (930 sq. metres) of laboratory space available, some of it on the roof of the Institute's main building, whilst the state governor placed a proportion of the National Guard hangar at Logan Field,[13] including a workshop, at NDRC's disposal. Initially entitled the 'Microwave Laboratory' of MIT but later changed to its more famous name of the 'Radiation Laboratory' or 'Radiation Labs', further to protect its anonymity, the new organisation was established during October 1940 with Dr Lee DuBridge[14] as its first director.[15]

By mid-November with recruitment in the hands of Professor Ernest Lawrence work at the new facility was well underway. On 6 November Bowen visited MIT to view the progress where he described the transformation as 'remarkable'.[16] DuBridge was established and hard at work organising the Labs and overseeing what Bowen describes as a 'veritable galaxy of talented people from all over the US', many of whom, in the years to come, would go on to win Nobel Prizes and some would join the *Manhattan Project* to build the world's first atomic bomb.[17] By the end of December the group at MIT comprising Professor Isidor Rabi who was responsible for magnetron research, Professor Louis Turner, modulators, Professor Alex Allen and Dr J.S. Hall, aerials and scanners, and Dr W.L. Barrow, waveguides, had assembled and run America's first 10 cm radar system.

1941

The Radiation Labs and Bowen's first task, designated 'Project No. 1', was the design and manufacture of a 10 cm AI set, suitable for installation in a night-fighter. Like AMRE before it, the people at MIT had to take the basic 10 cm set which comprised many parts weighing hundreds of pounds, and turn it into a set that could be installed in a twin-engined aircraft. To assist their progress the US Army Air Corps (USAAC) assigned the team a Douglas B-18 bomber, a machine that was the bomber derivative of the famous DC-3/C-47 transport aircraft, which was delivered to the National Guard hangar at Logan Field. Under the leadership of Professor Edward Macmillan the radar was installed in the B-18 towards the end of February 1941 and completed ready for flight testing in early March.

The 'B-18 radar' set employed a helical scanning system (see Appendix 1) that was driven electrically, with the R/O's presentation comprising a television-type display on a CRT some 5 inches square (130 × 130 mm), on which the target returns were shown in both range and direction, but not at the same time.[18] The first flight of the B-18 radar took place on 10 March when a number of clear echoes were received from the ground. Coincidently, this was the same date that George Edwards and Alan Hodgkin made the first flight of the prototype AIS at Christchurch.[19] By 27 March, with the set modified and improved, the team detected their first air-to-air target with Bowen sitting alongside Macmillan and two other scientists, Professor Luis Alvarez and Professor Ernest Pollard. It was, as Bowen describes, to be an epoch making flight:

> The target was a single engined machine borrowed from the National Guard and, to get clear air, we flew out over Cape Cod at about 10,000 feet [3,050 metres]. We did several runs on the target and got satisfactory ranges of detection of 2–3 miles [3.2–4.8 km]. We were delighted to see that this was not limited by the ground return – a prime reason for going to centimetre waves in the first place.[20]

Using a ship as a marker and with the scanner bore-sighted to the aircraft's line of flight, Bowen calculated the radar's range to be 10 miles (16 km). With such a good result and with sufficient fuel remaining the B-18 was flown down the east coast towards New London, Connecticut, which hosted one of the USN's bases, to see if they could detect a submarine. Thirty minutes flying time later they detected a fully surfaced submarine, alongside which they did several runs that confirmed a maximum range of 4–5 miles (6.4–8 km). With a second set of good results ASV was promptly added to the project list, which, as time went by, would subsequently attract as much research effort as did AI.[21]

On 29 April Bowen received a call from Air Chief Marshal Dowding who was visiting the US on behalf of MAP, and invited him to take a flight in the B-18 – a situation that was reminiscent of their first flight together from Martlesham Heath in 1939. With Macmillan operating the radar and accompanied by Bowen and DuBridge, Dowding was shown the set's capability against an aerial target. With the air-to-air range consistently showing returns at 5 or 6 miles (8–9.6 km) at 10,000 feet (3,050 metres), Dowding returned to the same question he had made in 1939 – what was the minimum range? With Dowding sitting next to him Macmillan guided the pilot to close on the target until the minimum range of slightly under 500 feet (150 metres) was reached. Dowding asked the pilot to hold the position, before climbing out of the nose compartment to see for himself. Whereupon 'he declared himself quite satisfied' and never raised the question of the minimum range again.[22]

Before leaving the US Dowding wrote to USAAC General Cheney to advise him of the equipment's performance and encourage him to press the US government to commit the system to production as quickly as possible. On his return to Britain Dowding expressed similar views to MAP, who in turn sent a telegram to the British Air Commission in Washington, requesting the despatch of ten of the NDRC Laboratory's 10 cm AI sets. In their turn, the Air Commission placed orders with the Western Electric Company (Western Electric) for a production version of the Radiation Labs' B-18 design, which in effect, was the first order for American radar placed in that country by Britain.[23]

To enable Western Electric to begin their task of engineering the 10 cm set for production, the Radiation Lab put in hand the manufacture of five updated copies of the B-18 prototype set, as the AI-10 radar (not to be confused with the later AI Mk.X), which incorporated all the modifications implementented during the set's airborne trials. One set was destined to replace the original prototype in the B-18, another to be shipped to Britain for evaluation, a third to NRC in Ottawa, the fourth to Bell Telephone and the final one to Western Electric as their working model.[24]

In order to have the British set delivered as quickly as possible Bowen proposed it be installed in a Douglas Boston for tests in the US, after which it could be flown across the Atlantic or dismantled for shipping. An alternative aircraft was the Douglas A-20 Havoc that was similar to the ones already in use by the RAF. Unfortunately, neither of these options met the Americans' aircraft delivery schedules. At this point the Canadians intervened and offered a Boeing 247D airliner,[25] then in use for experimental work at Rockcliffe. The offer was gratefully accepted and the aircraft was flown to Logan Field for the installation of the AI-10 to begin under the leadership of Professor Dale Corson. With the flight tests completed at Logan Field by early June 1941, the 247D was flown to Newark Airport for dismantling and cocooning as deck cargo for sea transport. Whilst the 247D was making its way across the Atlantic, Bowen and Corson flew to Britain in a Consolidated B-24 Liberator on 28 June to prepare for the Boeing's arrival at Liverpool Docks.[26]

Once reassembled and test flown, the 247D was delivered to FIU at Ford on 14 August and flown that day by Wing Commander Evans and Flight Lieutenant Ashfield. FIU's Operational Record Book (ORB) describes this and the other flights:

14 August 1941	Dr Bowen of TRE and Mr Heath of Canada appeared in their Boeing 247D equipped with American 10 cm AI. Wg Cdr Evans and Flt Lt Ashfield went solo in it forthwith – that is solo with about a dozen air advisors each.
15 August 1941	Practically everyone in the Unit flew in the Boeing, whilst it had its 10 cm AI tested.
16 August 1941	Flt Lt Ashfield undertook more flying in the Boeing, during which Sqn Ldr Hiscocks tried out the 10 cm AI.

17 August 1941   Flt Lt Ashfield and numerous others flew in the Boeing, whilst expert
observers, including Sqn Ldr Adams, examined the 10 cm AI.

As there are no further entries in the ORB, it is assumed the AI-10 completed is evaluation at FIU sometime after 17 August.[27] According to Bowen, who discussed the American 10 cm set's performance with Flight Lieutenant Ashfield:

At that moment [summer 1941], there were virtually no night attacks on Britain [true] and there was no pressing operational demand for anything better [than 1½ metre] radar. FIU tested the new equipment extensively and wrote a report about it which was full of approval; but I could not detect an urgent demand for production.[28]

Due to the rapid progress on the Americans' part with their 10 cm set, MAP was left with a situation whereby it had two AI radars, both of which appeared to have a comparable performance and employ similar technology. MAP therefore wished:

To obtain a direct comparison in performance with the British 10 cm set and to get a statement from FIU on the operational requirements so that American production could start in a form suitable to British needs.[29]

These issues needed to be resolved before the Air Ministry made a decision one way or the other. On the down-side the AI-10 radar employed helical scanning with which the British were unfamiliar, and from their viewpoint was untried in combat. Owing to the imminent delivery of the four Beaufighters to Christchurch to be fitted with AIS, TRE decided there was insufficient time to undertake a proper evaluation of the American system, since any delays incurred would jeopardise the timescales for the production of AI Mk.VII/VIII. This decision was undoubtly conveyed to MAP.

Two other factors also mitigated again the AI-10's adoption by the British; its displays and its bulk:

Although both range and direction could be determined from the position of the indication on the tube, they could not be read at the same time, and the range indication was of little use operationally. The whole equipment was too large and too heavy to be installed in a Beaufighter aircraft, but the necessary modification to reduce both bulk and weight were in hand. Despite the disadvantages mentioned, the equipment showed great promise and FIU reported that it would 'offer a weapon of the highest operational value against the night bomber'.[30]

Despite the American equipment having the better transmitter, the negative aspects of TRE's assessment, although a disappointment to Bowen did provide him and the Radiation Labs with sufficient encouragement to persevere and produce a set better suited to operational requirements. In retrospect Bowen nevertheless considered 'that both sides' had 'profited from the exchange'.[31] When Bowen returned to America in October the production of the AI-10 radar as the Western Electric SCR-520 had already begun (see Chapter 12). With the predicted delivery dates for the -520 being stated as May 1942, the decision was taken by the Air Staff in August 1941 to order 100 hand-built examples of the British AIS set as 'AIS Mk.I', on the grounds that the set would be the first system available to equip a small number of Beaufighter squadrons. AIS Mk.I was later retitled 'AI Mk.VII'. In the meanwhile, TRE was tasked with the development of an enhanced version that would include beacon working and IFF facilities. This system was designated AI Mk.VIII and was to be designed for large scale, quantity production.[32]

CONCLUSIONS

In his assumption that Britain's industry, particularly that associated with electronics, would quickly reach its maximum capacity under the demands of an increasingly technological war, Sir Henry Tizard was undoubtly correct. Having recognised that fact, he was also correct in proposing that America was the only country outside Europe in 1940 that had the technological base and the industrial capacity to meet Britain's increasing demands for war materiél. However, as he and the Prime Minister no doubt were aware, this dependence on the US came at a great price – the opening of Britain's scientific and technological 'treasure box' to American exploitation.

The Tizard Mission to Canada and the US in September 1940 took with it one of Britain's principal technological secrets, the magnetron valve, and the details of TRE's work on the development of centimetric radar. Once shown the device, the American's immediately realised its military potential and with the support of the President, the NDRC, the USN and the US Army Signal Corps, quickly established a research organisation based on British university lines. This, the Radiation Labs at MIT, was in turn backed by the research and manufacturing capabilities of US industry, who together produced the first US centimetric radar by Christmas 1940 and the first airborne equipment by the following March.

The transporting of the AI-10 radar in the Boeing 247 for trials in Britain beginning in the summer of 1941, and its pseudo 'competition' with TRE's AIS system, was decided in the latter's favour on practical grounds, rather than as Bowen suggests 'the not invented here' syndrome. AIS had a scanning system the British understood and trusted, it had a better display and was lighter and capable of installation in the RAF's standard night-fighter, the Beaufighter. However, in the longer term, Britain's gift of the magnetron valve and the subsequent development of improved AI radars (that are covered in subsequent chapters) by MIT and the US electronics industry, would, by the war's end, provide the Americans with a lead in the design and manufacture of airborne radar that persists to the present day. In this respect, it was Britain's defence electronics industry that would pay the price for Tizard's innovative thinking in the early months of 1940. Nevertheless, in retrospect, it was a reasonable price to pay to maintain Britain in the war.

Notes

1. E.G. Bowen, *op cit*, p. 150.
2. *Tizard, Sir Henry Thomas*, in The Dictionary of National Biography (DNB), 1951–1960, p. 979.
3. *Ibid*, p. 979 and E.G. Bowen, *op cit*, pp. 150 and 151.
4. The other members of the Tizard Mission were: Colonel F.C. Wallace (Army), Captain H.W. Faulkner (Royal Navy), Group Captain F.L. Pearce (RAF), Professor John Cockcroft (Army Research), Professor R.H. Fowler and Mr A.E. Woodward Nutt (Secretary).
5. These were, General Mauborgne, the US Army's Chief Signal Officer, Admiral Harold Bowen, the Director of the Naval Research Laboratory and others from the US Government's National Defence Research Council.
6. E.G. Bowen, *op cit*, p. 159.
7. *Ibid*, p. 159.
8. Dr Alfred Loomis trained as a lawyer before becoming an investment banker and a patron of the sciences. He served during World War One as an artillery officer, where he rose to become the Chief of R&D at the Army's Aberdeen Proving Ground and improved the Army's gunnery by the application of scientific methods. Post-war

he established his own laboratory at Tuxedo Park to study very accurate time keeping and the measurement of brain waves, for which he was recognised by the Franklin Institute.

9. Robert Buderi, *The Invention That Changed The World* (New York: Touchstone Books, 1997), p. 44.
10. E.G. Bowen, *op cit*, pp. 166–8.
11. *Ibid*, p. 168.
12. Robert Buderi, *op cit*, p. 44.
13. Now Logan Airport.
14. Lee A. DuBridge headed the Physics Department at the University of Rochester, where he helped build one of the first cyclotrons on the US east coast. He was also Rochester's Dean of the Faculty of Arts & Sciences and a proven administrator.
15. Robert Buderi, *op cit*, pp. 44–6 and E.G. Bowen, *op cit*, pp. 173–7.
16. E.G. Bowen, *op cit*, p. 177.
17. *Ibid*, p. 178.
18. *The Signals History, Volume 5*, p. 151.
19. E.G. Bowen, *op cit*, p. 183 and 184.
20. *Ibid*, p. 184.
21. *Ibid*, pp. 184 and 185.
22. *Ibid*, p. 185.
23. *Ibid*, p. 186.
24. *Ibid*, p. 186.
25. The Boeing Type 247D was an improved model of the all-metal, twin-engined airliner that entered service with United Airlines in 1932.
26. E.G. Bowen, *op cit*, p. 187.
27. *FIU ORB*, p. 26.
28. E.G. Bowen, *op cit*, p. 188.
29. *The Signals History, Volume 5*, p. 151.
30. Taken from *A Report on Science and War*, HMSO, 1945, and quoted in *The Signals History, Volume 5*, p. 151.
31. E.G. Bowen, *op cit*, p. 188.
32. *The Signals History, Volume 5*, p. 150.

CHAPTER ELEVEN

Centimetric Enters
the Battle

January 1942–April 1944

With the imminent delivery of the first batch of forty hand-built sets due sometime during the early part of 1942, TRE established two Service Liaison Sections at Christchurch in December 1941 to help manage the AI Mk.VII programme. One section was tasked with the supervision of the installation of the equipment in the Beaufighter, whilst the other dealt with the maintenance aspects. With an eye to the future, the RAF attached a number of radio mechanics to the Installation Section at TFU to work alongside them during the installation phase and prepare them for the forthcoming fitting of AI Mk.VIII, which was to be conducted at St Athan.

1942

The installation of the first Mk.VII sets began during the last week of February 1942, but proved more troublesome than had been expected and delayed the completion of the first two aircraft until 18 March. Thereafter, TFU delivered an average of three aircraft per week. The first squadron to receive AI Mk.VII Beaufighter Ifs, No. 29, was supported by a Service Liaison Party that accompanied the aircraft to West Malling to assist the Squadron's radio mechanics until they were competent to maintain the equipment themselves. Similar parties were despatched to No. 68 Squadron at Coltishall, No. 141 at Acklington and No. 604 at Middle Wallop. By May, all four squadrons were introduced to AI Mk.VII operations, but not wholly converted until later in the year.

As on previous occasions, it fell to FIU to claim the first enemy aircraft destroyed with AI Mk.VII on the night of 5/6 June, when Flying Officer Dickie Ryalls and his R/O, Flight Sergeant Owen, operating under the control of a coastal CHL station, detected a minelaying Do 217 over the Thames Estuary at a range of 4 miles (6.4 km). Obtaining a visual on the Dornier at 300 yards (275 metres), Ryalls closed and shot it down into the estuary below.[1]

Although hastily adapted to a semi-production standard to meet the pressing need to counter low flying enemy minelaying aircraft, AI Mk.VII proved to be an outstandingly successful radar by comparison with the early 1½ metre equipments. Before the first four squadrons were fully converted, seven enemy aircraft were claimed as destroyed before 15 May, in addition to several claimed as probably destroyed, or damaged. Until December

1942, when the first deliveries of AI Mk.VIII were made to the RAF, AI Mk.VII was the only centimetric radar in service with Fighter Command. However, the rush to get it into service coincided with a decline in *Luftwaffe* night operations, excepting for those aircraft engaged in coastal minelaying, where the majority of the Beaufighters' claims were made. Significantly, the motivation behind the introduction of centimetric AI, that of its ability to see long distances at low level was amply justified, and served to remove the *Luftwaffe*'s immunity to interception in this area. 'In the UK and later in the Mediterranean Theatre, AI Mk.VII was responsible for the destruction of just over 100 enemy aircraft, one for every set built'[2] – not a bad epitaph for what might be termed a 'lash-up'.

As has been said already, AI Mk.VII was delivered quickly to fulfil the requirements for low level interception, where its limited production was to serve as an interim solution until the arrival of the fully productionised AI Mk.VIII equipment. The AI Mk.VIII programme had as its objective the wholesale re-equipment of the night-fighter force with a state-of-the-art centimetric radar set, that incorporated AI Beacon and IFF facilities as standard. TRE's work on AI Mk.VII had identified a three stage production plan for Mk.VIII:

- Stage 1 – the delivery of 500 hand-built sets to an intermediate standard, designated AI Mk.VIIIA, and built by GEC, with manufacture beginning when the last Mk.VII had been completed. Although engineered in a similar manner to that of the Mk.VII equipment, this set was to have a higher output power and be able to interrogate centimetric AI beacons and the BABS Mk.I. The delivery dates for the first batch were given as 'the end of 1942', which GEC hoped to achieve by providing a pre-production prototype that TRE alone would approve. There were also to be no formal operational trials by FIU.
- Stage 2 – the main production version, designated AI Mk.VIII, was to be ordered in quantity, with a contract for 1,500 sets being awarded to E.K. Cole. This set would be similar to the Mk.VIIIA, but would be to an improved engineering standard with an even higher output power. It would also incorporate any modifications found necessary on the Mk.VIIIA production.
- Stage 3 – this was to complete the Mk.VIII production as AI Mk.VIIIB. It was to be a variant of the Mk.VIII with *Lucero*, AI beacon, BABS and IFF facilities that used the 1½ metre AI system.[3]

Orders for AI Mk.VIII therefore totalled some 2,000 sets, that were designed to replaced AI Mk.VII in the Beaufighter and AI Mk.IV and Mk.V in the Mosquito.

Up to this point only GEC and EMI were involved in the development of centimetric radar; however, with the need to expand the UK electronics industry's manufacturing capacity, MAP introduced the Southend-based E.K. Cole Company to the programme. E.K. Cole was not new to the radar industry having previously been involved in the manufacture of equipment for 1½ metre AI and ASV, but had no experience working with GEC, with whom, according to Alan Hodgkin, they did not get on:

It turned out that there was considerable rivalry between E.K. Cole and GEC and each firm was determined to engineer AI Mk.VIII in its own way, whereas the RAF rightly thought it essential to have identical sets of equipment. The reason why the two firms were involved was that the senior people at TRE, Dee, Skinner and Lewis, felt that GEC would always drag its feet because it hankered after its 25 cm project and that the only way to get things moving was to inject some competition into the system ….

I think the trouble was partly that in 1939 E.K. Cole had been suddenly evacuated from its home in Southend to the small town of Malmesbury where everyone, including the workforce, was thoroughly unhappy. In contrast, GEC's production factory in Coventry which had been badly bombed was relatively easy to deal with.[4]

The new set retained the spiral scanning technicque of the Mk.VII equipment and introduced a new version of the magnetron, the Type CV 64, or strapped magnetron. This device developed by Mr J. Sayers of GEC, was more stable and offered greater efficiency than the original E1198 magnetron. Operating in the 9 cm band, it demonstrated a peak output power of 50 kW, but required some 10 amps at 15 kV to drive it. From the installation designer's (Alan Hodgkin) viewpoint, this requirement caused a few headaches as there was little desire to pipe such high voltages around the aircraft. To overcome the problem, the designers chose to generate a 3 kV pulse in the modulator and route it to a pulse transformer, located close to the magnetron, where it was raised to the requisite voltage. Like the Mk.VII set, the scanner transmitter and receiver elements were duplexed using a reflex klystron.[5]

The Mk.VIII equipment was designed from the outset to work to IFF Mk.III. This set, which was trialled in December 1941, was produced in very large numbers and was destined to become the standard Allied IFF for the remainder of the war. It was capable of being interrogated by CH, CHL, GCI, GL, searchlight control (SLC) and 1½ metre airborne radars.[6] The procedure with the Mk.VIII AI for making an IFF challenge, was broadly the same as that for 1½ metre radar – the pressing of the IFF button would change the display to indicate a friendly aircraft. To accomplish this, the TRE team arranged that when the IFF button operated, the set would send out a stream of pulses at a PRF of 500 Hz, on a carrier frequency of 180 MHz. The response from a friendly aircraft was returned on the same carrier frequency and fed to the R/O's display, where it was displayed as a 'rising sun' ring around the screen at the same range as the target arc (see Figure 22).[7]

The designers chose to abandon the metric beacon system for night navigation and instead, proposed the installation of centimetric wide-band AI beacons at the principal

Figure 22.   IFF presentation for AI Mk.VIII.

(A) Shows the conventional display for an unidentified target aircraft.

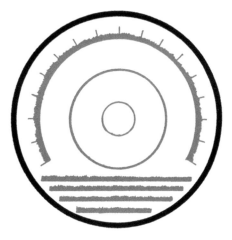

(B) Shows the same target with the radial IFF pattern that shows it to be 'friendly'.

night-fighter stations, 'which would receive, amplify and return an enlarged signal of frequency close to that received' …. 'Close to rather than equal, to help the return from the beacon stand out against the echoes from the ground'. When working on the beacon range, a multi-position switch converted the radial timebase to a vertical display that covered a range out to 95 miles (150 km), with deflections of this to the left or the right indicating the position of the beacon to port or starboard of the aircraft's track.[8]

The final refinement to the system concerned the incorporation of two radial timebases that were switchable between 8 miles (12.8 km) for the initial search and 2 miles (3.2 km) for the final stage of the interception. These comprised equi-spaced marker rings (concentric circles) at 2 mile (3.2 km) intervals on the 8 mile range and 2,000 feet (610 metre) intervals on the 2 mile range. On the 95 mile (150 km) beacon range, the marker rings were spaced at 10 mile (16 km) intervals.[9]

The AI Mk.VIIIA prototype set was tested by TFU at Christchurch sometime around March 1942, but does not appear to have been sent to FIU for service evaluation. Nevertheless, the prompt introduction of a production standard centimetric AI, would prove to be critical to the operations of Fighter and Bomber Commands in the coming months. With the German radar and night-fighter defences steadily increasing under the direction of their chief, *Generalleutnant* Josef Kammhuber, and Bomber Command's losses rising proportionately, TRE established a radio counter-measures (RCM) section to investigate the design of radar and radio jamming equipment. Led by Dr Robert Cockburn,[10] the new section began an investigation into the feasibility of jamming the *Luftwaffe*'s *Würzburg*[11] Flak and GCI radars by the use of reflecting dipoles. Allocated the codename *Window* by TRE, the task was handed to Mrs Joan Curran[12] who produced a radar reflector based on pieces of tin foil measuring 216 × 140 mm (8½ × 5½ inches) and 216 × 280 mm (8½ × 11 inches). Tests with these devices showed that if sufficient quantities were released, the resultant cloud could effect the screens of any radar operating at 200 MHz, or greater, and this included the British ones.[13]

With the potential greatly to reduce Bomber Command's losses, the CAS, Air Chief Marshal Sir Charles Portal,[14] ordered Bomber Command, on 4 April, to commence *Window* operations at the earliest opportunity. However, under pressure from Sholto Douglas, Portal was forced to rescind the order on 5 May, until a proper set of trials had been conducted to assess Fighter Command's vulnerability to *Window*-type jamming. At Lord Cherwell's[15] insistence the trials were to be conducted by Squadron Leader Derek Jackson, an officer on the staff at HQ Fighter Command and a former R/O, who had worked under Cherwell at Oxford prior to the outbreak of war.[16] Jackson's trials, which were undertaken at Coltishall, Norfolk, in co-operation with the station commander, Group Captain Ronald Lees, and the CO of the resident night-fighter squadron, No. 68, Wing Commander Max Aitken, took place between June and September 1942. Working with the local GCI station Jackson and his pilot flew some thirty flights against that station and other aircraft equipped with 1½ metre AI and AI Mk.VII. The results of the trials led Jackson to concluded that 'AI Mk.VII suffered severely' from *Window* jamming, whilst the older Mk.IV AI was affected 'to a lesser degree'.[17] The problem lay with the radial timebase presentation, which had difficulty in discriminating between a real target and a *Window* cloud. With these negative results now to hand, Sholto Douglas wrote to the Air Ministry asking that Bomber Command be prevented from using *Window* until such time as alternative air and ground-based radars were available to the RAF. Once again, with Cherwell's agreement, Portal deferred to his AOC-in-C Fighter Command.[18]

Cherwell was of the opinion that the employment of *Window* jamming would not bring about any substantial reduction in Bomber Command's losses to radar predicted Flak and searchlights. He calculated that only 1 in 1,000 of every aircraft despatched on a raid would be lost to these means – a statement he made towards the end of May 1942 in a minute to the VCAS, Air Chief Marshal Sir Wilfred Freeman – which he suggested was 'a poor justification for the premature release to the enemy of a device which is more effective against all our newer RDF than his, and to which we have no effective reply in sight'.[19] Unbeknown to Cherwell and the Air Staff, the enemy already had a *Window* device codenamed *Düppel*, which they had tested near Berlin in 1942 and later on trials over the Baltic. Fortunately for Britain, Göering for once appreciated its significance and ordered that all work on the system be destroyed to prevent the secret falling into Allied hands. Cherwell was not to know this, nor the importance of the *Würzburg* radars and the reliance placed upon them for Flak and night-fighter control, which would in reality have suffered greatly from *Window* jamming.[20] Whatever the reasons, the deployment of *Window* was now heavily dependent on the development of an AI radar that was not effected by a similar form of jamming.

The German jamming of the 1 ½ band during June and July provided a further incentive to get AI Mk.VIIIA into service as quickly as possible. With these objectives in mind, TRE was once again called in by the Air Ministry to take charge of the introduction to service of Mk.VIIIA. The diversion of TRE's resources to PDS duties was an unfortunate distraction from their primary responsibility for research, but with the knowledge and experience of centimetric radar, particularly in the field of maintenance, still limited within the RAF, there was little alternative. Consequently, TRE once more became involved in the work of instructing and supervising RAF personnel in the fitting and adjustment of radar sets.

To cope with the new workload the Service Liaison Sections were expanded, but with insufficient manpower available within the Special Installation Unit at Defford, the decision was taken to complete the installation of the first AI Mk.VIIIA sets in six Beaufighters at Defford with the help of RAF mechanics acting under instruction. When the first two aircraft were completed, one of each would be despatched to No. 32 MU at St Athan and No. 218 MU at Colerne, where they would serve as prototypes for all subsequent AI Mk.VIIIA installations. This approach worked well, for when the installation programme matured, it proved capable of delivering eighty aircraft per month to the operational squadrons – no mean achievement.

The modifications to the Beaufighter to accept AI Mk.VIIIA, like that of the earlier AI Mk.VII, were minimal and unlike that in the Mosquito (of which more anon) did not require a reduction in the aircraft's armament. The equipment boxes were installed in the cabin, with the receiver and display unit mounted in the R/O's position and facing backwards, with the scanner mounted on brackets that were bolted to the forward bulkhead in the extreme nose. A perspex radome, designed by the Bristol Aeroplane Company, that totally enclosed the scanner and its mechanism, completed the installation. In performance terms there was little to choose between the Beaufighter If fitted with AI Mk.IV or Mk.VIIIA.

The first pre-production AI Mk.VIIIA set was delivered from GEC's production line during July 1942, with flying trials to complete the set's type approval beginning at Defford on 25 August. These tests, which were conducted under TRE's supervision and in the presence of service representatives, lasted for two weeks and revealed a somewhat disappointing performance. Despite the Mk.VIIIA set having a more powerful transmitter output than that of AIS or AI Mk.VII, its effective range was measured at just 4 miles (6.5 km). Arcing was recorded as occurring at 22,000 feet (6,705 metres) which placed a restriction on the type's operational capability and tactical deployment. With the pressing need for the equipment to enter service, and following a series of recommended modifications, production approval was granted to GEC during October/November for the manufacture of 500 Mk.VIIIA sets.[21]

Technical difficulties on the production line at GEC delayed the delivery of the first ten Mk.VIIIA production models to the first week of December. The first two Beaufighter conversions went to FIU for operational testing, with the remainder being allocated to the squadrons selected for conversion: Nos 219, 68, 29, 125, 406, 604, 141 and 488 (in that order). FIU's first Beaufighter arrived from Defford on 23 December and flew its first test flight on 26 December, with Flight Lieutenant Davison at the controls and Flying Officer Mitchell as R/O. A Mr Barber accompanied the FIU crew on this flight but it is not known who he represented. Similarly, Mr Barber accompanied Flying Officer Crook, Pilot Officer Austin and Flying Officer Reece on a further test flight on 28 December. On both occasions the target aircraft was another Beaufighter. On 30 December Flight Lieutenant Ryalls flew the second AI Mk.VIIIA Beaufighter from Defford, to replace the first machine which was declared u/s on account of a 'smashed stern frame'. Whilst the operational testing continued at FIU, the first AI Mk.VIIIA Beaufighter was delivered to No. 219 Squadron at Scorton to begin their conversion to the new radar.[22]

1943

The first operational success with AI Mk.VIIIA occurred on the night of 20/21 January, when an FIU Beaufighter flown by Flight Lieutenant Davison with Fligt Lieutenant Clarke operating the radar, destroyed a Do 217, as the ORB records:

> They took-off in heavy rain and under pre-arranged radio silence and after reaching Beachy Head they were handed over to Worthing control. They were at once vectored onto a bandit. Since retaining contact they followed it through violent evasive manoeuvres and lost contact on three occassions. This chase lasted for approximately half an hour and the aircraft had never broken cloud since it had taken off. The last contact was lost when the pilot's gyro horizon spun. A new vector from the ground control was given and contact established on another, or the same bandit, travelling north at a range of 4 miles [6.5 km] to port. The bandit was made to pass in front and appeared at 4,000 feet [1,220 metres] on the starboard side. The pilot turned visually onto it and recognised it as a Dornier 217. Just before getting into a position astern he saw he was being attacked and dived to port, a short burst was given, resulting in a few strikes, but no apparent damage was done. AI contact was maintained and another burst was delivered onto the bandit as it climbed to starboard. The Dornier returned fire from its top turret, but this went wildy to starboard. His port engine caught on fire. He made an involuntary swing to port and caught another burst which set the fuselage on fire. He seemed to go almost on his back and dived vertically into the clouds.[23]

This was a hard battle which suggests the pilot was very experienced in executing a series of evasive manoeuvres and his aircraft was possibly equipped with some form of tail-warning radar.

Following the delivery of the first Mk.VIIIA equipped aircraft to No. 219 Squadron in December and January the R/Os quickly adapted to the new radar and became adept at its operation. Maximum ranges approaching 4½ miles (7.25 km) were regularly achieved, whilst the greater coverage of the Mk.VIIIA scanner gave the set an edge over the earlier Mk.VII equipment in its ability to follow 'jinking' targets. By the end of January No. 219 had completed its conversion, followed by No. 68 in February and Nos 125, 29 and 604 between March and May. During the conversion the opportunity was taken to equip Nos 219, 68, 29 and 604 with the more powerful Beaufighter Mk.VIf (see Appendix 3). Serviceability on these squadrons proved to be better than that experienced with AI Mk.VII, but on a number of occasions after running for a few hours the modulator broke down. The problem was eventually traced to overheating and arcing of the unit's CV57 valves which were being overrun, or the breakdown of the pulse transformer – problems already identified by Hodgkin and Edwards in the late summer of 1942.[24] As a palliative it was found that by carefully running-in the valves their lives were increased (there were three CV57s in the modulator) and by introducing modifications to the set to relieve the stress, that matters improved. The problem was finally overcome later in the year with the introduction of a more robust version of the CV57, which restored the modulator's operation to that of a 'reasonably high standard'.[25]

AI Mk.VIIIA's introduction to service coincided with a modest increase in conventional night-bombing and the introduction of nuisance raids by *Luftwaffe* fighter-bombers based along the French coast. During January the *Kampfgruppen* of *Luftflotte* 3 had been strengthened in numbers and in capability, with the introduction to KG 2 of the 'K' and 'M' variants of the Do 217 and the A-14 model of the Ju 88 to KG 6. The Do 217K-1 which entered service with KG 2 in the autumn of 1942, introduced a fully rounded nose, 1,700-hp BMW 801D radial engines, additional defensive machine-guns and a useful 20 mph (32 km/hr) increase in overall speed. The 'M' model differed from the 'K' only in its power-plants which were 1,850 hp Daimler-Benz DB603A in-line engines. The pressing need for night-fighters in the *Luftwaffe* meant that a relatively small number of 'M' models were delivered to the *Kampfgruppen*. The Ju 88A-14 was an updated version of the earlier A-4 variant that had been in service during and after the Battle of Britain and throughout the Winter Blitz. With increased armoured protection for the crew, balloon cable cutters as standard and other minor refinements, the A-14 was otherwise comparable in performance to the earlier model. By mid-January *Luftflotte* 3 had some sixty bombers of both types in its inventory.[26]

With its bomber strength thus increased and revitalised *Luftflotte* 3 assembled 118 aircraft to raid London in two waves, flying double sorties on the night of 17/18 January, making it the largest force to strike the capital since 10/11 May 1941. With the ground defences strengthened and an adequate supply of GL radars for the searchlight batteries, three of the five successful interceptions that night were attributed to searchlight illumination: a Ju 88A-14 of 1./KG 6 destroyed by Wing Commander C.M. Wright-Boycott and Flying Officer E.A. Sanders of No. 29 Squadron, another A-14 of *Stab* I./KG 6, again by Wright-Boycott and Sanders and a Ju 88A-4 of 2./KG 6 by Squadron Leader I.G. Esplin and Flying Officer A.H.J. Palmer, also from 29 Squadron and all flying AI Mk.IV Beaufighters.

Despite the problems with its modulator AI Mk.VIIIA became involved in the night battle during February, when the first squadron claim was submitted by No. 219 on the night of 3/4 February with Flight Lieutenant Willson and Flying Officer Bunch destroying a Do 217E-4 of 3./KG 2 over Yorkshire. This claim was quickly followed by 125 Squadron who claimed two on the night of 16/17 February when Pilot Officer Newton and Sergeant Rose despatched another Do 217E-4 from KG 2 off the Gower Peninsular and the CO, Wing

Commander Clerke, and Pilot Officer Spurgeon got a third, also from KG 2, near Beaminster, Dorset.

Although the increase in strength over that of 1942 was no doubt appreciated by the bomber crews of *Luftflotte* 3 their efforts by comparison to Bomber Command were almost insignificant. On the night of 1/2 March a force of 156 Lancasters, eighty-six Halifaxes and sixty Stirlings, supported by six Mosquitoes and a diversion force of forty-nine minelaying Wellingtons and Halifaxes, bombed Berlin and caused more damage than ever before – twenty factories badly damaged, 875 building destroyed and 191 people killed.[27] Hitler's response to this raid and the *Luftwaffe*'s feeble response on the night of the 3/4 March when only some 12 tons of bombs fell on London, was to castigate Goering and force him to examine the management of *Luftflotte* 3 by appointing an officer of proven ability to the post of *Angriffsführer* (Attack Leader) England. The choice for this rather grandiose title fell to a young, twenty-eight year former *Kommandeur* of II./KG 77 and Ritterkreuz (Knight's Cross) holder, *Oberstleutnant*[28] Dietrich Peltz. Rapidly promoted to *Oberst*,[29] Peltz took up his post later in the month and set about the not inconsiderable task of improving *Luftflotte* 3's *Kampfgruppen*. Short of resources and properly trained aircrews Peltz was under no illusions concerning his ability to affect the outcome of the war. Nevertheless, as a regular *Luftwaffe* officer he regarded it as his duty to obey his orders to the best of his ability. Aware that their bombing accuracy required some improvement by comparison with Bomber Command, Peltz began the slow process of forming I./KG 66 into a target marking unit along the same lines as that of KGr 100 during the Winter Blitz.[30]

His other initiative was to reorganise the *Luftflotte*'s fighter-bomber (*Jagdbombers* – Jabo) units into what he hoped would become a 'fast bomber force' – the *Schnellkampfgeschwader* (SKG). The potential use of Jabos with the consequent difficulty they presented to the defence in terms of the identification, tracking and destruction of such high speed targets had been established by *Generalfeldmarschall* Sperrle in March 1942. Using fighter-bomber conversions of the standard Focke Wulf Fw 190A-4 fighter,[31] Sperrle raised a 10th (Jabo) staffel within *Jagdgeschwaders* 2 and 26 (JG 2 and JG 26) and tasked them with attacks on coastal targets along the English Channel. Although usually only committed in sections of four aircraft, thirty Fw 190s drawn from JG 2, JG 26 and *Zerstoerergeschwader* 2 (ZG 2), supported by sixty escorts bombed Canterbury on 31 October causing little damage. A similar, but more audacious raid against London by twenty-eight 190s on 20 January 1943 initially confused the defence which later recovered and destroyed three of the raiders and six from the covering escort. Twenty bombs were dropped in Poplar and Deptford with little damage being recorded. The damage wrought by the defence on this and previous days would eventually lead to the Jabos being deployed at night, where they would initially prove very difficult to intercept despite the defences being the strongest of the war.[32] By April 1943 the cat's-eyes Defiants and the old Havocs in Fighter Command had been replaced by twin-engined Beaufighters or Mosquitoes, with many of the former Mk.If models being replaced by the Mk.VIf with AI Mk.VII or Mk.VIIIA (see Figure 23).

The *Schnellkampfgeschwader* were used for the first time at night on 16/17 April when II./SKG 10 sent twenty-eight 190s on a high-level bombing raid on London. Mixing themselves with Bomber Command aircraft returning from raids on Pilsen and Mannheim-Ludwigshafen, the 190s reached London but caused little damage and few casualties. Although admittedly experimental in nature, the raid would prove costly to the enemy who lost four aircraft when they became lost and force landed at various locations in Kent. The final one of these, an Fw 190A-4, flown by Feldwebel Bechtold, inadvertently alighted at

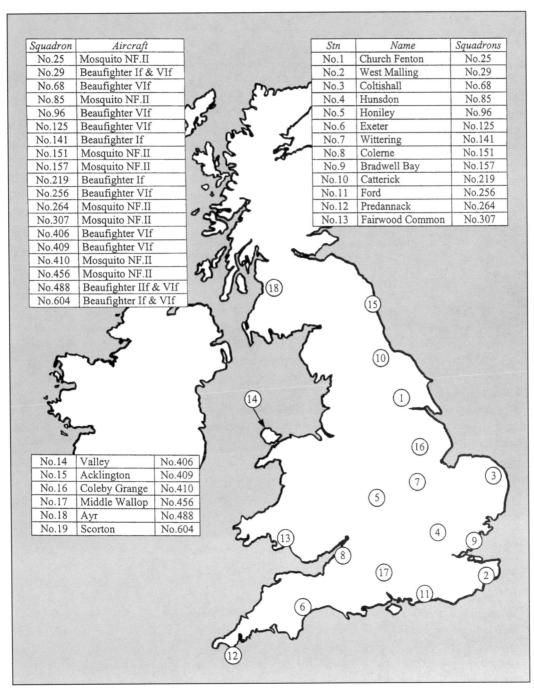

Squadron	Aircraft
No.25	Mosquito NF.II
No.29	Beaufighter If & VIf
No.68	Beaufighter VIf
No.85	Mosquito NF.II
No.96	Beaufighter VIf
No.125	Beaufighter VIf
No.141	Beaufighter If
No.151	Mosquito NF.II
No.157	Mosquito NF.II
No.219	Beaufighter If
No.256	Beaufighter VIf
No.264	Mosquito NF.II
No.307	Mosquito NF.II
No.406	Beaufighter VIf
No.409	Beaufighter VIf
No.410	Mosquito NF.II
No.456	Mosquito NF.II
No.488	Beaufighter IIf & VIf
No.604	Beaufighter If & VIf

Stn	Name	Squadrons
No.1	Church Fenton	No.25
No.2	West Malling	No.29
No.3	Coltishall	No.68
No.4	Hunsdon	No.85
No.5	Honiley	No.96
No.6	Exeter	No.125
No.7	Wittering	No.141
No.8	Colerne	No.151
No.9	Bradwell Bay	No.157
No.10	Catterick	No.219
No.11	Ford	No.256
No.12	Predannack	No.264
No.13	Fairwood Common	No.307

No.14	Valley	No.406
No.15	Acklington	No.409
No.16	Coleby Grange	No.410
No.17	Middle Wallop	No.456
No.18	Ayr	No.488
No.19	Scorton	No.604

Figure 23. Night-fighter Squadrons at April 1943. The conversion of the night-fighter force to the Mosquito was well underway by April 1943, with just under half the squadrons equipped with the NF.II. and the remaining Beaufighters having converted to the more powerful Mk.VIf fitted with AI Mk.VIII.

West Malling and presented the RAF with an intact example for evaluation.[33] A further two aircraft also failed to return to their bases in France.[34]

Further night-Jabo raids were conducted on 18/19 and 20/21 April which incurred few losses to the *Luftwaffe* and little damage to the citizens of London. As Peltz and the *Schnellkampfgeschwader* leadership expected, the early Fw 190 raids were not easily countered by the defences, with some *Luftwaffe* officers mistakenly believing they had found the solution to the night bombing problem. Although undoubtly fast and effective as a day-fighter, the Fw 190 in its Jabo-Rei configuration (the Fw 190A-4/U8 and A-5/U8) carrying a single 250 kg (550 lb) bomb and two 300 litre (66 gallon) fuel tanks under the wings, was much slower, less manoeuvrable and therefore more vulnerable to a new night-fighter recently deployed by Fighter Command.

Whilst the main line of development for AI Mk.VIIIA was centred on the Beaufighter the potential of the Mosquito was not ignored. During July 1942 a Mosquito Mk.II, DD715, was removed from the Hatfield production line and fitted with a thimble radome to accommodate one of the pre-production AI Mk.VIIIA radars. Trials with DD715 at Defford the following September led to the conversion of ninety-seven Mk.IIs that were taken from de Havilland's Leavesden factory and delivered to Marshall's Flying School at Cambridge. Here the four-gun Browning noses were removed and replaced with thimble radomes to accommodate the scanner mechanism. The first of these factory conversions, HJ945 and HJ946, were delivered to Defford on 13 February for evaluation, with the first 'production' aircraft reaching No. 85 Squadron at Hunsdon on 28 February. A productionised version of the NF.XII, the NF.XIII, that was based on the Mosquito FB.VI with Merlin 21 or 23 engines, stronger wings and greater fuel capacity (see Appendix 3), was introduced during the autumn of 1943.[35]

The Mosquito NF.XII/XIII represented a further improvement in capability, as one pilot, Flight Lieutenant Jack Meadows,[36] who flew the Beaufighter VIf and the Mosquito NF.XIII in 1944 recalls:

As I remembered sixty years later the cockpit [of the Beaufighter VIf] was overall the best I have ever used. Access was easy, up the ladder through the belly behind the seat. The back folded down and if in the rush of a scramble the ground crew failed to lock it back properly upright, the take off, when acceleration made it fall back again, could be interesting. There was plenty of room, good vision all round except astern, the controls all easily visible and at hand. Whilst lacking the Spitfire's light manoeuvrability [Jack's previous aircraft] it was nonetheless easy to throw around, not too heavy on the controls and easily able to cope with the manoeuvres of an enemy bomber. On endless practice interceptions, I never had any difficulty passing the time when on patrol, or felt incapable of doing anything I wanted. That the R/O was a disembodied voice from the rear was no problem to one unaccustomed to team work in the air. There was all that convenient space between us for passengers, observers and luggage (the 20 mm cannon were now belt fed, taking up less space and removing from the R/O the chore of fitting new magazines). And I knew baling out, if necessary, would, with the seat collapsed and through the hatch in the floor behind, be easier than in any other aircraft I flew. The only problem was that our Beau's were old and those sleeve-valve Hercules engines were worn. On three occasions, way out somewhere over the North Sea on patrol, I had a serious lack of oil pressure, or something else enough to make me return on one engine. But that was no problem, even landing. Yes, the Beau VI was a great aircraft, always remembered with affection.

By comparison:

Access [to the Mosquito XIII] was difficult, up a long flimsy ladder through a small door in the side of the nose. With side-by-side seating, the pilot had to go first, which slightly slowed up a scramble. Immediately obvious was the cramped cockpit space, despite the R/O's seat being set back 6 inches [15 cm] to allow more elbow room. And whilst most instruments, knobs and 'tits' were conveniently placed and visible, the fuel cocks were behind the pilot's armour-plated seat back. Some pilots let the R/O work them, I always did it myself, feeling behind my back. Visibility forward and around was good, particularly up ahead where it was needed for a night-fighter. It was better astern than in the Beau, but the engines blocked more of the side down view. Night vision was also impaired by the exhaust glow from the shrouds and the wings. Flying was a pleasure. It had the fighter 'joystick', Beau's had 'bomber spectacles', and felt almost like a Spitfire again. Aerobatics were discouraged as they upset the AI, but were sometimes secretly indulged in. It was so easy to fly, so tireless – it was almost the perfect aircraft. And that 60 mph [97 km/hr] more top speed meant we knew we could cope with anything – even day-fighters. It was less draughty than the Beau and had efficient heating. The thimble nose over the scanner slightly reduced forward vision when taxiing. The lack of the six 0.303s [7.69 mm] never concerned us – four 20 mm cannon were quite enough. Compared to the Beaufighter's feeling of great strength there was originally a slight feeling of frailty, but the wooden monocoque construction would prove to be stronger than any metal aircraft. Inspite of the proven reliability of the Merlin, I had three occasions when engine problems forced me to return from overhead the [Normandy] beachhead on one – but that did not cause the least concern, so good was the single engine performance. In conclusion, the Mosquito was a joy to fly and to operate, a really efficient killing weapon, and one of the world's great aircraft, which produced even more affection than the Beau.[37]

On the night of 16/17 May the Jabos of SKG 10 returned to bomb London in the face of opposition from the recently re-equipped No. 85 Squadron. Commanded from January 1943 by Wing Commander John Cunningham, with his Navigator (Radio), N/R,[38] Flight Lieutenant Jimmy Rawnsley as the Squadron's navigator leader, No. 85 claimed the destruction of five of SKG 10's Fw 190s: an A-4/A-5 by Squadron Leader Green and Flight Sergeant Grimstone that crashed into the English Channel, an A-4 from 3./SKG 10 by Flying Officer Thwaites that also fell into the Channel, a third A-4, again to Thwaites, crashed near Ashford, an A-4 from I./SKG 10 by Flying Officers Shaw and Howton that came down near Gravesend and, finally, an A-4 by Flight Lieutenant Howitt which crashed into the Channel off Hastings (see also Appendix 4). All these crews flew the Mosquito NF.XII equipped with AI Mk.VIIIA.

The CO claimed his first Focke Wulf on the night of 13/14 June, as Jimmy Rawnsley recalls:

We were one of the fortunate crews already in the air when the raid began. We had gone off on patrol just before midnight, and we were beating up and down the Channel off Dungeness at 20,000 feet [6,095 metres] when 'Skyblue' warned us that a fast customer was on the way in. The Controller timed our converging courses to a nicety, and the blip came scuttling across my cathode ray tube only a mile and a half [2.4 km] ahead and well below us. I had no fears about overshooting, only of being outdistanced. John opened up the engines as I brought him around in a tight diving turn, and we went howling down after the raider.

The range closed only very slowly, and the blip was as steady as a rock. This must be a new boy, I thought, one of those they-will-never-catch-me-at-this-speed characters I had been hoping to meet. The only thing to worry about now was the searchlights. If only they would leave us alone!

On the aircraft went, hell bent for London, and not the slightest sign of a [search]light broke the soft velvet of the summer night. And all the time we were creeping in.

… John saw the other aircraft against a patch of cloud. I looked up from the AI set, and there was no doubt about what we were after: it was an Fw 190 all right. The single exhaust flickered below the fuselage; the short, straight wings still had the drop tanks hanging from the tips; the big, smooth bomb was still clutched fiercely to its belly [identified as an Fw 190A-5/U8].

John very briefly touched the trigger, and the guns gave one short bark. The enemy aircraft reared straight up on its nose, flicked over and plunged vertically downwards. It all happened with incredible speed. Standing up and pressing my face to the window, I watched the blue exhaust flame dwindle as the aircraft hurtled earthwards.

Those watching from the [West Malling] aerodrome were apparently entranced with what was going on. They heard the one short bellow of the cannon and the echo from the surrounding hills. The note from the one-ninety changed its song and grew louder and louder and higher and higher until it was a tortured scream, which ended abruptly in a great red flash that silhouetted the trees to the west, followed by a crump that shook the ground and rattled the windows.

We had broken the spell, and had made the first kill of our second innings'.[39]

Surprisingly, the pilot Leutnant Ullrich from 3./SKG 10, escaped by being catapulted through the cockpit canopy when the aircraft flicked over into its dive. On recovering from the shock he opened his parachute and was made prisoner by the local searchlight crew when he landed near the village of Borough Green, Kent. His only injury was a broken arm.[40]

The first pre-production model of AI Mk.VIII was installed in a Beaufighter that arrived at FIU for operational trials on 23 December 1942, where it was subjected to some 120 hours of testing. With a transmitter that was ten times more powerful that that of AI Mk.VII, the maximum range results of just 4 miles (6.5 km), were disappointing. There were, however, occasions when this was stretched to 8 miles (12.9 km), but the equipment clearly needed modifying before it entered service. The first AI Mk.VIII production sets were delivered to No. 218 MU during May 1943, where they were immediately installed in a batch of Mosquito XIIs to re-equip No. 151 Squadron at Colerne from 11 June onwards. Throughout July and August the Squadron flew a series of intensive trials to compare the performance of AI Mk.VIII with that of the earlier Mk.VIIIA, but failed to find any appreciable difference between the two in terms of their maximum range. This would suggest that a mix of both radar types were installed in their Mosquitoes.[41]

The first success with AI Mk.VIII occurred on the night of 15/16 September, when Mosquitoes of No. 488 (New Zealand) Squadron destroyed a Ju 88A-14 of 4./KG 6 and a Do 217M-1 of 9./KG 2, both of which were shot down into the sea off the Kent coast.

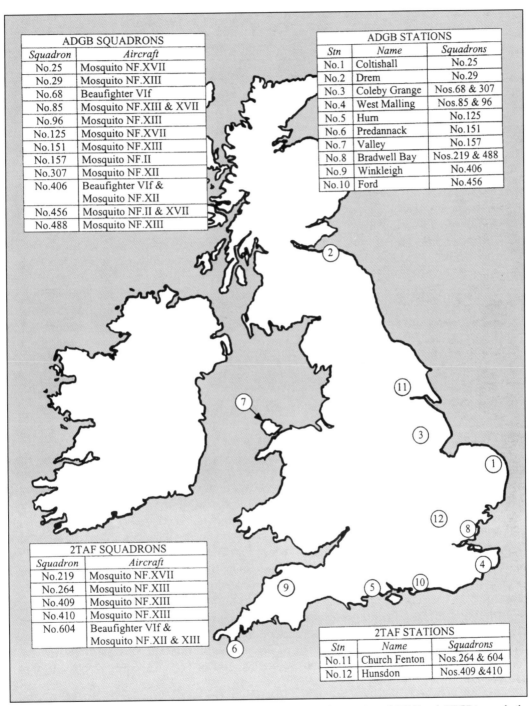

ADGB SQUADRONS	
Squadron	*Aircraft*
No.25	Mosquito NF.XVII
No.29	Mosquito NF.XIII
No.68	Beaufighter VIf
No.85	Mosquito NF.XIII & XVII
No.96	Mosquito NF.XIII
No.125	Mosquito NF.XVII
No.151	Mosquito NF.XIII
No.157	Mosquito NF.II
No.307	Mosquito NF.XII
No.406	Beaufighter VIf & Mosquito NF.XII
No.456	Mosquito NF.II & XVII
No.488	Mosquito NF.XIII

ADGB STATIONS		
Stn	*Name*	*Squadrons*
No.1	Coltishall	No.25
No.2	Drem	No.29
No.3	Coleby Grange	Nos.68 & 307
No.4	West Malling	Nos.85 & 96
No.5	Hurn	No.125
No.6	Predannack	No.151
No.7	Valley	No.157
No.8	Bradwell Bay	Nos.219 & 488
No.9	Winkleigh	No.406
No.10	Ford	No.456

2TAF SQUADRONS	
Squadron	*Aircraft*
No.219	Mosquito NF.XVII
No.264	Mosquito NF.XIII
No.409	Mosquito NF.XIII
No.410	Mosquito NF.XIII
No.604	Beaufighter VIf & Mosquito NF.XII & XIII

2TAF STATIONS		
Stn	*Name*	*Squadrons*
No.11	Church Fenton	Nos.264 & 604
No.12	Hunsdon	Nos.409 &410

Figure 24.   2TAF and ADGB night-fighter Squadrons at April 1944. The creation of 2TAF and ADGB towards the end of 1943 brought about a reduction in UK night defence and the transfer of the night-fighter force towards the east and south-east coasts to protect the invasion ports and provide the opportunity for field training for those squadrons transferred to 2TAF.

September and October 1943 witnessed an increase in *Luftwaffe* minelaying operations to the highest level throughout the year, during which all of Fighter Command's successes at night were made with AI Mk.VIII/VIIIA and thirty-one of the thirty-seven combats during October and November were made with the AI Mk.VIII/Mosquito XII combination. By this time centimetric AI was rapidly supplanting 1½ metre radar within the night-fighter community; however, there were still occasions when the old equipment had its successes. On the night of 10/11 December Flying Officer R.D. Schultz and his N/R, Flying Officer V.A. Williams, flying a Mosquito NF.II of No. 410 (Canadian) Squadron operating out of Hunsdon, destroyed three enemy aircraft off the Essex coast; all of which were Do 217M-1s from I./KG 2. Later that month No. 410 began its conversion to the Mosquito XIII.[42]

With eight squadrons scheduled to receive AI Mk.VIIIA during the winter and spring of 1943 and further ones to be equipped with AI Mk.VIII, training was expanded to cope with the increasing demand for centimetric expertise. The capacity of TRE's Service Liaison Group at Defford proved inadequate to deal with the demands of the MUs and the squadron's maintenance teams and recourse was made to a new organisation based within Technical Training Command. In mid-1943 No. 7 Signals School was established within the Science Museum in South Kensington, and, following the transfer of the instructional equipment from Defford the first course was run in January 1944. Likewise, the responsibility for supporting AI Mk.VIII was transferred from TRE's PDS section to Fighter Command's PDS Groups in the second half of 1943. Conversion would have been more rapid within Fighter Command had it not been for the need to provide night-fighter coverage in the other theatres of war (the Mediterranean, India and the Far East), which made increasing demands on the production of aircraft and equipment. Nevertheless, by April 1944 Nos 29, 96, 256, 264, 307, 406, 409 and 604 Squadrons had been converted to the Mosquito NF.XII/XIII with AI Mk.VIIIA/VIII, with which some would see out the war (see Figure 24).[43]

CONCLUSIONS

The introduction to service of AI Mk.VII was, according to FIU, one of the smoothest transitions of any set to that point in the war. Its operation was quickly learned by the N/Rs and when employed against low level minelaying aircraft it was found to be reliable and effective. Similarly, its fully productionised replacements, AI Mk.VIIIA and Mk.VIII, were assimilated into the Beaufighter force and easily adapted for use in the Mosquito, where, with its high performance, it performed useful work in the low-level role and against the high speed Jabos of SKG 10. Training with the new equipment was not ignored, with TRE supporting the installation work in the PDS phase at the MUs and on the squadrons, before handing over to Fighter Command's teams at the end of 1943. Conversion of the Command's night-fighters to an all centimetric force was not completed as quickly as some officers would have wished, due in part to the demands of the overseas theatres and the ability of industry to deliver the requisite number of fighters and radar equipment. Nevertheless, a substantial proportion of the UK-based squadrons were centimetric equipped by the spring of 1944.

The only disappointments with the new sets related to their poor maximum range at altitude and their susceptibility to *Window* jamming. Unfortunately these problems could only be resolved with the development of a completely new radar set that would, of necessity, be sourced in the US.

Notes

1. *The Signals History, Volume 5*, p. 153.
2. *Ibid*, pp. 153 and 154.
3. *Ibid*, p. 154.
4. Alan Hodgkin, *op cit*, p. 189.
5. Alan Hodgkin, *op cit*, pp. 190 and 191.
6. Ian White [2], *op cit*, pp. 15 and 16.
7. Alan Hodgkin, *op cit*, p. 192.
8. *Ibid*, pp. 192 and 193.
9. *Ibid*, p. 194.
10. Later Sir Robert Cockburn, KBE, CB, and the Chief Scientist at the Ministry of Aviation, 1959–1964.
11. *Würzburg*, which operated at a frequency of 560 MHz (0.53 metres), was introduced to *Flak* service before the outbreak of war.
12. Joan, later Lady, Curran, née Strothers, the wife of S.C. Curran, was the only lady scientist at TRE.
13. Alfred Price [4], *Instruments of Darkness* (London: Macdonald & Janes, 1977), pp. 114–16.
14. Later MRAF, the Viscount Portal of Hungerford, CAS from October 1940 to the war's end.
15. Frederick Lindemann was created the first Viscount Cherwell in 1941.
16. Dr D.A. Jackson, later Professor Derek Jackson, DFC, AFC, FRS, flew with 604 Squadron as an R/O from late 1940 to early 1942.
17. Alfred Price [4], *op cit*, pp. 116–17.
18. *Ibid*, p. 117 and H.G. Kuhn and Sir Christopher Hartley in *Derek Ainsley Jackson*, a Biographical Memoir by the Royal Society (London: The Royal Society), pp. 282 and 283.
19. Alfred Price [4], *op cit*, pp. 117 and 118.
20. *Ibid*, p. 118.
21. *The Signals History, Volume 5*, p. 155.
22. *The Signals History, Volume 5*, p. 37 and the *FIU ORB*, pp. 61 and 62.
23. *FIU ORB* p. 64.
24. Alan Hodgkin, *op cit*, p. 194.
25. *The Signals History, Volume 5*, p. 154.
26. Alfred Price [2], *op cit*, p. 145 and William Green, *The Warplanes of the Third Reich* (London: Macdonald, 1970), pp. 150, 151 and 456.
27. Martin Middlebrook and Chris Everitt, *The Bomber Command War Diaries* (London: Viking Penguin, 1985), p. 359.
28. A *Luftwaffe* rank equating to that of an RAF wing commander.
29. A *Luftwaffe* rank equating to that of an RAF group captain.
30. Winston Ramesy [3], *op cit*, p. 248 and Alfred Price [2], *op cit*, p. 149.
31. The Fw 190A-4/U3 (Jabo) was capable of carrying a single 250 kg (550 lb) bomb, or a 300 litre (66 gallon) fuel tank beneath the fuselage.
32. Heinz Nowarra, *The Focke Wulf 190, a Famous German Fighter* (Kings Langley: Harleyford Publications, 1965), pp. 46 and 47 and Alfred Price [2], *op cit*, pp. 142–9.
33. This aircraft was later serialled PE882 and passed to No. 1426 (Enemy Aircraft) Flight.
34. Winston Ramesy [3], *op cit*, p. 247 and Alfred Price [2], *op cit*, p. 149.
35. Sharp and Bowyer, *op cit*, p. 59 and AI in The Signals History, p. 38.
36. Later Wing Commander Jack Meadows, DFC, AFC, AE*, RAuxAF, who flew with No. 604 Squadron and later in 1944 as a flight commander with No. 219.
37. Ian White [1], *If You want Peace, Prepare For War* (London: 604 Squadron Association), pp. 157–9.
38. At some point in the late winter or early spring of 1942, the radar operators' title was changed from R/O, to Navigator (Radio), N/R, or simply 'navigator', which required the incumbents to complete a formal navigators course before putting up their brevet. The term N/R will be used from now on.
39. C.F. Rawnsley and R. Wright, *op cit*, pp. 245 and 246.
40. *Ibid*, p. 246.
41. *The Signals History, Volume 5*, p. 157.
42. *Ibid*, p. 157.
43. *Ibid*, p. 157.

CHAPTER TWELVE

Two Lines of Development

November 1942–April 1944

As with all things relating to warfare, the problems associated with night interception changed continuously throughout the course of the war. Although the introduction of centimetric techniques had overcome the low-level limitations of 1½ metre AI, new problems encountered during 1942 and 1943 raised the need for improved radars that were free from *Window*-type jamming and were capable of intercepting fast moving targets. Squadron Leader Jackson's trials in the summer of 1942 had concluded that an enemy bomber releasing the requisite quantities of *Window* foil would have little difficulty in avoiding interception by an AI Mk.VII or VIII equipped fighter. Furthermore, the GCI network that tracked the bombers and controlled the night-fighters was itself vulnerable to jamming. Another factor was the introduction by the enemy of faster and smaller aircraft in the form of the Fw 190 Jabo and the Messerschmitt Me 410 twin-engined light-bomber, with the possibility that the latter was fitted with a tail-warning radar. The availability of a tail-warning device required the fighter to operate outside that radar's coverage, which placed greater emphasis on approaches from the beam (port or starboard). In practise this approach proved very difficult to achieve as AI Mk.VIII's azimuth cover was restricted to a relatively narrow cone.

These and other factors were taken into account at the November 1942 meeting of the Interception Committee, when they considered the requirements for future AI systems:

- The best possible 'all round' coverage in plan view, but not necessarily spherical, to simplify ground control and enable the night-fighter to fly patrols independent of GCI assistance.
- Direction finding accuracy was to be within 10° of the dead-ahead position, ± ½°, with no lag in the display system in order to provide the pilot with a blind firing capability in cloud or in poor weather.
- The system's maximum range was to be of the order of 10 miles (16 km), with a minimum of no greater than 200 feet (60 metres), which were to be attainable at all altitudes between 500 and 5,000 feet (150–1,525 metres). The object of the low minimum range was to enable the system to meet the blind firing requirement.
- The provision of an immediate IFF response.[1]

In addition to these requirements, the Committee specified a set of 'desirable' features:

- Automatic target following (tracking).
- An accurately calibrated range meter.

- A visual indication from a beam approach system.
- A readable range of 100 miles (160 km) from an AI beacon.[2]

These requirements called for a review in the way that AI radars were developed, since the attainment of the specification was beyond the capability of AI Mk.VIII, particularly with respect to *Window* interference and maximum range. Three possible systems were assessed as having the potential to meet the specification: the first, a 10 cm (S-Band) radar with automatic lock follow (ALF), second, a 3 cm (X-Band) system broadly similar to the S-Band, but incorporating the necessary changes in wavelength and components, and third, a system based on the American television, or range-azimuth system then under development in the US. The S-Band system was the most attractive from TRE's viewpoint, since this could be operated and maintained from within the training and support organisations that were being created for AI Mk.VII and Mk.VIII. TRE therefore suggested a specification for what would become 'AI Mk.IX', that would operate on a wavelength of 9.1 cm (3.3 GHz). At a meeting called by the ACAS (Operations) held on 10 February 1943, it was agreed to proceed with the development of AI Mk.IX as a 'long term project', with the set being available for operations from the late summer of 1944.[3]

Whilst several variants of AI Mk.IX based on the S-Band radar with ALF were envisaged by TRE, they were persuaded to concentrate their efforts on producing a basic experimental set that could be installed in a night-fighter at the earliest opportunity. At broadly the same time, the Establishment undertook to modify an S-Band equipment to 3 cm X-Band working, in order that a comparative evaluation might be undertaken. At this juncture it is necessary to step back to 1941 to understand the techniques underpinning ALF and establish the starting point for the development of the Mk.IX equipment.[4]

1941

As Sir Bernard Lovell states in his *Echoes of War*, the essence of the lock-follow principle was relatively simple:

> By using a rotating offset dipole at the focus of the paraboloid [scanner] one could produce a split-beam. By comparing the strength of the received echo from both components of the beam [and converting these to voltages] the paraboloid could be driven so that it pointed continuously at the target. The pilot would then align his fighter with the axis of the paraboloid and, provided he was within destructive range of his armament, could fire blind with confidence.[5]

It was in March 1941 that Lovell began his investigations into ALF. Having previously been involved in the development of a magnetron-based GL set for the Army which employed the spilt-beam technique, Lovell set about the task of applying it to AIS. To keep things simple he began by constructing a ground-based ALF system on land in front of the stables at Leeson House; a site he had used previously to test a naval surface warning radar. By the month's end he had built an apparatus based on a rotating summer house that employed two cylindrical parabolas (one for transmitting and the other for receiving) and the other to produce a broad elevation beam that could be turned in azimuth by manual effort (he was at that time, the end of March, awaiting the delivery of a powered system from TRE's workshops). With his and Hodgkin's previous experience of working with Nash and Thompson in the design of the AIS scanner mechanism, Lovell opened discussions in April with Mr A. Tustin, a motor traction engineer working for the Metro-Vick Company in Manchester. Tustin appeared confident that an ALF steering mechanism, based on the Company's experience with their 'metadyne' system for automatic gun laying, could be adapted to AIS.[6]

Lovell's team of one was strengthened by the addition of Dr F.C. 'Freddie' Williams,[7] whom Lovell had known at Manchester University prior to the war, and Dr F.J.U. Ritson who had worked with Williams since joining AMRE at Dundee in November 1939. By the beginning of May TRE's workshops had delivered a motorised scanning lock mechanisum, which Lovell and Ritson fitted to the summer house and got working by 6 May, only to have it fail almost immediately when a bearing siezed. When a target echo was detected whilst the system was scanning in azimuth, an electronic gate, or strobe, was moved along the timebase by the operator until it coincided with the echo, whereupon the scanning ceased and the parabola would lock-on and follow the target automatically. At a later date Williams adapted the automatic strobe circuit he and Ritson had designed for AI Mk.VI, to hunt along the ALF timebase until it encountered an echo, whence the scanning would cease and the parabola would lock-on to the echo and track the target automatically in azimuth. On 25 May the system tracked a Blenheim and shipping in the Bay automatically for the very first time. All that remained was to change the apparatus to elevation working and confirm its operation, following which they would have the makings of a fully functional ALF system.[8]

When designing what became known as AI Follow (AIF),[9] Lovell had to overcome two significant problems; the first concerned the high pointing accuracy required of the scanner, and the second, the variation in amplitude, or fading, of the target echo. To enable the radar smoothly to track the target required a mechanism that was capable of aligning in azimuth and elevation with an accuracy of $\pm 1°$, with a consequent increase to $\pm \frac{1}{4}°$ for blind firing. Both requirements necessitated a mechanism that was free from the effects of inertia. The first ground versions of the ALF mechanism were based on an electronically controlled split-field drive motor developed by Metro-Vick, that incorporated circuits to limit the inertia effect. However, Lovell and Ritson were concerned that the sluggishness of the drive motor was such that it might prove unsuitable when working at the velocities encountered in operational flying (velocities of $10°$ per second and accelerations also of $10°/second/second$ – $10°/second^2$) and in the presence of signal fading. On a visit to the Ferranti Company in Manchester during mid-June with Williams and Dee, Lovell discussed the use of a magnetic friction clutch that employed two electro-magnets acting on two plates in proportion to the misalignment. Designating the Metro-Vick system as 'Class-A' and the Ferranti system as 'Class-B', Lovell returned to Leeson House to subject the former to a full scale trial.[10]

The TRE workshops completed the modifications to the summer house apparatus to enable it to work in azimuth and elevation and on 1 July Lovell recorded in his diary 'our automatic follow worked marvellously on aircraft and seemed very successful'. With the basics of the system proven, Lovell and Ritson set about the task of preparing the equipment for airborne trials in a Blenheim. Having decided on an installation similar to that employed in AIS, Lovell encountered another problem that affected the motor-drives. Unlike on the ground, where the echoes radiated by the target aircraft were of a relatively constant amplitude, those encountered in the air were anything but constant and caused the Metro-Vick Class-A system to break lock at low angular velocities. To better understand the problem Lovell instituted a programme to provide a detailed analysis of signal fading. In a paper on the subject dated 19 July 1941, Lovell showed that the Class-A system gave an output of 80 volts per degree of misalignment of the target up to a displacement of 3–4° off the scanner from its centre line, after which it remained constant. These values were in turn effected by the speed of rotation of the dipole at the scanner's centre. Lovell considered that if the dipole could be spun to 6,000 rpm (it was currently set at 600), the scanner mechanism would be provided with fresh information once every 1/100th of a second and would even out the echo variations. This figure was eventually proven by photographic means, using a system devised by Andrew Huxley, that employed a modified 35 mm Leica camera focused on the CRT, with the film pulled through by hand to produce adequate images of the individual echo pulses. By October Huxley's method proved the need for the dipole to rotate at 6,000 rpm to provide a smooth output that was free from violent amplitude variations.[11]

Whilst Lovell and Ritson were grappling with the fading problem, Freddie Williams was devising a solution to ensure the scanner's servo-motors would smoothly follow the target. Using a system that employed a combination of velocity and acceleration feedback which induced an inertia in the scanner movement that was proportional to the angular velocity, a smooth motion was achieved. According to Sir Bernard:

> Williams achieved this brilliant solution ... by mounting a tachometer on the motor shaft from which he derived a voltage proportional to the speed of rotation ... which was applied to the feedback network together with the error signal. A development of this idea by Williams eliminated the ... tachometer ... and enabled the necessary feedback voltages to be derived from the armature of the driving motor. Williams named this versatile low-power variable speed drive the *Velodyne*.[12]

From this point on the Class-A and B solutions were abandoned in favour of Williams' Velodyne system.

Lovell was allocated a Rootes Securities-built Mk.IV Blenheim, T1939, for the trials of AIF. T1939 was delivered to Christchurch for the installation of the equipment, much of which was based on the AIS prototype installation in Blenheim N3522. The task of building the lock-follow scanner mechanism was allocated to Metro-Vick's Trafford Park factory, who were tasked with building a properly engineered version of Williams' prototype. Whilst the fitting party at Christchurch installed the cables and connectors in T1939, Lovell shuttled between Swanage and Manchester throughout late August and on into early September to encourage the Company to deliver the scanner as quickly as possible. The scanner was finally delivered to Leeson House on 19 September, where it remained for formal testing before being installed in T1939 on 14 October. From then onwards, Lovell moved to Christchurch to nurse the system through its flight trials.[13]

The AIF system installed in T1939 operated on a wavelength of 9.1 cm using a split-beam that was generated by an offset dipole spinning at 800 rpm, whose position was switched synchronously with the receiver by means of a commutator. The resultant four voltages, one from each of the four quadrants, were used to equalise the scanner mechanism and keep it aligned with the target. The tracking of the target was by means of a manual range strobe. During the search phase the scanner swept 120° in ten seconds, before reversing and sweeping back another 120°, with the observer able to elevate or depress the sweep between +60° and −20°. The observer was provided with two CRTs, the one on the left being the Spot Tube which provided azimuth and elevation information, whilst that on the right, the Range Tube, displayed a conventional linear timebase. When a target was detected during the scan phase an echo was displayed on the CRTs and the observer switched from SCAN to AUTO. The strobe was then driven along the timebase trace on the Range Tube by the observer until it was aligned over the echo, following which the system would automatically track the target in azimuth and elevation. The scanner positions were extracted as voltages taken from potentiometers mounted on the mechanism to drive the observer's display and a wings display for the pilot that acted very much like the pilot's indicator on AI Mk.V. Whilst the observer maintained the strobe over the target return, range data was fed to the pilot's indicator to increase the width of the wings as the distance between target and fighter closed. It only remained for the pilot to steer the aircraft until the wings centred on his display, to effect the interception. In the final stages the pilot switched from COARSE to FINE, which changed the range lines on his indicator from $\pm 60°$ to $\pm 5°$.[14]

Ground testing was completed by 21 October and showed promising results, with the first flight being scheduled for that day, but later cancelled due to a fault on one of the Blenheim's engines. The next attempt was made a week later on 28 October, but this too was cancelled ten minutes into the flight due to the failure of the aircraft's air speed indicator (ASI). Events continued to conspire on the following day when the radar itself failed and the radome had to be removed for a close inspection. Finally on the morning of Friday, 30 October, with Lovell flying as the observer, the system was given its first air test. The system work reasonably well, but the range was a disappointing 3–3½ miles (4.8–5.6 km) and Lovell had difficulty in finding the target. Simlar results were obtained by Willams and Ritson during a second flight in the afternoon.[15]

Testing continued throughout the autumn of 1941. With Christchurch airfield being close to the coast another hazard presented itself in the form of interceptions by friendly fighters. The Blenheim's 'odd looking nose' frequently attracted the attention of Fighter Command's patrolling Hurricane and Spitfire squadrons and the possibility that T1939 would be mistaken for a 'hostile' was ever present for Lovell and his pilot. Nevertheless, despite this unwanted distraction, steady progress was made and by early November T1939 had completed ten hours of testing with promising results. In a report to Dee dated 4 November, Lovell stated that despite the poor performance of the scanner mechanism, good results were being obtained in the case of (low-angular) tail-on approaches to the target, where the system delivered an accuracy of $\pm \frac{1}{4}°$ and a lag of 1½–2° for angular velocities of 8° per second. He also confirmed the need to increase the dipole speed to 6,000 rpm to counteract fading, but wished to retain the slow speed design for the first Beaufighter models.[16] In conclusion, Lovell appeared confident that improvements to the mechanical design of the scanner mechanism, coupled with greater torque in the drive motors would provide sufficient accuracy for blind firing.[17]

By Christmas the results with T1939 had improved further, with the lock-on facility now being 'extremely good' and holding 'securely in Rate 2[18] turns at a range of 1–2 miles (1.6–3.2 km)'. Only the acquisition of the target in continuous tight turns was causing the system some problems, when it became confused and locked onto the ground returns. This was overcome by searching at a low level with the scanner elevated, a procedure that was also adopted for the interception of low-flying aircraft. The first blind firing trials had also been conducted using a G22 camera gun to confirm the system's accuracy against Blenheim, Gladiator, Battle and Wellington targets. The interception was made in the 'normal way' with the observer acquring the target before lock-on and closing until the pilot's indicator showed a range of 1,000 feet (305 metres), at which point the camera gun was switched on. An analysis of the results by statistical means showed an error of 0.41° in azimuth and 0.32° in elevation. Interceptions from dead astern were precluded as the turbulance generated by the target aircraft upset the system. This too was overcome by the fighter approaching the target from below before bringing up the nose at an angle of 5° and opening fire. The G22 camera was replaced with a G45 cine gun to access the effects of bursts of fire. With the system modified to enable the pilot to pre-select his firing point and the camera gun enabled to operate automatically when the target was in line and within a predetermined range, the trials were continued. The results of the 'movie' trials showed that with burst durations of one half to greater than four seconds, 42 per cent of all bursts showed an error in azimuth and elevation of less than ± ½°. When extrapolated to the closely spaced forward firing cannon armament of the Beaufighter, these figures showed that at least 50 per cent of the rounds would strike the target.[19]

At the year's end with Williams, Ritson and Lovell heavily involved in the testing and preparing for the installation of AIF in the Beaufighter, Lovell was posted within TRE to take charge of the development of the *H2S* navigation and bombing radar. To replace Lovell, Rowe moved Dr A.C. Downing, who had previously worked with Hodgkin and Edwards on AI Mk.VII, to continue the testing of AIF.[20]

1942

Prior to the Tizard Mission in September 1941, the American electronics industry had 'cut its teeth' on the manufacture of airborne radar using an example of AI Mk.IV that was shipped across the Atlantic by the US Army Signal Corps the previous May. This set was handed to Bell Labs for examination and testing and later to the Army's Aircraft Radio Laboratory at Wright Field, Dayton, Ohio, for flight testing. A further flight test model was constructed by the Labs and installed in an A-20 bomber at Mitchell Field, Long Island, New York, in September. Testing was completed on 5 December, following which the set was put into production by Western Electric[21] as the SCR-540 for use in the Douglas P-70 Havoc night-fighter.[22]

The two examples of the first AI-10 radars issued to Bell Labs and Western Electric formed the basis for America's first centimetric production radar, the SCR-520A. This S-Band set was designed from the outset to be capable of installation in the Beaufighter and a heavy night-fighter, the P-61 Black Widow, that was under development for the USAAC. With their experience in the manufacture of the SCR-540, Bell Labs and Western Electric were awarded a contract to build the SCR-520 in August 1941, whilst the AI-10 was undertaking its evaluation in Britain (see Chapter 10). Despite their decision to adopt AIS/AI Mk.VII as the first centimetric AI, the Air Staff had not given up on the Americans and instructed MAP to order 200 SCR-520 sets to a British specification as SCR-520(UK) for installation in Beaufighters. To help the Americans, a Beaufighter was shipped to the US in December to test the (UK) prototype. However, in March 1942, when the bulk and weight of the equipment and its excessive power consumption was presenting the Americans with installation problems, the order was cancelled.[23]

Although overweight and with a scanner that was too large to be installed comfortably in the Beaufighter, Western Electric nevertheless began their deliveries of the SCR-520A in May 1942. Whatever their allies might have felt, the new radar offered a considerable technical advance on the SCR-540 and the USAAC planned its deployment as an interim system pending the introduction of an improved model. However, with the requirement for a good sea search radar having a greater priority than AI, many of the SCR-520A units were adapted, along with a 7 inch (180 mm) PPI, to the SCR-517 ASV set for use by the USN. Altogether, over 2,000 SCR-520/517 sets were delivered to the US forces by the war's end.[24]

Whilst it was appreciated that the SCR-500 series of radars had been developed and manufactured in the shortest possible time to meet the needs of the USAAC and the USN, the weight and size of the units (in particular the RF units, the modulators and power supplies) were difficult to fit in current aircraft. With these shortcomings to the fore, Bell Labs-Western Electric set about the task of building smaller and lighter radars that incorporated more sophisticated circuitry and higher output powers. The term 'small package' was used to differentiate these sets from the 500-series. With large contracts for the SCR-520/517 radars waiting to be fulfilled, Western Electric proposed these be amended on an ad hoc basis to cover the development of the small package radars. Subsequently, small package equivalents of the 500-series were developed, from which America's next AI set, the SCR-720, gradually emerged.[25]

Designated SCR-720-T2 in US service, the new radar, like its predecessors, worked in the S-Band and employed the same helical scanning techniques as the AI-10 and SCR-520 sets. The N/R was provided with two CRT displays, a azimuth/range (C-Scope) presentation and an azimuth/elevation (B-Scope). The maximum range of the equipment above 5,000 feet (1,525 metres), was 6–8 miles (9.5–12.8 km), with a minimum of 300 feet (90 metres). Below 5,000 feet the range reduced proportionally to height. D/F accuracy was approximately 5° at any angle within the field of scan. The equipment was well suited to high altitude operation as many of the units were pressurised.[26]

At this point we need to return to the problems associated with *Window* during July 1942, when the CAS ordered that the use of the counter-measure be banned until such time as less vulnerable AI and GCI radars were available to Fighter Command. In October Dr R.V. Jones,[27] the Head of Scientific Intelligence at the Air Ministry, received a report from a Danish agent who overheard a conversation between two female *Luftwaffe* auxiliaries on a train. The report highlighted a number of factors concerning the use of women auxiliaries in the enemy's night-fighter control stations, but more importantly, stated that the night-fighters were wholly dependent on those stations with 'the pilot almost [being] a passenger' and the fighters carried radar.[28] The auxiliaries had also heard that British bombers in raids over the Rhineland, had confused the fighter control stations by releasing 'aluminium dust' that fooled the fighter crews into firing on the resultant 'cloud'.[29] Be it an intelligence 'plant' or not, the conversation drove a coach and horses through the argument that the *Window* principle was unknown to the enemy and needed guarding to prevent them using it as a counter-measure over the British Isles.

As a result of Jones' subsequent report, the CAS, Sir Charles Portal, called a meeting on 4 November to discuss the matter with the AOC-in-Cs of Fighter and Bomber Commands, Air Marshals Douglas and Harris, Cherwell, Watson Watt in his role at the Air Ministry's Scientific Advisor on Telecommunications (SAT) and the by now Wing Commander Derek Jackson, who attended at Cherwell's request. On the day Air Vice Marshal Robert Saundby, Harris' deputy, represented Bomber Command. Jackson informed the meeting of the development of AI Mk.IX, the title by which AIF was then known, and its automatic target following capability, that would potentially be able to distinguish between a target and a *Window* cloud. However, Saundby did not push for the introduction of *Window* as they were happy with the *Mandrel*[30] jammer that was recently introduced in his Command. The meeting ended with the CAS asking Jackson, Jones and Sir Robert to ascertain precisely how much *Window* would be required to be dropped to wipe out the German defences.[31]

The following month Jackson and Downing used the prototype AI Mk.IX Beaufighter which was made available to them following its initial flight testing at TRE that autumn, to assess AI Mk.IX's vulnerability to *Window* interference. In the process of thirteen flights undertaken during November and probably into early December, Jackson was able to show that the radar unfortunately locked onto the *Window* cloud and not to the target. Following a series of modifications to the locking circuits at Downing's behest, the system was to be re-evaluated by Jackson on 23 December. What followed next was a tragedy both for the people concerned and the future development of British AI radar.

It had been intended that Jackson and Downing would fly together in the TRE Beaufighter to evaluate the radar, with a second Beaufighter acting as target and discharging the bundles of *Window* foil. However, at the last minute it was discovered there was no observer available to launch the foil, so, rather than delay the trial, Jackson took his place. Operating once again from Coltishall the pair were airborne, when, in Jackson's words:

> I heard the control order two Spitfires to scramble: they were to be ready to intercept an unidentified plane. I was somewhat dismayed at hearing this, particularly because the Controller gave no indication of how far away the unidentified aircraft was, and he appeared to be about to ignore my request that the fighters stationed at Coltishall should be kept well away from the two Beaufighters making the *Window* trial. When we were over the sea I saw one of the Spitfires coming towards us in the most sinister manner, and at once said to [the pilot] Squadron Leader Winward on the intercom that a Spitfire was approaching us in a manner which I did not like. Winward immediately started a very sharp diving turn; almost simultaneously were were hit by cannon fire from the Spitfire. Although neither the pilot nor I were aware of it, this had severed the intercom leads, so each thought the other had been killed, as there was no reply over the intercom. It seemed to me that the only thing for me was to bale out: but I was in error; a few hundred feet above the water, the Beaufighter levelled out: I saw the fault in the intercom and remedied it just in time to hear Bill Winward say on the R/T '... and he has killed my observer!' Our aeroplane was damaged so we had to return to Coltishall immediately: but as we turned Bill saw another aeroplane burning on the suface of the sea; our fears that this was the other [TRE] Beaufighter and that both the pilot and Dr Downing had been killed were well founded.[32]

Downing's death and the loss of the prototype equipment in the depths of the North Sea, were a severe set-back to the release of *Window* and the development of the lock-follow programme.

Aware that AI Mk.IX would not be in widespread use for some considerable time and conscious of the need to supply a radar free from *Window* interference, the Chiefs of Air Staff encouraged the Americans in their development of an improved AI set.

1943

Shortly before the fatal flight on 23 December, Western Electric delivered the first example of the SCR-720 radar (SCR-720B) for RAF service trials and evaluation. To save time the equipment was installed in the nose of a Vickers Wellington bomber and undertook its first *Window* trials in the hands of Wing Commander Jackson during the first week of January 1943. The ability of the B-Scope display to be switched in various ranges down to 2 miles (3.2 km) enabled Jackson to differentiate a target against the background of the *Window* cloud. Jackson's trial removed the last objection to the deployment of *Window* by Bomber Command, who were now able to proceed with the preparation and planning of its

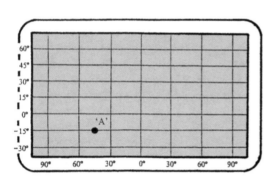

C-Scope (Azimuth/Elevation)

Showing single Target 'A' from the strobe band
at 45° to port and an angle of 15° below the fighter.

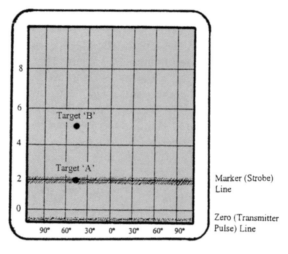

B-Scope (Range/Azimuth)

Showing two targets on a bearing of 45° to port
at ranges of 2 and 5 miles respectively.

Figure 25. Navigator's displays for AI Mk.X (SCR-720). The diagram shows two targets, 'A' and 'B', on the B-Scope at ranges of 2 miles (3.2 km) and 5 miles (8 km) on a bearing of port 45°. The navigator has moved the strobe control to overlay Target 'A', which now appears on the C-Scope display on the same port 45° bearing, but with an elevation bearing of 15° below. His objective now will be to instruct his pilot to bring the target echo to 0° in azimuth and 0° elevation, or slightly lower depending on the pilot's preference.

deployment in the coming spring.[33] This they accomplished with telling effect on the night of 24/25 July, when 791 bombers deploying *Window* foil[34] raided Hamburg, crippling the German radar defences and creating a fire storm that killed 40,000 of its inhabitants and forced two thirds of its population to flee the city.[35]

Further trials of the Wellington installation by TFU were successfully completed before the set was transferred to a specially adapted Mosquito NF.II, DZ659, between January and April.[36] However, although TRE were generally satisfied with the set's performance, they did recommend a number of modifications before it was introduced to service: changes to the scanner's tilt limits better to suit the RAF's interception techniques, the replacement of a poor RF feeder cable, changes to the range scale to indicate 3, 5, 10 and 100 miles (4.8, 8, 16 and 160 km) and the marker width and the design of a new visor for the B-Scope display. These were incorporated into the Mosquito installation in good time for its testing by TFU and FIU. These trials were regarded as being 'highly successful'[37] and cleared the radar for introduction to RAF service as AI Mk.X. A comparison of the SCR-720B with AI Mk.VIII showed the former to have better discrimination between target and *Window* cloud, better coverage and increased maximum range. The Director of Radar Development (DRD) was thoroughly pleased with the set and thought it:

> a 'grand job' and recommended that a large programme of re-equipment of the night-fighter force with AI Mk.X should be undertaken with high priority, particularly with regard of its effectiveness against *Window*.[38]

The re-equipment programme was heavily dependent upon the Western Electric production schedules, which naturally had priority to provide the USAAC with a home-built night-fighter in the form of the P-61. Until this fighter came into service the Americans were dependent on the supply of British Beaufighters fitted with AI Mk.VIII to fulfil its night defence responsibilities in the European theatres of operation.[39] AI had also not been accorded a high level of priority during 1942, which compounded the problem and delayed SCR-720B deliveries until at least July 1943, with a small batch being predicted for May. After July Western Electric's production schedules optimistically showed a total of 100 sets being delivered in July, rising to 200 per month thereafter. At the June meeting of the Air Interception Committee, Britain's initial order was stated as comprising 2,900 sets, of which some 250 SCR-720B models were expected by the end of 1943, followed by 120 each month until deliveries were completed. Following this meeting the Prime Minister was given an assurance, presumably based on US forecasts, that following the installation of the first 100 sets in Mosquito night-fighters, sixty aircraft per month would be delivered to Fighter Command. This statement made by the members of the Committee, was one of the principal factors in clearing *Window* for Bomber Command, for which the Prime Minister directed 'that the most strenuous efforts must be taken to increase the fitting programme and to have the largest number of sets in operational use as soon as possible'.[40]

With the RAF firmly committed to the American equipment, the continued development of AI Mk.IX was reduced to a much lower priority. However, there was still some interest in the development of AI Mk.IX and the possibility that it might be adapted to 3 cm working. In a memo dated 23 September 1943 from Mr W.C. Cooper, a DDCD in MAP, the present position on AI Mk.IX and the possibilities for 3 cm radar were laid out. The Ministry proposed four versions of the Mk.IX system:[41]

- AI Mk.IX – the basic system with windscreen projection of the pilot's indicator information.
- AI Mk.IXB – windscreen projection with roll stabilisation.
- AI Mk.IXC – a gyro gun-sight using a version of the Automatic Gun Laying Turret (AGLT) radar proposed for Bomber Command.
- AI Mk.IXD – an electrical prediction system similar to Mk.IXC with windscreen projection and an electrical computer to work out the firing calculations.

Somewhat optimistically, TRE proposed that the first elements of these systems could be air tested as early as October 1943, with the experiments on the Mk.IXD beginning in January 1944 and the first units being delivered by the following October. Needless to say, these dates were not met.

The Air Ministry where hopeful that the 9 cm AI Mk.IX equipment could be adapted to 3 cm, with many of its boxes being capable of adaptation to the shorter wavelength. Some of this work was in the hands of E.K. Cole Ltd who were awarded a development contract to investigate the adaptation of the Mk.IX indicating unit, receiver and control unit, all of which were common to both wavelengths. Working with 3 cm offered a number of advantages in relation to smaller scanners, sharper beam angles and improved tolerance to *Window* jamming. On the negative side the power output was calculated at a mere 50 Watts (against 200 Watts for 9 cm), which would reduce the range and there was already a 3 cm development programme underway in the US (see also Chapter 13).

In conclusion, Mr Cooper recommended 'that work on 3 cm and 9 cm should proceed in parallel with the same degree of priority until such times as flight trials could prove the definitive advantage of one or other of the two systems'.[42] We will return to the development of AI Mk.IX later.

Having taken the decision to adopt AI Mk.X as the standard radar for the foreseeable future, the problem arose of the integration of this and other Allied radar systems into an Allied IFF scheme. Following the acceptance of IFF Mk.III as the standard inter-service scheme for British forces in early 1942 (see Chapter 11) it was agreed at an inter-Allied meeting in January 1942, that the Mk.III system would also be deployed by US forces. The scale of the programme to equip US, British and Commonwealth forces with a single IFF system was immense, for not only did the British have to contend with converting their air and ground radars to the Mk.III system, they also had to accommodate those of their Allies – one of which was SCR-720B/AI Mk.X.[43]

In order that night-fighters might properly identify US and British bombers, Fighter Command's Beaufighters and Mosquitoes required the installation of IFF Mk.III interrogator/responders, the TR3171 for AI Mk.IV and V, and SCR-729 for AI Mk.X. The SCR-729 equipment also supplied a navigation capability by means of a built-in *Rebecca/Eureka* radar beacon system.[44] A total of 400 SCR-729 equipments were scheduled for delivery in 1943, with the first six being due in June. Similarly, there was a need to identify Allied night-fighters on the new generation of centimetric GCI stations then coming into use in Britain. For this the Beaufighters and Mosquitoes were required to carry an IFF Mk.IIIG interrogator/responder, which was built and supplied from the US as the SCR-695. This operated in the IFF G-Band (200–210 MHz) for centimetric GCI working and on the old A-Band (157–187 MHz) to enable night-fighters to be recognised by all the other radar systems, both land and air.[45]

Because of the size of its scanner and the commencement of the run-down of the Beaufighter force, AI Mk.X was only installed in the Mosquito. Once again a Mk.II airframe, HK195,[46] was adapted to take the new radar, a task that was completed at Defford by 12 July, after which it was evaluated by TFU and delivered to FIU for operational testing on 11 August. Whilst at FIU, HK159's radar was evaluated by Wing Commander Jackson on 13 August with Flight Lieutenant Davidson as his pilot and again on 5 September with Squadron Leader Christopher Hartley.[47,48] Fitted with what became known as the 'bull-nose' on account of the shape of its perspex radome, the resultant Mosquito NF.XVII had single-stage Merlin 21 or 23 engines and no provision for wing tanks. A small batch of Mosquito IIs were converted by Marshalls shortly after they had completed the work on the Mk.XIIs, after which three batches of Mk.XVIIs were built by de Havilland's Leavesden factory.[49] The main AI Mk.X production version was the NF.XIX which was based on the Mk.XIII airframe with Merlin 25 engines, the bull-nose radome and the FB.VI's stronger wing with attendant extra tankerage. As a protection again delays in the delivery of AI Mk.X, this version was also capable of carrying AI Mk.VIII – a wise decision, as the late deliveries of the American radar would soon prove. This, and the installation for the Mk.XVII, were based on the development work undertaken by Mosquito DZ659 at Defford in April.[50]

Despite the high priorities set by the Air Staff and the Prime Minister, the deliveries of AI Mk.X were subjected to considerable delays. With the Western Electric production line firmly committed to delivering the US version of the SCR-720B equipment, there was little time, nor effort, available to incorporate the modifications recommended by TRE. Consequently, when the first forty Mk.X units were received in the late autumn of 1943, the modifications had to be undertaken by TRE before the equipment was installed in the Mosquito. This work was carried out at Defford by TRE's Special Installation Unit,[51] itself a part of Establishment's PDS Group, who then installed the first twelve sets of a batch of ninety, in Mosquito NF.XVIIs and Wellington Mk.XVIIs.[52] The latter were standard Wellington Mk.X bombers with all armament removed and a radome installed in the position occupied by the nose turret to accommodate the scanner. They were used in the role of 'travelling conversion' for N/R training. A further version, the Wellington Mk.XVIII, again an adaptation of the Mk.X airframe, had the W/Op's and navigator's compartments redesigned to accommodate four pupils, their instructor and the 'boxes' for AI Mk.X. The scanner was once again enclosed in a radome in the extreme nose of the aircraft. Following the delivery of the first twelve Mosquito NF.XVIIs, the responsibility for the remaining aircraft in the batch passed to No. 218 MU at Colerne.[53]

The Mk.X re-equipment programme suffered a further setback following the completion of the first Mosquito installations, when it was discovered that RF radiation from the modulator unit was interfering with the aircraft's VHF R/T set. This fault was regarded as being 'serious' by DRD, who also asked why it had not been identified during the trials phase.[54] The RF source proved difficult to eradicate, but after a series of investigations and much trial and error the problem was eventually solved by a series of modifications that included: improvements to the screening of a number of connectors, the fitting of special filters and chokes to the modulator and RF units, the repositioning of the VHF aerials and the addition of further suppression components to the VHF set. By the time these had been completed and tested, the programme was seriously delayed, with the first five fully fitted aircraft not being delivered to the RAF until the end of January 1944.[55]

1944

Whilst TRE struggled to overcome the interference problem, a flight was established under the auspices of Fighter Command to train the N/Rs in the intricacies of the new equipment. Using the Wellington XVII as a basis, the flight visited the first squadrons scheduled to convert to the Mosquito NF.XVII, No. 85 at Hunsdon and No. 25 at Church Fenton. No. 85 had already received its first Mosquito XVIIs during November 1943, with No. 25 following closely behind in December. By the end of January the conversion of their N/Rs was nearly completed, enabling No. 85 Squadron to fly the first operational AI Mk.X sortie on 12 February, when one aircraft was rostered on the night flying programme. The first enemy aircraft destroyed by the AI Mk.X/Mosquito NF.XVII combination occurred on the night of 20/21 February, when Pilot Officer J.R. Brockbank with Pilot Officer D. McCausland destroyed a Ju 188E-1 of 5./KG 2 that crashed near Wickham St Paul, Essex. With only two squadrons operating the Mosquito NF.XVII, the results for AI Mk.X interceptions for the month of February were sixty AI contacts on enemy aircraft, of which seven were destroyed, one probably destroyed and a further five damaged (see also Appendix 4). The high percentage of lost contacts was later attributed to the speed of the Fw 190 Jabos and to the 'vigorous evasive tactics' employed by many of the German pilots.[56]

With the radio mechanics being trained on the new equipment, first at TRE's School and later at No. 7 Signal School, the re-equipment of further squadrons gathered pace. With Mk.X serviceability after the first two weeks of operations being regarded as 'satisfactory', at an average of twenty-five to thirty flying hours per fault, and the equipment delivering maximum ranges varing between 5½ and 6 miles (8.8–9.6 km) and minimum ranges of 400 feet (120 metres), confidence in the new set quickly grew. A large proportion of the faults were centred on the electrical drive mechanism on the scanner unit, which never performed satisfactorily from the tactical viewpoint due to its inability to provide adequate cover in the maximum down position. A number of modifications were proposed, but that by the Reyrolle Company, which required the unit to be re-balanced, proved the most effective. To further understand the problem a party of RAF 'technicians' visited the US to assist Western Electric and the USAAC in the formulation of a new scanner design that would improve its coverage and reliability. Sets which employed the new design, designated SCR-720D, or AI Mk.XA in RAF parlance, were delivered to Britain shortly before hostilities ceased and the UK/US Lend-Lease Agreement was ended.[57]

It had been the Air Staff's intention to install AI Mk.X in all of the new types of Mosquito then coming into service with Fighter Command – the NF.XIX, NF.30 and NF.36, of which more later. However, with the uncertainty surrounding the AI Mk.X supply situation at the end of 1943, they were forced to reconsider their decision and fit the first fifty of the Mk.XIX Mosquitoes with AI Mk.VIII in anticipation of it being cleared to operate over enemy territory in time for the invasion of northern Europe. AI Mk.X, however, would remain 'restricted' as it was the only AI radar capable of reasonable discrimination in the face of *Window*-type jamming. Consequently, the number of Mk.VIII equipped Mosquitoes XIXs was increased to 100.[58]

It will be recalled that the primary reason for the introduction of AI Mk.X was its ability to operate in the face of *Window* jamming. To this end, a comparison of the performance of AI Mk.VIII and Mk.X in the period January to June (D-Day) in the defence of UK air space may be of interest. The comparison, which was conducted by TRE and published in their house journal for July 1945, shows there was very little difference in the operational efficiency of both types in standard GCI interceptions – both returning interception rates of 12 per cent.[59] Overall, the level of *Window* jamming was much lighter than had been anticipated and was in the main deployed as a self-protection counter-measure for bombers. There was no systematic assault on the UK's ground and air radar defences of the type employed over Germany during the bomber support campaign of 1943/45.[60,61]

CONCLUSIONS

The design of AI Mk.IX provided another step in the development of centimetric AI radar towards the concept of an automated fire control system. Its proven ability to lock-on to a target, at first with the aid of the N/R and later fully automatically, and provide sufficient information to the pilot to engage the enemy without actually seeing him, made AI Mk.IX the most sophisticated AI radar of its period (1943). Its inability to operate in the face of *Window* jamming was unfortunate and unforeseen when it was being designed by Dr Lovell and his team. The equally unfortunate loss of Dr Downing and his pilot in December 1942, along with the recently modified prototype equipment (which may or may not have countered the jamming), put back the development of British AI radar for several months and provided an opening for the Americans to field one of their own, the SCR-720/AI Mk.X.

For their part the Americans learned well from the British after the Tizard Mission and successfully pooled their academic and commercial research facilities to develop the SCR-720 equipment and make a prototype available to TRE for evaluation. By great good fortune the equipment arrived shortly before the loss of the Mk.IX equipment to provide the British with a fully functioning, *Window* resistant radar, which cleared Fighter Command's objections to the use of the device by Bomber Command. The resultant raid on Hamburg on the night of 24/25 July 1943 undoubtly proved the effectiveness of this decision.

With the decision to commit to the SCR-720 radar and the placing of Mk.IX development on the 'back burner', the Americans were committed to deliver the system in sufficient quantity to have it installed in the Mosquito by the middle of 1943. This they failed to achieve, in part due to their concentration of production to meet the needs of the USAAC, and also to the inevitable delay in production deliveries that afflicts most industrial programmes in time of war. Ironically, the slip in the delivery schedules required the Air Staff to back-track somewhat and equip the Mosquito force with further deliveries of AI Mk.VIII, thanks to someone in the Air Ministry or the de Havilland Company making the Mk.XIX Mosquito compatible with both types of radar.

However, the principal outcome of the loss of Dr Downing's life and his Mk.IX radar was its effect on the UK electronics industry. From 1944 onwards the RAF would be wholly dependent on the US for the supply of its front line AI radar. This situation would extend into the post-war period when all RAF and naval night-fighters were be equipped with American radars, whose production was lost to UK industry, and even more importantly, provide an inertia against the development of home-built systems until well into the 1950s.

Notes

1. *The Signals History, Volume 5*, p.163.
2. *Ibid*, p.163.
3. *Ibid*, p.164.
4. *Ibid*, p.164.
5. Sir Bernard Lovell, *op cit*, p.69.
6. *Ibid*, p.70.
7. Later Professor Sir Frederick Williams, FRS, Professor of Electrical Enginering at Manchester University.
8. *Ibid*, p.71.
9. The system was also referred to as AIS Follow, or AISF.
10. Sir Bernard Lovell, *op cit*, pp.72 and 73.
11. *Ibid*, pp.74 and 75.
12. *Ibid*, pp.75 and 76.
13. *Ibid*, pp.76 and 77.
14. *Ibid*, p.77.
15. *Ibid*, p.77.
16. The decision to engineer the AIF installation for use in the Beaufighter was made one month before the prototype flew in T1939.
17. Sir Bernard Lovell, *op cit*, p.78.
18. An aircraft turning through 360° in two minutes is said to have undertaken a Rate 1 turn. Similarly, an aircraft completing a 360° turn in one minute is undertaking a Rate 2 turn, i.e. twice that of a Rate 1.
19. *Ibid*, pp.78 and 79.
20. *Ibid*, pp.79 and 80.
21. Western Electric was the production arm of the Bell Telephone Company.
22. AT&T, *Radar*, in Engineering & Science in the Bell System (New Jersey: AT&T), pp.93 and 94.
23. *The Signals History, Volume 5*, pp.151 and 152, and AT&T, *op cit*, p.95.
24. AT&T, *op cit*, pp.96 and 97.
25. *Ibid*, pp.97 and 98.
26. AP1093D, *Notes on the History of AI, IFF & Radar Beacons*, sections 79–102.
27. Later Professor Reginald Jones, CB, CBE, FRS, the post-war Director of Intelligence at the Air Ministry and Professor of Natural Philosophy at Aberdeen University from 1946.
28. R.V. Jones, *Most Secret War* (London: Hamish Hamilton, 1978), pp.293.
29. *Ibid*, p.293.
30. *Mandrel* was a jamming device that targeted the enemy's *Freya* early warning radars.
31. R.V. Jones, *op cit*, pp.294 and 295.

32. H.G. Kuhn and Sir Christopher Hartley, *Derek Ainslie Jackson, a Biographical Memoir* (London: the Royal Society), p. 285.
33. *The Signals History, Volume 5*, p. 164, Alfred Price [4], *op cit*, p. 128.
34. By July 1943 *Window* foil had been refined to strips measuring 1–2 cm in width by 25–30 cm in length. Quoted in V.A. Pheasant, *op cit*.
35. V.A. Pheasant, *The Sixtieth Anniversary of Window, 1943–2003* (Chemring Group, 2003).
36. Sharp and Bowyer, *op cit*, p. 59.
37. *The Signals History, Volume 5*, p. 165.
38. *Ibid*, p. 165.
39. The USAAC's 414th Night Fighter Squadron (NFS) was equipped with Beaufighter VIfs from January 1943 in the Mediterranean Theatre, before converting to the P-61 in December 1944. Similarly the 415th NFS and the 417th NFS operated Beaufighters from February 1943, with the 415th converting to the P-61 in March 1945 and the 417th in April/May of the same year.
40. *The Signals History, Volume 5*, p. 165.
41. Memo from W.C. Cooper, DDCD2, dated 25 September 1943 and filed in AVIA10/66.
42. *Ibid*.
43. Ian White [2], *op cit*, p. 15.
44. *Eureka* was the codename for a ground based radar beacon that was designed by TRE in 1941/42 to provide a homing capability by means of a *Rebecca* interrogator mounted in an aircraft. It was designed for the accurate landing of airborne troops and for aircraft employed in dropping supplies to agents in occupied Europe. In good conditions the beacon could be detected at ranges out to 50 miles (80 km).
45. Ian White [2], *op cit*, pp. 17–19 and the minutes of the SCR-720 Panel held at Church House on 24 July 1943 and filed in AVIA10/66.
46. This aircraft is mentioned in the minutes of the SCR-720 Panel held at Church House on 24 July 1943 and filed in AVIA10/66.
47. *FIU ORB*, pp. 89 and 98.
48. Later Air Marshal Sir Christopher Hartley, KCB, CBE, DFC, AFC and the CO of FIU from November 1943 to July 1944.
49. HK363–HK382, HK728–HK327 and HK344–HK362.
50. Sharp and Bowyer, *op cit*, pp. 59 and 60.
51. The Special Installation Unit grew out of the TRE Fitting Party formed for the purpose of installing AI Mk.VIIIA in Mosquito XIIs at Christchurch during the spring of 1942.
52. *The Signals History, Volume 5*, pp. 165 and 166.
53. Andrews and Morgan, *Vickers Aircraft Since 1908* (London: Putnam, 1988), pp. 358 and 359.
54. Loose minute signed by Mr G.M. Brown, DRD, dated 26 October 1943 and filed in AVIA10/66.
55. *The Signals History, Volume 5*, p. 166.
56. *Ibid*, p. 166.
57. *Ibid*, p. 167.
58. *Ibid*, p. 167.
59. This is the percentage of all GCI controlled interceptions that resulted in a 'visual'.
60. The use of AI in the Bomber Support role falls outside the terms of reference of this book; however, for completeness, No. 100 (Bomber Support) Group was formed within Bomber Command in November 1943 to support the night-bombing campaign by attacks upon the enemy's air and ground radar and night-fighter defences by means of jamming and/or deception. It was disbanded in December 1945 after barely twenty-five months of very successful service.
61. *The Signals History, Volume 5*, p. 168.

CHAPTER THIRTEEN

The Final Push

January 1944–May 1945

The continuing decline in Germany's fortunes in Russia, North Africa and Italy during the first half of 1943 and the transfer of forces to support Operation *Citadel*, the armoured offensive at Kursk, alongside the air defence of the Reich, did not impede the *Luftwaffe*'s continuing efforts to improve the quality and quantity of its aircraft. With production geared to the manufacture of fighters with which to equip the home defence units and those in Russia, bomber production was accorded a lower priority. Nevertheless, new *Kampfgruppen* were scheduled to be formed during the coming summer, whilst others were re-equipped with new or improved aircraft. Of these *Luftflotte* 3s II./KG 40 received their first Heinkel He 177A-5s at Bordeaux-Mérignac in October for anti-shipping operations over the Atlantic, whilst V./KG 2 received Messerschmitt Me 410A-1s in June to operate in the *Schnellbomber* role over England. V./KG 2 was joined by I./KG 51 which converted to the Me 410A-1 in the late summer. Throughout the second half of the year *Luftflotte* 3 retained the services of SKG 10, KG 2, I and II./KG 6 and I./KG 66 to maintain the night offensive over England.

The only new aircraft to enter the inventory was the He 177 'Greif' (Griffon) heavy-bomber. First tested in prototype form by KG 40 in the summer of 1941, the He 177 was destined to fail by nature of its unreliable double-powerplant that married two Daimler-Benz DB 601 in-line engines to form the DB 606. The He 177A-5 was a capable aircraft when things went right. With the 3,100 hp DB 606A-1/B-1 it possessed a maximum speed of 303 mph (485 km/hr) at 20,000 feet (6,100 metres) and an economical cruising speed of 210 mph (335 km/hr) at 20,000 feet. Its defensive armament comprised a 20 mm cannon in forward ventral and tail positions, a 7.9 mm (0.3 inch) machine-gun in the nose and two 13 mm (0.5 inch) machine-guns in the forward dorsal barbette and one in the aft dorsal turret. Offensively, the A-5 could carry bombs of various calibres to a maximum of 2,200 lb (1,000 kg), or externally two Henschel Hs 293 air-to-surface guided missiles, or two FX 1400 'Fritz' guided bombs. The range was proportional to these loads, but 3,417 miles (5,470 km) could be achieved carrying two Hs 293s.[1]

The updated or improved aircraft comprised the Me 410, the Ju 88S and the Ju 188 light-bombers. Designed as a replacement for the failed Messerschmitt Me 210 programme and employing many of its parts, the Me 410 was regarded as a 'formidable adversary' when it was introduced to service over England in June 1943. Fitted with DB 603A in-line engines rated at 1,625 hp, the Me 410A-1 *Schnellbomber* could reach speeds up to 388 mph (620 km/hr) at 21,980 feet (6,700 metres) and accommodate bombs up to 2,200 lb (1,000 kg) in its internal bomb bay, although when operating over England this was usually restricted to a single 1,100 lb (500 kg) weapon. Defensive armament comprised two fixed forward firing 20 mm cannon and two 13 mm (0.5 inch) machine guns and a further two 13 mm machine-guns in remotely controlled rearward facing barbettes.[2]

Aware that the standard Ju 88A series was by 1942 obsolescent, the Junkers design staff replaced the A-4s BMW 801A radial engines with the more powerfull 801D, added a nitrous oxide (GM 1) boost system to increase the engine power and reduced the defensive armament, armour and the crew from four to three (pilot, navigator/bomb-aimer and W/OpAG). The resultant Ju 88S-1 with two BMW 801G-2 engines of 1,730-hp with GM 1 boost, could reach 379 mph (605 km/hr) at 26,250 feet (8,000 metres) – making it quite a challenge for a Mosquito! Possessing a cleaner cabin, extended wing tips to improve altitude performance and BMW 801D-2 engines of 1,700-hp for take-off, the Ju 188E-1 retained the four seats of the Ju 88A-4 series and its heavier defensive armament. Consequently its maximum speed suffered, just 310 mph (495 km/hr) at 19,685 feet (6,000 metres), but it could lift a respectable 4,400 lb (2,000 kg) of bombs.[3]

1944

The light night-bombing of England during the autumn and early winter of 1943/44 was nothing compared to the pounding dealt out by Bomber Command to the cities and towns in Germany's heartland. It also cost *Luftflotte* 3 dear in respect of its losses to night-fighters based in the south and east of England and intruder operations over its French, Belgian and Dutch airfields. Nevertheless, the OKL hoped that by drafting in bomber units from across the Reich and equipping them with the latest aircraft and weapons, the scales might once more be tipped in Germany's favour. On 28 November 1943 *Reichsmarschall* Goering summoned *Generalmajor* Dietrich Peltz, his *Angriffsführer England*, to a high level conference at Neuenhagen, near Berlin, to outline his plans for yet another offensive with London as the principal target. Using the codename Operation *Steinbock* (Ibex), Goering informed the meeting he had promised Hitler the *Luftwaffe* would be ready for its first attack in fourteen days time (12 December). To reach the required number of sorties, some 550–600 were planned, Goering proposed that *Luftflotte* 3's aircraft fly three separate operations during the night, a figure immediately reduced to two following discussions between Peltz and his C-in-C.[4]

To support *Steinbock* the OKL assembled just short of 500 aircraft under Peltz's command, made up from the regular units of *Luftflotte* 3 and two *Gruppen* withdrawn from KG 30, KG 54 and KG 76 (six in total) in Italy's *Luftflotte* 2 and unit's re-equipping in Germany: the Ju 88s of II./KG 6, He 177s of I./KG 100 and Me 410s of I./KG 51. The force by 20 January is shown in Table 6. Together these units totalled 491 aircraft of which 431 were serviceable – a remarkable 88 per cent availability rate.

In order to achieve the maximum destruction Goering ordered that each He 177 be able to carry two of the *Luftwaffe*'s heaviest bomb, the SC 3500 of 5,500 lb (2,500 kg), whilst medium bombers were to deliver the AB 1000 bomb container holding 620 1 kg (2.2 lb) incendiary bombs. To improve navigation and bombing accuracy the Ju 188s of the force's pathfinder unit, I./KG 66, were fitted with the *Egon* electronic bomb release system and the Y-beam system.[5,6] Another system codenamed *Truhe*, a German copy of *Gee* that used British *Gee* network, was also carried by the pathfinders to improve their navigation.[7] The RAF were alerted to *Steinbock* through the *Ultra*[8] system in December, when radio intercepts detected

TABLE 6. *LUFTWAFFE* UNITS ASSEMBLED FOR OPERATION *STEINBOCK*,
JANUARY 1944.[9]

Luftflotte 3		Seconded Units	
Unit	Aircraft	Unit	Aircraft
Stab./KG 2	Do 217M-1	II./KG 30	Ju 88A-4
I./KG 2	Do 217M-1	1./KG 40	He 177A-3
II./KG 2	Ju 188E-1	Stab./KG 54	Ju 88A-4
III./KG 2	Do 217M-1	I./KG 51	Me 410A-1
V./KG 2	Me 410A-1	I./KG 54	Ju 88A-4
Stab./KG 6	Ju 88A-4	II./KG 54	Ju 88A-4
I./KG 6	Ju 188E-1	I./KG 66	Ju 188E-1
II./KG 6	Ju 88A-4	Stab./KG 76	Ju 88A-4
III./KG 6	Ju 88A-4	I./KG 100[10]	He 177A-3
I./SKG 10	Fw 190A-5		

the moves of the *Kampfgruppen* from Italy and further decrypts in January confirmed the move of the six bomber *Gruppen* to bases in Germany, and in one case, a second move to Belgium. The need for secrecy that was stressed in the signals, provided further evidence to the RAF's intelligence officers of the possibility of impending heavy attacks on England.[11]

During the preparations for Operation *Overlord*, the Allied invasion of Europe that was planned to take place in 1944, the Air Staff created the 2nd Tactical Air Force (2TAF) under the command of Air Marshal Sir Arthur Coningham. By June 1943 2TAF had absorbed Army Co-operation Command, Fighter Command and Bomber Command's No. 2 Group and re-established the old Air Defence of Great Britain (ADGB) to control all defensive fighter operations, which retained the old group structure. To command ADGB the Air Staff chose the former AOC No. 12 Group, Air Marshal Sir Roderick Hill. A First World War fighter pilot who was blessed with a sensitive nature and was an accomplished artist, Hill, unlike his predecessor Air Marshal Leigh Mallory, approached the problems of air defence from a more scientific and humane viewpoint, but was nevertheless firm in his ideas and actions.

The strength of ADGB in 1944 is a complex equation, since aircraft that were allocated to 2TAF were deployed in support of ADGB and transferred to the Continent from June to cover the D-Day landings. Similarly, squadrons on rest from operations in France and Belgium were allocated to Home Defence on their return to Britain. Consequently the composition of ADGB at any one time from D-Day onwards is a 'snap-shot' of its true strength.

ADGB strength at the beginning of the *Steinbock* campaign comprised sixteen squadrons scattered over fifteen airfields, from Predannack in Cornwall to Drem in Scotland and as far west as Valley on the Isle of Anglesey. This strength requires some explanation, since six squadrons (Nos 85, 151, 264, 307, 410 and 456) were in the process of converting to one or other marks of the Mosquito or working-up to an operational level, whilst one, No. 29, had just arrived back in Britain and was awaiting its turn to convert to the NF.XVII. Of the remainder five (Nos 68, 125, 406, 409 and 604) were still operating the Beaufighter VIf. The key sectors as far as *Steinbock* is concerned were those covered by Nos 11 and 12 Groups. Here the situation is different with only four squadrons (Nos 68, 85, 96 and 488) in the front line, of which all excepting No. 68 were equipped with the Mosquito carrying centimetric radar. In total, a maximum of approximately sixty-five night-fighters[12] were available to oppose an enemy force approaching 500.

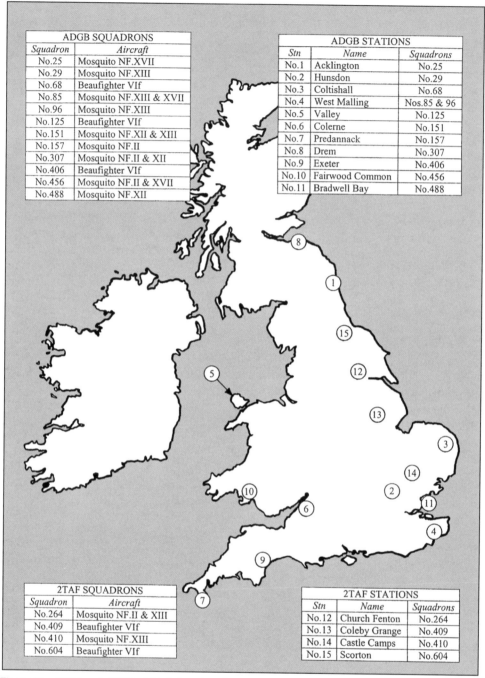

ADGB SQUADRONS	
Squadron	*Aircraft*
No.25	Mosquito NF.XVII
No.29	Mosquito NF.XIII
No.68	Beaufighter VIf
No.85	Mosquito NF.XIII & XVII
No.96	Mosquito NF.XIII
No.125	Beaufighter VIf
No.151	Mosquito NF.XII & XIII
No.157	Mosquito NF.II
No.307	Mosquito NF.II & XII
No.406	Beaufighter VIf
No.456	Mosquito NF.II & XVII
No.488	Mosquito NF.XII

ADGB STATIONS		
Stn	*Name*	*Squadrons*
No.1	Acklington	No.25
No.2	Hunsdon	No.29
No.3	Coltishall	No.68
No.4	West Malling	Nos.85 & 96
No.5	Valley	No.125
No.6	Colerne	No.151
No.7	Predannack	No.157
No.8	Drem	No.307
No.9	Exeter	No.406
No.10	Fairwood Common	No.456
No.11	Bradwell Bay	No.488

2TAF SQUADRONS	
Squadron	*Aircraft*
No.264	Mosquito NF.II & XIII
No.409	Beaufighter VIf
No.410	Mosquito NF.XIII
No.604	Beaufighter VIf

2TAF STATIONS		
Stn	*Name*	*Squadrons*
No.12	Church Fenton	No.264
No.13	Coleby Grange	No.409
No.14	Castle Camps	No.410
No.15	Scorton	No.604

Figure 26. 2TAF and ADGB night-fighter Squadrons at January 1944. The deployment of night-fighter squadrons in preparation for the invasion of Europe, provided a defence for southern England and in sufficient strength to deflect the weak *Luftwaffe* bomber forces participating in the 'Baby Blitz'. At this juncture Fighter Command was deploying Mosquito squadrons equipped with AI Mk.VIII and Mk.X in increasing numbers and had very nearly completed the removal of the Beaufighter.

Although scheduled to begin in December, the first raid of the *Steinbock* offensive, otherwise known in Britain as the 'Baby Blitz', occurred on the night of 21/22 January when the first elements of a force of 227 bombers that included, for the first time, He 177s drawn from I./KG 40 and I./KG 100, Do 217Ms from I./KG 2, Ju 88s from II./KG 54 and I./KG 76, based in France and from Holland more Ju 88s from II./KG 6 and II./KG 30 and the Me 410s of V./KG 2. Using every 'trick in the book' and led by white marker flares dropped by the pathfinders from I./KG 66, the main force crossed the English south coast between Hastings and Folkestone at 2040 hours and headed straight for London. Whilst on course to the target the crews threw out large quantities of *Duppel* foil, which caused a number of the controllers at the older metric GCI stations some difficulty in tracking their targets. If this was not bad enough, the sheer number of enemy aircraft present that night overwhelmed the controllers of the un-jammed centimetric sets, forcing the fighter crews to fall back on the searchlights to indicate their targets.[13]

Considering the relative strength of the first wave, remarkably few crews found their way to London and consequently little damage was recorded. The second phase was equally weak. British records show that whilst 200 aircraft crossed to the landward side of the coast that night, only forty reached Greater London to drop just 32 tons (32.5 tonnes) of bombs. Fighter Command flew ninety-six sorties and claimed sixteen enemy aircraft destroyed. By comparison the *Luftwaffe* claimed a total of 447 sorties and the loss of twenty-five aircraft over Britain and a further eighteen to other causes[14] – an overall loss-rate of 9.6 per cent. The failure of Peltz to secure a more worthwhile result may be attributed to a number of factors that were outside his control to correct, namely:[15]

- The jamming of the Y-beams that placed many of the pathfinder markers in the wrong place.
- The failure to create a concentration that would saturate the defences due to the small numbers involved.
- The poor training and experience of the bomber crews.
- The obsolescence of the bombers and the unreliability of their equipment, particularly that of the He 177.
- Poor weather conditions during the second attack wave.
- The strength and technical superiority of the British ground and air defences.

This trend was set to continue throughout the remainder of the *Steinbock* offensive.

The excessive loss-rate and the need to conserve his force as a viable entity, persuaded Peltz to abandon the two wave method of attack and instead to concentrate his forces in a single wave. Poor weather intervened to delay the second attack to the night of 29/30 January, when 285 bombers departed their airfields, of which only some 130 crossed the English coast and penetrated inland. Of these only thirty reached London, whilst others released their bombs over Kent and Essex, killing fifty-one people (forty-one in London), seriously injuring a further 124 (116 in London) and causing 343 fires. Fourteen bombers failed to return.[16]

Events in the Mediterranean during January conspired to rob Peltz of aircraft and crews and further reduce his strength. On 22 January the American 5th Army landed on the beaches at Anzio on the Italian western coast, putting considerable pressure on German forces in the region and forcing the C-in-C, *Generalfeldmarschall* Kesselring, to request the return of four of his *Kampfgruppen*, three of Ju 88s and one of He 177, from France and

Holland with immediate effect. By this means *Steinbock* was deprived of some 100 aircraft and their crews.

Peltz returned to the offensive on the night of 3/4 February striking at London, but causing little damage, and again on 13/14 February when 230 sorties managed to release just 16 tons (16.25 tonnes) of bombs on the capital. The bomber's performance on the following three nights improved dramatically and brought about the heaviest raids since May 1941. On the night of 18/19 February sorties by 200 bombers succeeded in delivering 139 tons (141.25 tonnes) on Greater London (see Table 7), in an attack that lasted barely thirty minutes and caused 480 fires in fifty-six of the capital's boroughs. Widespread damage to the rail network, bridges, telephone exchanges, hospitals and civil defence buildings was recorded and 643 people were killed or seriously injured. The *Luftwaffe* returned on the 20/21 February when once again 200 bombers led by I./KG 66 dropped 118 tons (119.90 tonnes) that started 606 fires in Central London and others in sixty-three of the boroughs. As on the previous raid casualties were heavy, with 216 persons being killed and 417 seriously injured.

Lesser raids were carried out during the remaining nights of February and on into March (see Table 7), when much of the bombing was scattered and on some occasions no bombs fell on the intended target. On the night of 19/20 March London earned a respite for one night whilst Peltz concentrated his forces for an attack on the east coast town of Hull, along with its docks and port facilities. 131 bomber sorties were flown, but because of poor target

TABLE 7.   SUMMARY OF *STEINBOCK* RAIDS OVER BRITAIN, JANUARY–MAY 1944.[17]

Date	Target	Bomber sorties	Bombs on target (tons)	Bombers lost (to all causes)	Loss rate (%)
21/22 Jan	London	447	32.5	43	9.6
29/30 Jan	London	285	36.5	14	4.9
18/19 Feb	London	200	139	9	4.5
20/21 Feb	London	200	118	8	4.0
22/23 Feb	London	185	75	7	3.8
23/24 Feb	London	161	49	4	2.5
24/25 Feb	London	185	89	9	4.9
1/2 March	London	165	65	4	2.4
14/15 March	London	187	81	13	7.0
19/20 March	Hull	131	0	9	6.9
21/22 March	London	144	87	10	6.9
24/25 March	London	143	52	15	10.5
27/28 March	Bristol	139	0	13	9.4
18/19 April	London	125	53	13	10.4
20/21 April	Hull	130	0	8	6.2
23/24 April	Bristol	117	0	8	6.8
25/26 April	Portsmouth	193	1	6	3.1
26/27 April	Portsmouth	78	1	4	5.1
27/28 April	Portsmouth	60	1	2	3.3
28/29 April	Portsmouth	58	1	1	1.7
29/30 April	Plymouth	101	8	3	3.0
14/15 May	Bristol	91	3	11	12.1
15/16 May	Portsmouth	106	1.4	4	3.8
22/23 May	Portsmouth	104	1.5	7	6.7
27/28 May	Weymouth	28	13	0	0
28/29 May	Falmouth	51	0	2	3.9

marking, bombing was scattered over Lincolnshire and Norfolk, with not a single bomb hitting the town.

From the end of March the serviceability rate amongst the *Kampfgruppen*, which was as high as 89.6 per cent at the beginning of the offensive, began to decline. Of the 441 bombers and fighter-bombers in *Luftflotte* 3's service, only 252 were available for operations – an overall serviceability rate of 57 per cent.[18] This combined with losses to the defences and operational causes (navigation, intruders and landing and take-off accidents) that had by 20 March accounted for some 110 aircraft, seriously affected Peltz's ability to maintain a reasonable force for night raiding. From April onwards the sortie rate began a gradual decline, excepting for the night of 25/26 April when 193 raided Portsmouth, but only a few managed to hit the target with one or two bombs and lost six aircraft in the process. With the targets shifting to the south coast *Overlord* invasion ports towards the end of April and on into May and with Fighter Command continuing to take a steady toll of the raiders, eleven on the night of 14/15 May when the enemy raided Bristol, the *Steinbock* offensive finally collapsed on the night of 28/29 May.

Operation *Steinbock* was the last conventional bombing offensive of the war, during which the *Luftwaffe* in a desperate bid to seek some compensation for Britain's bombing of Germany and improve the country's morale, killed and injured several thousand civilians and damaged a great deal of their property. Fortunately for the nation few targets of any military or industrial significance were struck and the flow of men, munitions and equipment for the forthcoming invasion of Europe was not impeded in any way. The enemy on the other hand lost significant numbers of aircraft and experienced crews and failed to fulfil Hitler's demands for retaliation against Britain. *Steinbock* did, however, serve as the marker that would define the beginning of the terminal decline of the *Kampfgruppen* in the west.

By the time of the execution of Operation *Overlord* at the beginning of June 1944, the number of night-fighter squadrons available to ADGB and 2TAF were at their peak in terms of their numbers and the quality of their equipment. With eight squadrons allocated to ADGB and six to 2TAF (see Figure 27), with all but one, No. 68, equipped with the Mosquito, the defence was able to provide sufficient night-fighters to protect the embarkation ports and secure the invasion beaches. However, ADGB did see a small diminution in its overall strength compared to that in January, with the transfer of two of its squadrons, Nos 85 and 157, to strengthen No. 100 Group's bomber support operations over the Continent.[19]

The choice of Normandy as the site of the invasion, because it was less well defended than the Pas de Calais and had better beaches, took the air battle outside the range of the western coastal GCI stations and necessitated the deployment of fighter direction tenders (FDT) to provide GCI cover for the invasion fleet. Three of these vessels, FDT Nos 13, 216 and 217, were converted from landing ship tanks (LST) and provided with two 1½ metre AMES Type 15 GCI sets, a 50 cm Type 11 GCI, an AI beacon and R/T and D/F equipment. With the control equipment manned by RAF personnel, two of the FDTs were positioned off the beaches and one in mid-Channel, to provide GCI cover over the invasion front and air warning for the Allied fleet. It was planned that the two beach tenders would be relieved of their air warning responsibility, but not the control of air interception, when a mobile GCI had been landed on D-Day in each of the British and American sectors.[20]

Responsibility for the air defence of the beachhead and inland towards German positions in and around Normandy rested with No. 85 Group, 2TAF, with the invasion ports and the south coast of England falling to ADGB. With all 2TAF squadrons, including those of No. 85 Group, being capable of deployment under field conditions and with a high level of

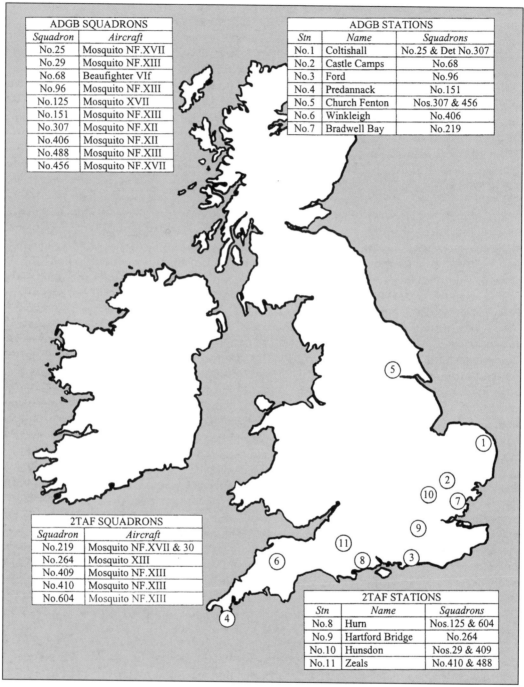

ADGB SQUADRONS	
Squadron	Aircraft
No.25	Mosquito NF.XVII
No.29	Mosquito NF.XIII
No.68	Beaufighter VIf
No.96	Mosquito NF.XIII
No.125	Mosquito XVII
No.151	Mosquito NF.XIII
No.307	Mosquito NF.XII
No.406	Mosquito NF.XII
No.488	Mosquito NF.XIII
No.456	Mosquito NF.XVII

ADGB STATIONS		
Stn	Name	Squadrons
No.1	Coltishall	No.25 & Det No.307
No.2	Castle Camps	No.68
No.3	Ford	No.96
No.4	Predannack	No.151
No.5	Church Fenton	Nos.307 & 456
No.6	Winkleigh	No.406
No.7	Bradwell Bay	No.219

2TAF SQUADRONS	
Squadron	Aircraft
No.219	Mosquito NF.XVII & 30
No.264	Mosquito XIII
No.409	Mosquito NF.XIII
No.410	Mosquito NF.XIII
No.604	Mosquito NF.XIII

2TAF STATIONS		
Stn	Name	Squadrons
No.8	Hurn	Nos.125 & 604
No.9	Hartford Bridge	No.264
No.10	Hunsdon	Nos.29 & 409
No.11	Zeals	No.410 & 488

Figure 27.   2TAF and ADGB night-fighter Squadrons at June 1944. The map shows the night-fighter force deployed at or about D-Day, 6 June 1944, when the 2TAF squadrons came under the command of the Allied Air Commander, Air Chief Marshal Sir Trafford Leigh-Mallory.

mounted in the rear cockpit. Having a beam that was broad in azimuth, but narrow in elevation, the ASH set was designed for low-level operations against ships and, to a lesser extent, aircraft. Its maximum range was assessed as being 4 miles (6.5 km) in the air-to-air mode, with a minimum range of 250 feet (75 metres) in the hands of an experienced operator – in other words, just what was required for hunting V-1 carrying He 111s over the North Sea. In part to exploit the radar's potential and to provide the Navy with operational experience in night-fighting, NFIU approached HQ Fighter Command during October for permission to join in the battle on an occasional basis.[31]

With the Command's agreement NFIU undertook its first detachment to Coltishall on the night of 25/26 October, when Firefly MB419 flown by Lieutenant Kneale, with Lieutenant Harrison operating the radar, was scrambled to intercept a suspect Heinkel operating in the North Sea. Unfortunately the R/T failed and the sortie was abandoned. The Navy tried again two nights later, but this time the enemy failed to put in an appearance. Kneale and Harrison flew their next sortie on 14/15 November, when poor weather forced them to return to Coltishall before making contact with the enemy. On the evening of 22/23 November MB419 was flown to Coltishall by a new crew comprising Lieutenant Mike Howell and observer Lieutenant Lester. On this occasion the crew were able to track, but not catch, a released V-1 which disappeared off Lester's screen and most probably fell into the sea. From then on it was apparent to the crews that the Firefly did not have the performance to maintain contact with the missile, forcing NFIU on subsequent occasions to attempt to intercept the Heinkel carrier aircraft before the V-1 was launched. Lieutenant Kneale returned to Coltishall in DT933 on the afternoon of 24/25 November, but despite being scrambled in good time they were not put on to a target by the GCI controller. On the evening of 12/13 December with observer Lieutenant George Davies,[32] operating the radar, Kneale was directed towards a target on which they closed following a protracted AI chase in low cloud. Kneale opened fire blind and without visually identifying the target, but to no avail. The Navy's last patrol was carried out two nights later on 14/15 December, with no claim submitted.[33]

Like the RAF, the Navy found chasing low-flying He 111s over the North Sea very difficult and in an operation where they had to compete with ADGB's night-fighters, they were allocated few targets. Despite their lack of success NFIU's pilots and observers did prove the Firefly's worth as a night-fighter and confirm its suitability in the night/all-weather fighter role, albeit requiring a radar with a blind-fire capability.

By November with the Allied armies threatening to break out into northern Germany and the Ruhr, all three of KG 53 *Gruppes* were withdrawn to airfields in the vicinity of Oldenburg and Bremen, but nevertheless managed to launch 316 V-1 sorties during that month. Despite the cover of darkness and low altitude flying, casualties amongst KG 53's crews began to mount and with them a shortage of carrier aircraft after the Red Army overran Heinkel's Rostock factory. To ensure the supply of replacement airframes, KG 27 was disbanded in September and its 111s converted to V-1 platforms and delivered to KG 53.

Although pulled back to airfields in Germany, KG 53 was not quite finished with the air launched missile programme. Firings continued throughout December and drew in another innovation as a counter to their attacks on Britain. Throughout 1941 and on into 1942 and 1943, TRE had conducted a series of operational experiments using a Wellington bomber in the 'Air Control of Interception' (ACI) role, or what today would be called 'Airborne Early Warning' (AEW). First mooted in August 1941 by Watson Watt in his capacity as DCD, as a means of directing fighters onto the Focke Wulf Fw 200 Condor patrols in the North Atlantic,

TRE had equipped Wellington Ic, R1629, with a rotating Yagi dipole array, an ASV Mk.II receiver, a special high powered transmitter and a nine inch (23 cm) PPI display. Successfully trialled several times in 1942 and 1943, but dismantled in April 1943, R1629 was written off in a ground accident the following October. During January 1945 the Fighter Interception Development Squadron (FIDS), a part of the recently created Central Fighter Establishment (CFE)[34] which was raised to supersede FIU in October 1944, undertook a series of trials at Ford and Manston in the ACI role under the codename Operation *Vapour*.[35]

Using an ex-Coastal Command Wellington fitted with an ASV Mk.VI radar and PPI display, these trials were practised in the Channel during the hours of daylight in early January 1945 and flown operationally off the Dutch coast at very low level, in the company of five Mosquito night-fighters. Using the *Rebecca/Eureka* beacons system to maintain a formation, the set-up was controlled by a New Zealand civilian scientist, Mr E.J. Smith, from the Wellington, who provided vectors to likely targets for the accompanying Mosquitoes. Overall, the results proved disappointing since the sea returns restricted the ACI's maximum range when flying at low level, but ranges of 14 miles (22.5 km) were recorded at higher altitudes – which were twice as good as the Mosquito's AI Mk.X. However, at the critical point the *Luftwaffe* ceased air-launching operations on 14 January and the ACI project was allowed to fall into abeyance.[36]

From early September 1944 to mid-January 1945, 1,012 air launched V-1s had been fired at the UK by KG 53, at the cost of seventy-seven He 111s destroyed along with the majority of their crews.[37] In response the defence destroyed 404 V-1s, or just under 40 per cent of those launched, with 72½ being credited to the RAF (7.2 per cent) and 320 to the guns (31.6 per cent). Of the remainder, 220 failed for a variety of reasons (21.7 per cent) and 388 (38.3 per cent) penetrated the defences to land in East Anglia or on London. In terms of capability, the AA guns proved to be by far the most effective means of defence.[38] In retrospect it is not known how many casualties may be attributed to the air-launched V-1 programme, but it must extend into the thousands.

1945

The deteriorating military situation in January 1945 forced the retreat of the air units of *Luftflotte* 3 into the western portion of the *Reich*, for which it then resumed responsibility in co-operation with *Luftflotte Reich*. With Soviet forces in the east having reached the German-Polish border and the boundaries of East Prussia, all available bombers, with the exception of those occupied in V-1 operations by KG 53 or allocated to reconnaissance, were transferred to the Eastern Front. When KG 53 ended its V-1 operations in mid-January, all manned bomber flights against Great Britain effectively ceased. Reconnaissance continued, with small *Staffel*-sized units flying sorties as far north as Hull with jet-powered Arado Ar 234B Blitz aircraft. Based at Rheine in western Germany, these machines which flew as high as 32,800 feet (9,755 metres) at speeds up to 435 mph (695 km/hr), were virtually immune to interception and passed unnoticed by the general public. Similarly, *Aufklärungsgruppe* 122 based with *Luftflotte* 2 in Norway and Denmark continued to fly reconnaissance sorties over the north of England and East Anglia with Ju 88 and Ju 188s. On the night of 20/21 March, Flight Lieutenant Kennedy and his N/R, Flying Officer Morgan, of No. 125 Squadron flying a Mosquito NF.XVII, shot down an unidentified Ju 188[39] into the North Sea 10 miles (16 km) to the north-east of Cromer, Norfolk, and with it marked the destruction of the last enemy aircraft in UK air space.

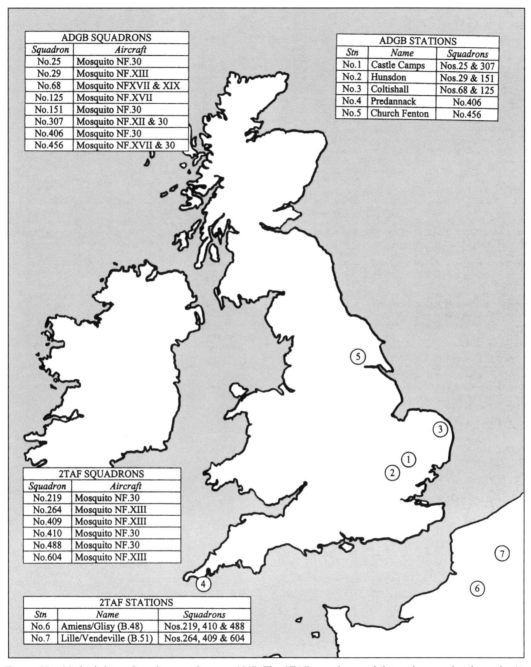

ADGB SQUADRONS	
Squadron	*Aircraft*
No.25	Mosquito NF.30
No.29	Mosquito NF.XIII
No.68	Mosquito NFXVII & XIX
No.125	Mosquito NF.XVII
No.151	Mosquito NF.30
No.307	Mosquito NF.XII & 30
No.406	Mosquito NF.30
No.456	Mosquito NF.XVII & 30

ADGB STATIONS		
Stn	*Name*	*Squadrons*
No.1	Castle Camps	Nos.25 & 307
No.2	Hunsdon	Nos.29 & 151
No.3	Coltishall	Nos.68 & 125
No.4	Predannack	No.406
No.5	Church Fenton	No.456

2TAF SQUADRONS	
Squadron	*Aircraft*
No.219	Mosquito NF.30
No.264	Mosquito NF.XIII
No.409	Mosquito NF.XIII
No.410	Mosquito NF.30
No.488	Mosquito NF.30
No.604	Mosquito NF.XIII

2TAF STATIONS		
Stn	*Name*	*Squadrons*
No.6	Amiens/Glisy (B.48)	Nos.219, 410 & 488
No.7	Lille/Vendeville (B.51)	Nos.264, 409 & 604

Figure 28. Night-fighters Squadrons at January 1945. The 2TAF squadrons whilst no long under the authority AOC-in-C Fighter Command, are included in this disposition as they were the first line of defence against attacks on the UK by *Luftwaffe* bombers or overflights by reconnaissance aircraft.

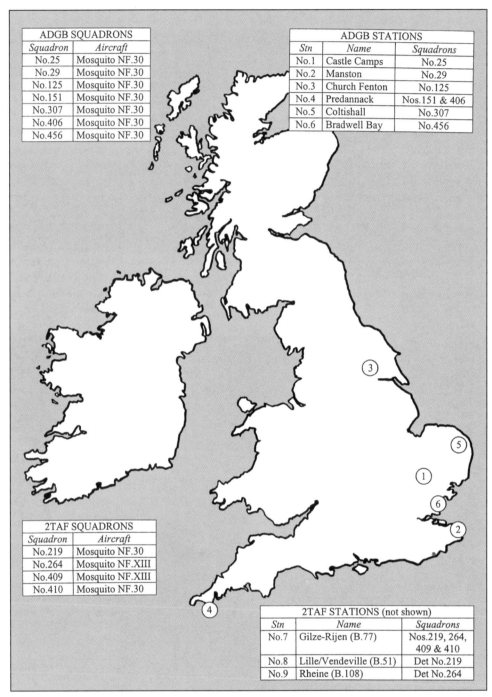

ADGB SQUADRONS	
Squadron	*Aircraft*
No.25	Mosquito NF.30
No.29	Mosquito NF.30
No.125	Mosquito NF.30
No.151	Mosquito NF.30
No.307	Mosquito NF.30
No.406	Mosquito NF.30
No.456	Mosquito NF.30

ADGB STATIONS		
Stn	*Name*	*Squadrons*
No.1	Castle Camps	No.25
No.2	Manston	No.29
No.3	Church Fenton	No.125
No.4	Predannack	Nos.151 & 406
No.5	Coltishall	No.307
No.6	Bradwell Bay	No.456

2TAF SQUADRONS	
Squadron	*Aircraft*
No.219	Mosquito NF.30
No.264	Mosquito NF.XIII
No.409	Mosquito NF.XIII
No.410	Mosquito NF.30

2TAF STATIONS (not shown)		
Stn	*Name*	*Squadrons*
No.7	Gilze-Rijen (B.77)	Nos.219, 264, 409 & 410
No.8	Lille/Vendeville (B.51)	Det No.219
No.9	Rheine (B.108)	Det No.264

Figure 29.   Night-fighter Squadrons at May 1945. The night-fighter squadrons at the war's end were severely diminished by comparison to those required to defend the UK during the Blitz and the *Luftwaffe*'s major campaigns against England in 1942 and 1944. Once again it was the squadrons of 2TAF who had help push the *Luftwaffe* back over its borders and rendered England beyond the enemy's reach.

CONCLUSIONS

The night air defence situation in May 1945 bore little relationship to that in January of the previous year. During that period Fighter Command, which had metamorphosised into ADGB, had fought off and defeated the threat posed by *Generalmajor* Peltz's bombers during the period of the Baby Blitz. At the same time, ADGB's night-fighters sought to protect the *Overlord* assembly ports and the invasion forces further inland, from the prying eyes of the *Luftwaffe*'s reconnaissance aircraft, whilst 2TAF's fighters provided a twenty-four hour air defence umbrella over the Normandy beaches. The appropriation of part of ADGB's strength to 2TAF, who were to continue the night-battle in Europe, saw its squadrons diminished in the light of a reduced threat as the Allied armies pushed the *Luftwaffe* out of France and Belgium and back into the Germany.

However, although not unexpected by the Government, the opening of the V-1 campaign early in June 1944 did provide the defence with a number of problems, particularly at night. These small, high speed and for their day, relatively sophisticated weapons, were difficult and dangerous to shoot down, whilst the German's tactics after September 1944 of air launching them at night over the North Sea increased the problems fourfold. In reality, the He 111 V-1 carrier aircraft flying just above the wave tops were very difficult to detect, which makes the destruction of seventy-seven of them that more creditable. Nevertheless, the object of the exercise was the destruction of the V-1, and in this task the RAF failed to achieve decisive results.

By the end of January 1945, Fighter Command had total domination of UK air space and that over northern Europe and stood above all other countries in the techniques and practice of night air defence. The country was covered by a comprehensive collection of CH, CHL, GCI, GL, SLC and light AA radars and nine night-fighter squadrons,[40] including four that had been re-equipped with the latest and the last wartime fighter, the Mosquito NF.30.[41] However, it would not be long before this system, which had been so carefully built-up in the light of experience gained over six hard fought years, would be rendered obsolete in a short space of time by two inventions, the atomic bomb and the jet-engine, then followed a marked deterioration in the east-west political climate.

Notes

1. William Green, *op cit*, pp. 344 and 345.
2. *Ibid*, p. 662.
3. *Ibid*, pp. 477 and 489.
4. Alfred Price [2], *op cit*, p. 159
5. This system employed two *Freya* early warning radars, the first to measure the range to the pathfinder very accurately and track it along an arc of constant distance towards the target and the second to measure the distance along the arc to the pathfinder. When the two crossed the pilot was ordered to release his marker flares. The British *Oboe* system employed a similar technique.
6. Alfred Price [5], *Operation Steinbock, The Baby Blitz* (London: IPC Media, 2002), p. 74.
7. Alfred Price [2], *op cit*, p. 160.
8. *Ultra* was the codename accorded to intelligence derived from the interception and analysis of German *Enigma* codes that were broken at Bletchley Park by the Government's Code and Cipher School.
9. Based on Winston Ramsey [3], *op cit*, p. 318.
10. Less the 3rd *Staffel*.
11. Alfred Price [5], *op cit*, p. 74.
12. This figure assumes the squadron strength as being sixteen aircraft, of which two or three would be unserviceable. By the war's end squadron strengths had increased to eighteen aircraft.

13. Alfred Price [2], *op cit*, p. 162.
14. Navigation errors and landing accidents on poorly lit airfields.
15. Winston Ramsey [3], *op cit*, p. 321 and Alfred Price [2], *op cit*, p. 164.
16. *Ibid*, p. 321.
17. *Ibid*, pp. 318–66.
18. *Ibid*, p. 324.
19. As a consequence of the disastrous losses suffered by Bomber Command during its raid on Nuremburg on the night of 30/31 March 1944, the Air Staff transferred Nos 85 and 157 Squadrons to No. 100 Group during April 1944.
20. Derek Howse, *Radar at Sea* (Macmillan Press Ltd, 1994), pp. 217 and 218.
21. Bowyer and Sharp, *op cit*, p. 175 and Howse, *op cit*, p. 219.
22. When released from 20,000 feet (6,095 metres) the Fritz-X was capable of penetrating the armoured deck of a battleship. On 9 September 1943 the surrendering Italian battleship *Roma* was struck by two Fritz-X bombs and sunk, with the battleship *Warspite*, the cruiser HMS *Uganda* and the US cruiser *Savannah* being struck and seriously damaged a few days later.
23. *The Signals History, Volume 5*, p. 168.
24. Winston Ramsey [3], *op cit*, pp. 378 and 379.
25. Chaz Bowyer, *op cit*, pp. 135 and 136.
26. C.F. Rawnsley & Bob Wright, *op cit*, pp. 347 and 348.
27. *Ibid*, pp. 349 and 350.
28. *Ibid*, p. 351.
29. William Green, *op cit*, p. 308.
30. Major Skeets Harris, DSC, OBE, was the officer responsible for the creation and training of the Royal Navy's, carrier-based, night-fighter force and a test pilot for Westland Aircraft in the post-war period.
31. W. Harrison, *Fairey Firefly* (Shrewsbury: Airlife Publishing Ltd, 1992), p. 53.
32. Later Lieutenant Commander G.L. Davies, DSC, RNVR.
33. W. Harrison, *op cit*, pp. 53 and 54.
34. CFE was formed at Wittering on 1 October 1944, where it absorbed Fighter Command's Fighter Leaders School and FIU, which became the Fighter Interception Development Squadron (FIDS) of CFE.
35. Lawrence Hayward, *Radar Pioneer, The story of Wellington R1629*, in Aviation World, Spring 2004, pp. 24–6.
36. *Ibid*, pp. 26 and 27.
37. William Green, *op cit*, p. 308.
38. Winston Graham [3], *op cit*, p. 379.
39. Although unidentified, it is most probable that the aircraft was a Ju 188F-1 from 3.(F) 122 based with *Luftflotte* 2 in Norway or Denmark.
40. This total includes those night-fighter squadrons attached to 2TAF, which provided an air defence barrier to the UK.
41. Nos 219, 410, 456 and 488 Squadrons.

CHAPTER FOURTEEN

A Defence Looking for an Enemy

May 1945–1949

At the war's end in August 1945, Great Britain possessed the best air defence system in the world, bar none, and no natural enemy that might oppose it. However, unlike its predecessor at the end of the First World War, the system was not dismantled with the coming of peace, but did require restructuring to meet the demands imposed by new technology, particularly that of jet-powered aircraft. The extent of the restructuring would, however, be heavily dependent on a number of external factors concerning the country's finances, the demobilisation of its armed forces and the technological lead in radar design achieved by the US.

Further, Britain's replacement as the principal super-power in the West by the US and its reduction in status to that of a regional (European) power, albeit one with global 'reach', was, surprisingly, a situation that was not readily accepted by the incoming Labour Government. Notwithstanding the fact the country had insufficient money[1] to fund a large military establishment and undertake a whole raft of social policy initiatives in the fields of nationalisation and social welfare, seems to have escaped many in the political elite.

In Europe, Britain's neighbours were hemmed-in by strong Soviet forces to the east, who contrived, by fair means or foul, to annex a number of countries to provide a buffer zone of satellite states as a block to western influence. Determined to prevent any further invasions of the Motherland by European countries, the Red Army maintained some 12,000,000 men and women under arms in the occupied territories and satellite states in May 1945. The Soviet Union, therefore, had no desire to see an economically strong America spread capitalism and its version of democracy throughout Europe. Consequently, with Germany defeated and the reason for Allied co-operation now gone, the wartime alliance began to crumble. The subsequent withdrawal of inter-government co-operation and a deterioration in relations between the US and Russia, created an atmosphere of mutual mistrust that would eventually lead to the realignment of Europe into Eastern and Western Blocs. All of this Britain viewed as a relative bystander, for by 1945 her influence and economy were in decline and her Empire was in retreat.

1946

In an atmosphere of severe financial constraint, the Labour Government set about the task of demobilising the armed forces and bringing home large numbers of servicemen and women from Europe, the Empire and the Far East. By May 1945 the RAF had an establishment of very nearly 55,500 aircraft, of which 9,200 were in the first line, and a manpower strength in excess of 1,000,000 British, Commonwealth and Allied airmen, of whom 193,300 were aircrew. The Air Staff's Plan 'E' anticipated the RAF's front line strength by mid-1947 to be 1,500 aircraft, comprising fifty-one fighter squadrons, forty-one of bombers, thirteen of maritime aircraft, forty-two of transports and twelve reserve squadrons.[2] These being deployed in Germany as part of the British Air Forces of Occupation (BAFO) to supervise the dismantling of the *Luftwaffe* and support the British Army, in the Far East to protect Hong Kong, Malaya, Singapore and Aden, in India to restore imperial rule and in the Middle East to ensure the flow of oil from Persia (now Iran), Iraq and Saudi Arabia and support our troops in Palestine. This left very few aircraft, squadrons and personnel for Home Defence. As was to be expected, RAF strength as proposed by Plan 'E' would not be achieved, since there were insufficient airmen to man it and the Cabinet was more interested in getting industry back on its feet and placing the national finances on a more firm footing.

Fortunately for the RAF, and for the Cabinet, there were few, if any, countries capable of mounting an air attack on the UK. The two principal bomber forces in the world, those of Bomber Command and the USAAF, were even then being downgraded to a peacetime establishment and the only potential protagonist, the Red Air Force, had no bomber forces worthy of that name. However, as we shall see later, this was set to change before the end of the decade.

From 1943 onwards the air defence of the UK, which primarily meant night air defence, decreased in importance as the *Luftwaffe*'s assets were sucked into the battle in the east and the Allied armies crossed Europe, pushing the enemy further away from the British Isles. Whilst the air defence system, which had been devised in the late 1930s and expanded in the light of experience, performed well through to the war's end, its structure was by 1945 overly complicated and heavily dependent on manpower. The introduction of jet-powered aircraft in the latter stages of the war and the consequent need to be able to track targets flying at speeds up to 600 mph (960 km/hr), or 10 miles (16 km) per minute, mitigated against a manual control and reporting (C&R) system and the myriad of operations rooms and radars that supported it.[3]

Late in 1945 the AOC-in-C Fighter Command, Air Marshal Sir James Robb,[4] addressed the issue of the air defence organisation in his Command's report 'A Memorandum on the Raid Reporting and Control Aspects of the United Kingdom Air Defence Organisation'. This work, which was largely the effort of an Auxiliary officer on Fighter Command's staff, Group Captain John Cherry, and as such now bears his name, identified the failings in the organisation and suggested a series of improvements. The Cherry Report identified the delays in the tracking system and highlighted the large numbers of radars that were required to provide cover in the vertical and horizontal planes. He also took issue with a system that provided the lower level radar stations, particularly the GCIs, with a local picture that was provided in real time (i.e. straight off the screen), when their commanders in the Sector Operations Rooms (SOR), who were responsible for scrambling the defensive fighters in good time, were observing a plot that failed to show the presence of any enemy aircraft! Although aware of the financial and political constraints likely to be imposed by government, and being an Auxiliary and having no career to worry about, Group Captain Cherry and his AOC-in-C did not hold back when recommending what they saw as the best solution to Fighter Command's problem.[5]

Cherry's recommendations impacted on the command structure by separating the AOC-in-C and the Group AOCs from the battle. Previously, the AOC-in-C and his Group AOC's had been active participants in fighting the air battle down to sector level. However, with the introduction of jet-bombers and their high approach speeds, it was no longer possible to provide these gentlemen with sufficiently accurate up-to-date information on which to base their response. In order to overcome the problem Cherry recommended the separation of senior command from active participation in the air battle and the transfer of that responsibility to the GCI controllers.

To this end he argued the current structure of multiple sectors into which UK air space was divided, be reduced to six regional sectors (see Figure 30) and that the functions of 'tactical control' and 'interception control' be separated within them. Tactical control applied to the need to scramble fighters in order to counter a specific threat, or raid, a function which was the responsibility of the SOR, whilst interception control pertained to the guidance of those fighters, or groups of fighters, to an enemy aircraft, or a group of enemy aircraft, by a GCI controller.[6]

With the controllers fighting the battle from the screens of their radar sets, the AOC-in-C, in his role as the nation's Air Defence Commander (ADC), and his Group AOCs were now largely responsible, by means of proper administration, effective logistical support and good training, for providing the best possible system and placing that in the hands of the Sector Controller and his team to get on with.

Fighter Command adopted the new structure and created six regional sectors, with each being allocated a Sector Operations Centre (SOC) that was collocated on one of the GCI sites, to provide the tactical control function and a number of GCI stations within the sector to fulfil the interception control. However, the implementation of this programme required significant changes to the existing facilities and a major re-routing of the communications links provided by the General Post Office (GPO).[7] As an interim measure Fighter Command nominated a Master GCI station in each of the sectors that would fulfil the responsibility of the SOC and connect it to the other GCIs in the sector. By this means the Master GCIs, later Master Radar Stations (MRS), undertook the tactical control role and their subordinate GCIs the interception responsibility.

The system was trialled in March 1946 when four MRSs at Sopley, Hope Cove, Treleaver and Wrafton were brought into service for a period. Unfortunately, due to a lack of maintenance and operating personnel Treleaver and Wrafton were forced to close after only a month's service. Nevertheless, the trial was judged a success and in due course led to the opening of four permanent MRSs at Sopley in the Southern Sector, Trimley Heath in the Metropolitan, Neatishead in the Eastern and Patrington in Yorkshire. Each was equipped with a Type 7 GCI and two height finding Type 13s, with the low overland cover provided by a Type 11.

During 1947 great efforts were made to implement the system throughout the remaining sectors, but the shortage of skilled manpower in No. 90 (Signals) Group overcame the planners efforts and forced the Fighter Command staff to accept a lower standard by fully protecting only the eastern and south-eastern seaboard of England. In the other sectors, mainly in Nos 10, 12 and 13 Groups, manning was reduced to a part time basis.[8]

The Cherry Report did not redress the situation with regard to the operation and deployment of night-fighters, nor the way in which they should be manned and maintained. Britain's incoming Labour Government found it necessary to retain sufficient manpower in the armed forces to counter the growing effects of Communism in Europe and the Far East and maintain law and order in its overseas colonies in the face of growing nationalism and

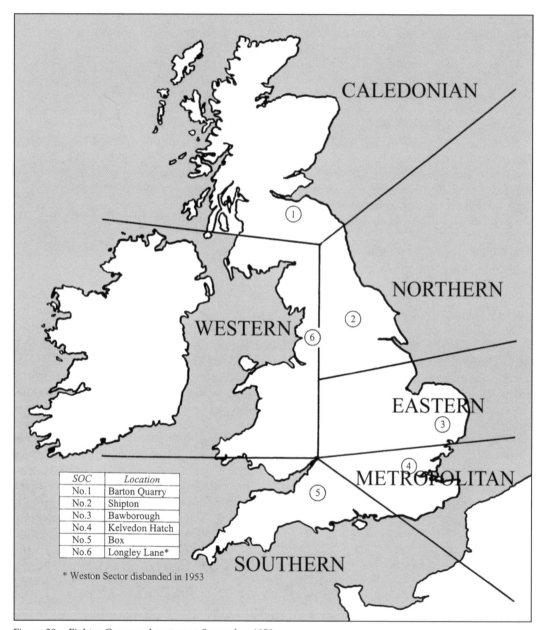

SOC	Location
No.1	Barton Quarry
No.2	Shipton
No.3	Bawborough
No.4	Kelvedon Hatch
No.5	Box
No.6	Longley Lane*

* Weston Sector disbanded in 1953

Figure 30.   Fighter Command sectors at September 1950.

demands for self determination. Some means, therefore, had to be found to maintain manning at the required levels and at reasonable cost. One method of accomplishing these conflicting requirements was to retain the system of conscription and allocate some of the manpower demand to the reserve and auxiliary forces.

By May 1946 the Government's thinking had crystallised around the idea of an Air Force Reserve to support the regular RAF in, amongst other fields, the air defence of the UK.

Whilst the AAF had participated as equals in Fighter Command during the war, the RAF's senior officers felt constrained by the type of aircraft the reservists could be expected to fly and operate effectively. However, on the positive side of the argument, there existed a large pool of aircrew who had returned to their civilian roles with their expertise and experience in fighters almost intact. If employed in the Home Defence role there would be no requirement for them to serve outside the UK, except on exercise, and hence their pre-war conditions for flying from fixed bases in the locality of their recruitment, could be re-established.[9]

Authority was granted later that month (May) for the establishment of twenty AAF squadrons, each with one flight of nine, later ten aircraft, that would by December 1947 be expanded to two, operating the later marks of the Spitfire (F.Mk.21 and Mk.22)[10] and the Mosquito NF.30. In reality the 'later' marks were required by the Regulars and the Auxiliaries were issued with wartime Spitfire Mk.14 and Mk.16. Unfortunately for the Auxiliaries, one senior officer did not appreciate their returning to operate in the night-fighter role. Air Marshal Sir Basil Embry,[11] the ACAS (Training), expressed the view that the pool of wartime veterans would quickly be depleted and require replacement with fresh crews who would have to be trained at the Government's expense. He was also of the opinion that the Auxiliaries would be incapable of maintaining the necessary standards required of night-fighter crews. Sir Basil's criticisms were backed by the Air Staff, who in turn doubted the Auxiliary's ability to absorb sufficient training to become fully proficient. In other words, the Auxiliaries might be able to fly the aircraft, but they could not, given the few hours each week available to them, operate it effectively as (what today we would call) a 'weapon system'.[12] The outcome of these senior officers' decision was to see those Auxiliary squadrons (Nos 500, 504, 605, 608, 609 and 616) that had been equipped with the Mosquito when they were re-formed in 1946, have their aircraft and N/Rs taken away and replaced by Spitfires by 1948.

As with the other Commands at the war's end, Fighter Command's establishment was drastically reduced and with it the size of the night-fighter force. The May 1945 strength of Fighter Command and 2TAF's seven night-fighter squadrons (see Figure 29), plus a further eight operating in the offensive bomber support role under the auspices of No. 100 Group, were dramatically reduced at the war's end. By the late summer of the following year, the UK night-fighter strength comprised eight squadrons (see Figure 31), with the majority operating the final variant of the Mosquito fighter to enter service, the Mk.36. A development of the Mk.30 with late-model AI Mk.10 radar and a higher powered version of the Merlin, the Mk.113/114 (see also Appendix 3), the NF.36 entered service with No. 85 Squadron at Tangmere in January 1946. It saw service with six fighter squadrons in the UK and was destined to be the last piston-engined night-fighter in Fighter Command's inventory, when No. 23 Squadron exchanged its Mosquitoes for Vampires in June 1952.

One of the N/Rs who flew in the Mosquito NF.36 was Flight Lieutenant Ken Wright[13] of No. 29 Squadron, who makes the following observations on his training and the Mosquito/AI.10 combination:

> I was first introduced to AI Mk.10 at RAF Leeming in the autumn of 1950 and was to spend the next four years using this equipment in a variety of aircraft. The training of navigators/radar was done at [No. 228 Operational Conversion Unit], Leeming, after they had completed a series of aptitude tests to assess their ability to interpret the signals on their CRTs. The ground training was pretty basic and included the use of a

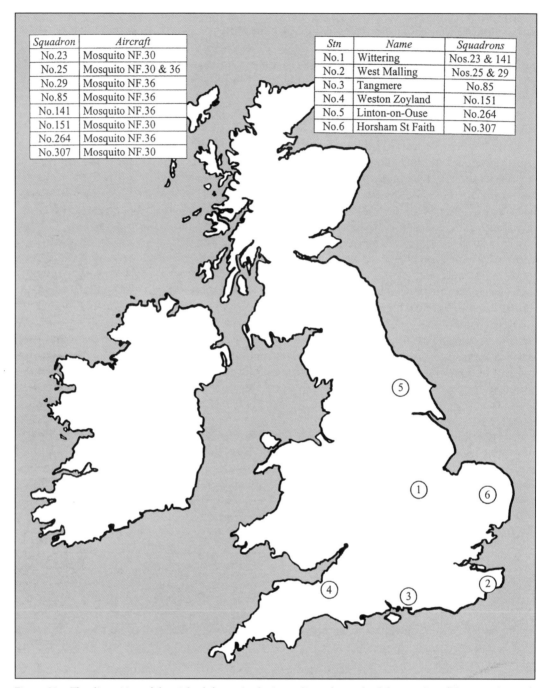

Squadron	Aircraft
No.23	Mosquito NF.30
No.25	Mosquito NF.30 & 36
No.29	Mosquito NF.36
No.85	Mosquito NF.36
No.141	Mosquito NF.36
No.151	Mosquito NF.30
No.264	Mosquito NF.36
No.307	Mosquito NF.30

Stn	Name	Squadrons
No.1	Wittering	Nos.23 & 141
No.2	West Malling	Nos.25 & 29
No.3	Tangmere	No.85
No.4	Weston Zoyland	No.151
No.5	Linton-on-Ouse	No.264
No.6	Horsham St Faith	No.307

Figure 31.   The disposition of the night defences in the immediate aftermath of the war. It will be seen that with there being no recognisable 'enemy' and in the light of the demobilisation of the RAF and the need to reduce expenditure on the armed forces, the night defences are significantly reduced. However, their strength is still oriented towards an attack from the east or south-east.

simulator to give the prospective operator a good appreciation of the blip movements. The final training in the air was done in modified Wellington [T.Mk.XVII and T.Mk.XVIII] bombers. Having flown them in wartime at the [Operational Conversion Units] OCUs and on navigator refresher training, their performance in this role as a 'fighter' was stretching it a bit. The 'crew' was a pilot, nav/rad instructor and two students. One student would sit in the second pilot's seat, while the instructor and the other student were seated at what had been the old navigator's station. Both faced the radar set, looking towards the port side of the aircraft.

The benefits of the Wellington were that it was slow to manoeuvre and therefore didn't quite give the instantaneous response that we would later get in the Mosquitoes, Vampires, Venoms and Meteors. Miles Masters were used as targets and the combination was quite satisfactory for training purposes. Most of the sorties were flown at around 10,000 feet [3,050 metres]. All likely scenarios were faithfully carried out from head-on attacks, to normal quarter approaches, usually from slightly below.

Low level (500 feet [150 metres]) sorties were carried out from the head-on acquisition to simulate catching an enemy aircraft approaching to land at his own airfield. This called for hard work on the part of the pilot as he had to get the old Wimpey round through 180° for the navigator to position him behind the target for the final 'shoot-down'. Also, the ground returns on the set would swamp half the picture and called for the navigator to work with as low a 'gain' as possible to keep track of the target.

The degree of serviceability of the equipment was quite good. Out of the twenty-seven training flights I flew, we only had equipment failure three times (11 per cent). When I crewed-up with my pilot and flew the Mosquito NF.30, it wasn't quite so good. Out of eleven sorties, we had the set fail five times (45 per cent). Most of the practise interceptions in the Mosquitoes were done between 20,000–30,000 feet [6,100–9,145 metres], usually at the request of the GCI operators. The ceiling on the Mossie was around 35,000 feet [10,670 metres], which was adequate to meet the piston-engined bomber threat at that time [represented by the Tu-4].

Arriving at my first squadron, No. 29, which was flying the Mosquito NF.36, the serviceability factor improved dramatically to 100 per cent and when we switched to the Meteor NF.11 the same high percentage was maintained. I left the Squadron in 1951/52 and went north to re-form No. 151 Squadron on Vampire NF.10s, at Leuchars.[14]

The reduction in aircrew numbers went hand-in-hand with a reorganisation of the night-fighter OTUs and the technical training establishments. At the beginning of 1946 a reassessment of the post-war training requirements saw No. 54 OTU at Leeming merge with the Middleton St George based No. 13 OTU (light-bombers and intruders) to form No. 228 OCU at Leeming on 1 May 1947. Of the other night OTUs, No. 60, at Finmere, was absorbed by No. 13 OTU in April 1945 and No. 51, the old Beaufighter and Havoc OTU, at Cranfield was closed down and disbanded in June 1945. No. 228 OCU was charged with the training of tactical light-bomber and night-fighter crews, for which it was divided into three squadrons, No. 1 for AI training, No. 2 for night-fighters and No. 3 for light-bombers. Administration was the responsibility of No. 12 (Fighter) Group, with the initial establishment being stated as: three Mosquito T.3s, six Miles Martinets, nine Mosquito FB.6s, seven Mosquito NF.30s, an Oxford and seven Wellington T.18s.[15]

Although deprived of the services of many of its senior scientists and with no requirement to replace AI.10, TRE did, nevertheless, assign some time to examining the feasibility of

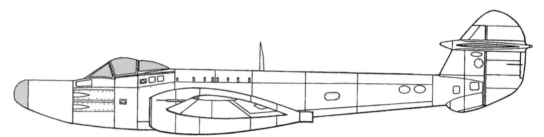

Figure 32. The AN/APS-4 installation on Meteor EE348. EE348 was an early Mk.3 Meteor, hence the original bulbous nacelles, that was modified by the replacement of the standard symmetrical nose section, with one of harsher profile carrying the AN/APS-4 radar.

installing radar in a jet-fighter and continuing the development of AI Mk.9. During September 1945 Gloster Meteor F.Mk.III, EE348, that had previously seen service with Nos 222 and 616 Squadrons, was delivered to the Central Radar Establishment (CRE) at Defford, Worcester, for the installation of an AN/APS-4 radar. The choice of the APS-4, known as AI Mk.XV in RAF service, was a practical decision based on its known performance in naval service (see Chapter 13), its light weight and physical dimensions, enabling it to be mounted in the Meteor's nose. The task of adapting the Meteor airframe to take the APS-4 was assigned to CRE's chief designer, Mr David Henderson,[16] who was also responsible for the installation programme.[17]

Taking the nosewheel framework as a starting point, Henderson designed a pick-up frame for the radar and a fairing to blend the nose profile to that of the forward fuselage. When these were fabricated in Defford's workshops a transparent acrylic radome was added to cover the front of the ASH bomb. The cannon ports on the fuselage sides were plugged and the guns removed to compensate for the weight of the radar. It is, however, possible that weight was added to the rear fuselage to balance the aircraft. When completed, the new nose section was painted in standard RAF medium sea grey, but the camouflage pattern was not extended forward to cover it. Other than these modifications, EE348 was a standard Meteor F.3 when it was delivered to CFE at Tangmere, Kent, sometime in 1946. It was also the first British jet-fighter to carry an AI radar and the ugliest Meteor ever flown.

During its time at CFE EE348 was used to assess the potential of radar as a gun-ranging device that could be used to supply ranging data to a conventional 'head-up' glass reflector, lead computing, gyro gunsight (GGS) Mk.II, with which all RAF fighters were fitted. Provision of the ranging data saved the pilot having constantly to adjust the sighting graticule during the approach to the target. In this respect it proved successful. EE348 remained in service with TRE throughout the remainder of the decade and on into the 1950s and is known to have been used by de Havilland Sea Hornet pilots for ASH experience prior to the aircraft's service entry.[18] In October 1951 it was still in service with an experimental homing head installed for guided weapon research.[19]

1947–1948

The year 1947 was to prove something of a watershed in the development of night air defence, with the appearance of the Soviet Union's first long-range, strategic bomber and the release of a specification for an advanced, all-weather, day and night fighter, to replace the

Mosquito. During 1946 the RAF's Directorate of Operational Requirements began the task of drafting a specification for a new, jet-powered, all-weather fighter, that would incorporate the lessons learned during the war in relation to the advances made in bombers, concerning blind bombing techniques (*H2S*) and long-range navigation systems (*Gee*, *Oboe* and *Knickebein*). The improved performance of bombers and their sheer size required the study of 30 mm cannon batteries, single-barrelled recoilless guns and air-to-air guided missiles (AAM), all of which would require development. In order that the new fighter be capable of intercepting high speed targets at the greatest possible distance, a powerful AI radar (AI Mk.9 was the only candidate) would also be required.

During December 1946 the Air Staff released Operational Requirement (OR) 227 and OR228 for a day-fighter and a two-seat night-fighter, which were issued to the British aircraft industry as Air Ministry Specifications F.43/46 and F.44/46 the following January. F.44/46 called for an aircraft capable of taking-off and climbing to 45,000 feet (13,715 metres) in ten minutes, with a level speed at that altitude of not less than 605 mph (970 km/hr) and an endurance of two hours. The operational equipment was to comprise four 30 mm cannon, AI radar, a comprehensive communications and navigation suite, cockpit pressurisation and ejection seats for the two-man crew.[20] Three companies showed interest in the new aircraft, Gloster Aircraft, the de Havilland Company and Hawker Aircraft who prepared formal responses to the Ministry.[21]

Following the release of the specification, the project was to be marked by muddled thinking on the part of the Air Staff and confusion within the Air Ministry and the Ministry of Supply (MoS),[22] all of which resulted in the late delivery of a much needed interceptor and the subsequent need to build a 'stop-gap' replacement. Nevertheless, in response to F.44/46 de Havilland proposed their DH.110, whilst Gloster responded to F.43/46 with the P.234, which was a large aircraft with a delta wing and 'V' tail, and to 44/46 with a Meteor derivative, the P.228. Hawker produced their P.1057 to cover F.44/46, but were really interested in developing a day-fighter to 43/46. In January 1948 the Air Ministry was persuaded to let Hawkers proceed with the day-fighter, which eventually became the Hunter, whilst Glosters and de Havilland competed for the 44/46.[23]

By the time this decision had been made Glosters had refined their thinking following a tendering conference with the MoS in April 1947, at which the Ministry appeared to view the P.234 as the superior design. The Company received confirmation it would be given the go-ahead, but by June no order had been issued. At this juncture the Air Staff changed their requirement and the MoS altered its specification to include *Red Hawk* AAMs in place of the guns, a large AI radar, plus other changes, which resulted in Gloster re-drafting their design as the two-seat P.259. De Havilland's DH.110 was a large twin-boomed aircraft that was intended to meet the requirements of the proposed day and night-fighters. With the F.43/46 and 44/46 specifications being altered to the improved F.3/48 and F.4/48 in April 1948, the MoS awarded development contracts to Glosters to build four of what had by then been recast as the G.A.5 night-fighter and de Havilland for four DH.110s, whilst Hawkers got the F.3/48 contract to build the P.1067 Hunter.[24]

Whilst all this too-ing and fro-ing was going on in the corridors of the Air Staff and the MoS, the Russians were taking the first steps towards the creation of a strategic air force – the Long Range Aviation division of the Red Air Force, or the *Aviatsiya Dal'nevo Deistviya* (ADD). During the war the Military Air Forces of the Red Army, the *Voenno-vozdushnye sily* (V-VS), neglected the development of long-range bombers in favour of the production of large quantities of tactical aircraft, which were better suited to supporting the Soviet field army. Apart from a few operations by four-engined Petlyakov Pe-8 bombers early in the war, the Peoples Commissariat of the Aircraft Industry, the *Narodnyy komissariaht aviatsionnoy promyshlennosti* (NKAP), showed little interest in the development of long-range, multi-engined, strategic bombers. Equally, the Soviet intelligence services were unaware of developments in the US, where the Boeing Company were building the B-29 Superfortress for the USAAF, until a casual remark by presidential advisor, Eddy Rickenbaker,[25] exposed the existence of the aircraft.[26]

Whilst not having a requirement to develop a strategic air force (that decision rested with the Politburo) the Soviet leadership was nevertheless impressed by the bombing campaigns of the USAAF and Bomber Command and 'dabbled' with the drafting of a specification for a long-range bomber. Pre-warned of the B-29's existence, the Soviet Military Mission in Washington made an official request for supplies of the bomber, which the Americans subsequently refused. The leadership's prayers were nevertheless answered at the end of 1944, when a B-29 of the 770th Bombardment Squadron (BS) landed at a naval airfield close to the eastern city of Vladivostok on 20 July, following battle damage sustained during a raid in Manchuria. The process was repeated in November when a B-29 of the 794th BS force landed in Soviet territory following a raid on the Japanese homeland. The Russians were 'gifted' a third B-29 on 21 November, when another aircraft from the 794th landed after sustaining battle damage.[27]

In May 1945 the NKAP proposed to the Soviet leadership that the bomber be copied and placed in service with the V-VS. Approval was granted by the State Defence Committee and the design bureau (*Opytno konstruktorskoe byuro* – OKB) headed by Andrei Tupolev (OKB-156) was assigned the task of reverse engineering the design. Using the second B-29 as their model, Tupolev's team succeeded in completing its drawings by April 1946 and building the prototype 'B-4', later the Tupolev Tu-4, at Plant No. 22 in time to roll it out on 28 February 1948. Serialled '220001', the prototype was ready for flight testing by May and undertook its first flight in the hands of test pilot Nikolay Rybko on 19 May. On the successful completion of basic handling tests, 220001 was flown to Stakhanovo for a thorough examination, where it was joined by the second and third prototypes in June and August. By the autumn of 1947 Plant No. 22 had completed ten B-4s which were used for the type's State Acceptance Trials in 1948.[28]

The three prototypes were shown to the Soviet people and the assembled mass of western air attachés at Tushino airfield outside Moscow in August 1947. Not convinced the Russians were capable of building an aircraft as complicated as the B-29, the majority of the attachés reported back to their governments that the B-4s were repaired American machines. It was not until 1949, when the recently formed North Atlantic Treaty Organisation (NATO) Co-ordinating Committee realised the truth, that the, by then Tu-4, was allocated the reporting name *'Bull'*.[29]

Powered by the indigenous 2,400 hp Shvetsov Ash-73TK, turbo-supercharged, 18-cylinder radial engine, the Tu-4 showed itself capable of achieving a range of 3,343 miles (5,350 km) with a 4,400 lb (1,995 kg) bomb load. With an operational load of 9,000 kg (19,842 lb) of bombs, the production Tu-4s were capable of flying at 33,630 feet (10,250 metres), at 346 mph (555 km/hr) over a range of 2,225 miles (3,560 km). These performance figures enabled the Tu-4 to strike targets in North America 'over the pole' and bases in Europe and Great Britain from airfields in the northern and western Soviet Union.[30] At altitude, therefore, the Tu-4 was not that much slower than the Mosquito NF.36.

As was mentioned in Chapter 13, the development of AI Mk.IX did not end with the loss of Dr Downing and the TRE Beaufighter in 1942. Whilst TRE concentrated on introducing the SCR-720B set into RAF service in 1943/44, they did not neglect the development of what they and the Air Ministry saw as its post-war successor, AI Mk.IX. With its lock-follow capability already in development, the scientists pursued another facility which they believed would further lighten the night-fighter crew's workload – windscreen projection.

Windscreen projection was not a new innovation in the night-fighter world. The idea for the projection of radar information directly onto the windscreen of a fighter became a reality with the design and testing of AI Mk.V in 1941 and its introduction to service during the spring of 1942 (see Chapter 7). It is difficult to say who first suggested the image from the AI Mk.V's pilot's indicator might be projected directly onto the windscreen. Whoever it was, TRE had taken up the idea in the summer of 1942 and by October had built what they termed the 'Automatic Pilot's Indicator' (API), whose output was projected directly onto the windscreen.

API took the output of the Mk.V's pilot's indicator CRT, on to which was inserted an electronically generated artificial horizon, along with the directional gyro 'pip' and combined these with the graticule of a standard GGS Mk.II and projected them onto the windscreen, as shown in Figure 33. The API Beaufighter was flown from Defford to FIU's base at Ford, who were then tasked with undertaking a 'quick and dirty' evaluation of the device. The aircraft was flown for the first time by Wing Commander Chisholm on the evening of 13 October, with Flight Lieutenant Clarke, FIU's Special Signals Officer, and Wing Commander Adams from HQ Fighter Command, monitoring the system in the rear.[31] Overall, FIU thought the idea worth pursuing, but highlighted the poor light levels and the obstruction of some of the flight instruments by the gunsight unit on the cockpit coaming.[32]

A year later Mosquito NF.II, DZ301, fitted with windscreen projection was delivered to FIU sometime in September 1943 for an operational assessment. This unit, still based around

Figure 33. Windscreen projection for AI Mk.V, October 1942.

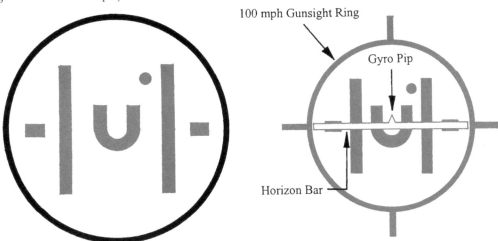

(A) Conventional Pilot's indicator for AI Mk.V showing a target above and to starboard of the fighter at a range in excess of 7,500 feet.

(B) Pilot's Indicator information combined with an artificial horizon and gyro pip of GGS Mk.II gunsight and projected onto the windscreen of an FIU Beaufighter.

the AI Mk.V/pilot's indicator arrangement, was, like the previous article, a bulky item which obscured the air speed indicator (ASI) and required it to be repositioned in a more appropriate location. The image projected by the 1943 version comprised the GGS Mk.II's 100 mph (160 km) gunsight ring, that was bisected by a horizon bar in green that rolled in sympathy with the aircraft and along which the directional pip was able to travel. The target spot operated in the same way as on the Mk.V (as the target approached the spot grew wings in relation to the distance from the fighter), but had the vertical and horizontal marker bars removed, as shown in Figure 34.[33]

Experience with the earlier model showed the need for a cut-off switch to remove the 100 mph ring, leaving just the target spots visible on the windscreen. However, as one of the FIU pilots, Flight Lieutenant Jeremy Howard-Williams, recalls:

> in the absence of the ring, the dot was sometimes hard to find, and in any case was useless for deflection shooting or, of course, for range estimation, [whilst] sensitivity in the horizon was satisfactory, with no tendency to premature toppling in steep turns. [It] was slightly sluggish in pitch but was acceptable for interception purposes[34]

However, by the end of 1943 AI Mk.V was obsolescent for night interception purposes as the superseding centimetric AI Mk.VIII and Mk.X sets were already in use by Fighter Command. The single display of AI Mk.VIII was better suited to windscreen projection and a model of this set was adapted and installed in a Mosquito NF.XII, HK419, before being delivered to FIU in January 1944. As far as can be ascertained, the complete Mk.VIII picture was projected through the GGS Mk.II, straight on to the windscreen. Subjected to intensive trials by FIU, the installation in HK419 was unanimously praised by all those who flew with it.[35]

However, with both TRE and the Air Ministry intending that AI Mk.IX be the next British AI to see service, the set was adapted to incorporate windscreen projection, as AI Mk.IXB.

Figure 34.   Windscreen projection for AI Mk.V, September 1943.

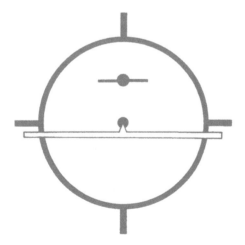

(A) Conventional Pilot's Indicator display for AI Mk.V, showing a target above the fighter and dead ahead at a range of 1,000 feet.

(B) The same information viewed through the GGS Mk.II gunsight and projected onto the windscreen of Mosquito DZ301.

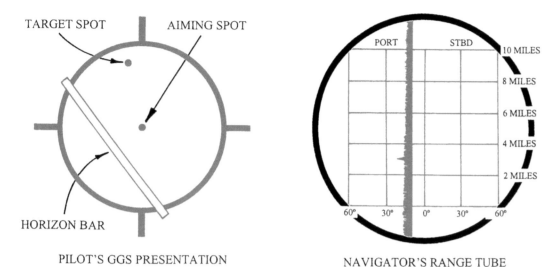

PILOT'S GGS PRESENTATION   NAVIGATOR'S RANGE TUBE

Figure 35.   Windscreen projection for AI Mk.IXB. The navigator's display shows a target at a range of 3 miles (4.8 km) and 15° to port off the aircraft's nose. With the target strobed by the navigator the pilot's display shows the same target to be well above the fighter which is banked and climbing to port to close the range.

The N/R's displays for AI Mk.IX comprised a Spot Tube, similar in concept to the display in AI Mk.VIII, and a Range Tube that showed range/azimuth information. In the Mk.IXB set the pilot was provided with a spot display projected through the GGS Mk.II and on to the windscreen, as shown in Figure 35. A pre-production Mk.IXB set was installed in an NF.XVII Mosquito, HK311, and delivered to CFE's FIDS on 22 December 1944. This was not the first time that CFE had evaluated the set. An earlier version had been installed in Mosquito HK946, a Mk.XII, and presented to FIU some twelve months earlier, but had been returned to FIU with a recommended list of modifications. The pre-production Mk.IXB set tested throughout the early months of 1945, therefore, incorporated those modifications recommended in the 1944 tests, in addition to the windscreen projection system.[36]

Between the end of December 1944 and April 1945, HK946 flew sixty-one day sorties and eighteen by night, totalling 168 flying hours. The tests were conducted to assess the radar's general performance over land and sea and evaluate the windscreen projection system. Following these tests, further tests to evaluate the radar's tactical performance were conducted, with the results being divided accordingly. The trial showed the overall performance of the radar to be 'good', except when operating at low levels (2,000 feet [610 metres] above the sea and 5,000 feet [1,525 metres] over land) where the lock-follow system was ineffective and coverage inferior to that of AI Mk.X. In respect of the windscreen projection facility, CFE considered it to be 'a highly successful method of following' an evading target, whilst 'enabling controlled visuals to be obtained, even in steep diving turns'. However, good though this was, the system fell short of that required for operational use on account of the GGS Mk.II obscuring the upper half of the blind flying panel and problems with lighting contrast between the target spot and the ranging ring. By comparison with AI Mk.X, the Mk.IXB set was superior at those heights where the lock-follow system could be employed (i.e. at medium and high altitudes) and would present few problems in converting experienced Mk.X navigators. However, more time would be required to train *ab initio* navigators on the Mk.IXB system, as a greater number of hours would need to be flown at night, since the pilot's display could not easily been seen in daylight (where most of the initial training was normally conducted).[37] HK946 was returned to Defford along with a whole raft of suggestions and recommendations.

1949

With East/West tension increasing through the Soviet imposed blockade of the German capital, Berlin, in June 1948 and the Anglo-American airlift to feed its population underway the same month, the question of Europe's security quickly rose to the top of government agendas. To counter Soviet expansionism and provide a common defence bond between Europe and North America,[38] NATO was established in April 1949, with its HQ, the Supreme Headquarters Allied Powers, Europe (SHAPE), located in Paris. This organisation was to have a profound influence on British defence policy, as its various military committees took over the responsibility for formulating aircraft equipment policy, strategic planning and command and control.

The national defence situation in both Great Britain and America changed again in 1949 with the explosion of Russia's first atomic device in September and the potential, not though the actuality, for a nuclear attack against both countries, and/or their NATO allies. This in turn required the UK air defence commanders to look towards the protection of their ground radar stations, as these would inevitably become prime targets in any nuclear strike. To counter the threat the Government and the RAF instituted a programme, under the code-word *Rotor*, that would see the SOCs and GCI radar stations rehoused in underground concrete bunkers to protect them from the effects of nuclear blast and fall-out. Because of their approaching obsolescence and their employment as a simple warning barrier, the decision was taken to phase out the CH stations and replace them with the new AMES Type 80 radars that would provide seamless cover to 40,000 feet (12,190 metres). These high powered radars had a range of 200 miles and were provided with PPI displays of high discrimination for accurate fighter direction. The Type 7 GCIs were retained at some, but not all sites, as an adjunct to the Type 80s to provide additional capacity should the need arise. The *Rotor* programme also made provision for the SOC's to be relocated in underground accommodation and remained the primary structure until the next reorganisation of air defence in 1955.[39]

By 1949 over two years had elapsed without any metal being cut on the F.4/48 fighters and with a development programme stretching towards 1953, something had to be done to provide the RAF with an interim jet-powered night-fighter that was capable of tackling the Tu-4 and its successors. With their experience of the radar design in EE348 and David Henderson's work, which proved the forward section of the aircraft could be adapted to carry a substantial load, the MoS considered an adaption of the Meteor would provide the best solution. Subsequently, in the spring or summer of 1948, Glosters were approached with a view to them developing the interim night-fighter. The company responded by releasing a design brochure in October 1948 based on the Meteor trainer, the T.Mk.7, powered by R-R Derwent 5 engines of 3,500 lb st (1,590 kg st). With Glosters fully committed to production of the day-fighter variants, the night-fighter design was passed to another company in the Hawker Siddeley Group, Armstrong Whitworth Aircraft (AWA), at Bagington, near Coventry, where it became the responsibility of their Chief Designer, Mr H.M. Watson. In discussions with the late Bryan Philpott, in 1984, Mr Watson, describes the changes necessary to convert the Meteor T.7 to a night-fighter:

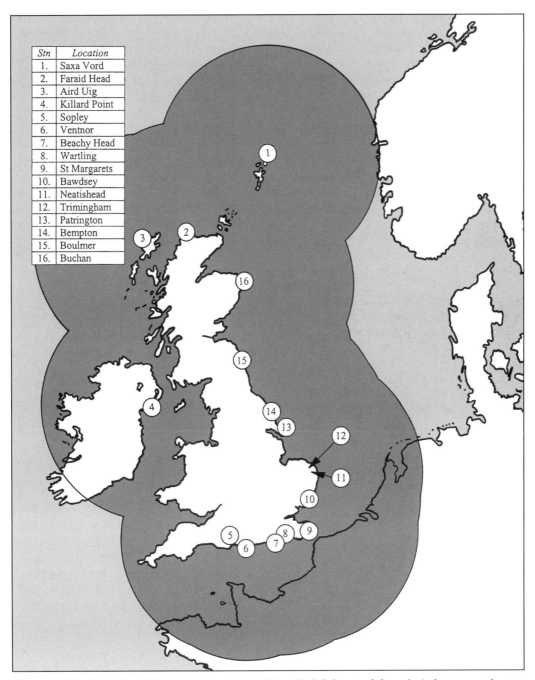

Stn	Location
1.	Saxa Vord
2.	Faraid Head
3.	Aird Uig
4.	Killard Point
5.	Sopley
6.	Ventnor
7.	Beachy Head
8.	Wartling
9.	St Margarets
10.	Bawdsey
11.	Neatishead
12.	Trimingham
13.	Patrington
14.	Bempton
15.	Boulmer
16.	Buchan

Figure 36. High level Type 80 Radar Cover at 1957. Although slightly out of chronological sequence, the map shows the high level cover provided by the Type 80 radars when they were probably operationally most numerous. With the RAF undergoing what might be described as a 'period of change' that was brought about by the reductions in National Service manpower and a contracting defence budget, radar stations were being opened and closed on an almost daily basis. This diagram, therefore, does not purport to represent the true position on any particular date in 1957, but hopefully will provide a flavour of the Type 80 radar cover at its best.

The radar scanner is the major feature from which most of the other changes arise. For a relatively small fighter aircraft the positioning of the scanner in the fuselage nose is practically 'Hobson's choice'. Housing the equipment means lengthening the fuselage, and of course fitting a dielectric/fibreglass nose forward of the scanner. Operating the [AI] equipment required a second crew member, and here extensive use was made of the Meteor trainer, with its long canopy using a fore and aft hinge along the starboard side. This arrangement is satisfactory for a trainer but not for a fighter with ejection seats [although perhaps specified, ejection seats were never fitted in night-fighter Meteors]. The guns and ammunition for the day fighter were fitted into the fuselage, and they had to be transferred to the wings. This is not an easy task and I will explain why. The ammunitions boxes must go between the spars and the doors form part of the upper wing surfaces, the requirements for these are therefore; they must be quickly detachable for rearming, capable of carrying aerodynamic loads without appreciable deformation, and joined along all four edges by some shear carrying device to maintain the torsion rigidity of the wing.[40]

To prove the viability of the new fighter, the fourth production T.7, VW413 was set aside for conversion as the aerodynamic prototype. Fitted with long span wings taken from the F.3 to provide greater lift, VW413 was flown in a series of tests beginning on 23 December 1948, before returning to the workshops at Bitteswell where a mock-up nose and appropriate ballast to simulate the radar were installed. On 28 January 1949 VW413 made its first flight in the hands of AWA's test pilot Bill Else in the guise of the 'long-nosed' Meteor T.7. Flight tests showed the longer nose had little or no effect on the aircraft's handling. The aircraft was then returned to Bitteswell for the fitting of a more angular tailplane taken from the experimental E.1/44 fighter.[41] This tailplane, which was first fitted to the F.Mk.8 fighter to cure pitch-up problems, was subsequently adopted as standard for all night-fighter Meteors and reconnaissance variants. It was flown in the new form on 8 April 1949 by AWA's chief test pilot, Eric Franklin, who reported no difference in rudder effectiveness compared to the old version. With the new tailpane, VW413's length was again increased to 48 feet 6 inches (14.78 metres).[42]

Whilst the work continued to define the final configuration of the Meteor night-fighter, the future of AI Mk.9 was already sealed. During a trial undertaken by CFE and concluded in February 1948, AI Mk.9B was pitted against AI Mk.10. The AI Mk.9B radar was installed by TRE in a Mosquito NF.38, VT658, and flown on thirty-seven occasions for a total of forty-two hours and forty minutes, under the watchful eye of Wing Commander Bob Braham, the OC of the Fighter Interception Development Squadron (FIDS). Four navigators took part in the trial, the majority of whom were rated 'above average' in terms of their proficiency and it is largely on the basis of their expertise that the final report rested. The trial was broken into three parts:

- Part 1 – to make a recommendation to introduce AI Mk.9B as the RAF's next AI radar or retain the Mk.10.
- Part 2 – to evaluate the windscreen projection system.
- Part 3 – to assess the capability of the Mosquito NF.38 as a night-fighter, bearing in mind the performance of the B-29/Tu-4.

As has already been mentioned, the NF.38 was the last development of the Mosquito, which if it proved superior to the Mosquito 36 and showed a marked performance superiority to the B-29, could be the country's last piston-engined night-fighter. This solution would have saved the Exchequer hundreds of thousands of pounds by not having to develop an interim fighter, nor purchase a new radar set from the US. Unfortunately for them and the Air Staff, the trials highlighted a shortcoming in both radar and fighter capability.

In a situation not dissimilar to that in their April 1945 report, CFE concluded that AI Mk.9 had a poorer performance than AI.10 at altitudes below 3,000 feet (915 metres) over land and a poorer azimuth cover than AI.10 at altitude (120° compared to AI.10's 180°). The lock-follow minimum range of 600 feet (180 metres) was 'unreliable' and the target could escape if the lock was broken. Similarly, AI.9B did not perform well in the presence of *Window* jamming due to the auto-follow facility failing to achieve lock. Turning to the windscreen projection system, CFE reported that under normal daylight conditions the gunsight graticule, the projected horizon bar and the directional blip were almost invisible, whilst at night they were far too bright and rendered the target invisible at minimum range. If this was not enough, TRE had failed to overcome the problem of the bulk of the GGS Mk.II, which blotted out the Mosquito's airspeed indicator, artificial horizon and vertical speed indicator.[43]

The Establishment's overall recommendation was damning and most probably brought about the cancellation of the AI.9 programme:

> The opinion of this Establishment is the AI Mk.9B is operationally unacceptable in free-lancing, broadcast control, or bomber support operations. It is, therefore, recommended that AI Mk.9B should not be accepted for service use.[44]

Having dealt with AI.9B, the CFE report turned its attention to the Mosquito NF.38. The flying trials showed that under normal conditions and with normal fuel loads, the take-off run for the NF.38 was 'considerably longer' than that of the Mosquito NF.36, whilst both aircraft took the same time to reach their flight safety speeds. The Mk.38 was heavier to handle and less manoeuvrable than the Mk.36 and over the altitude range 2,000–25,000 feet (610–7,620 metres) the Mk.38 was generally slightly faster by a couple of knots (3.5 km/hr) than the Mk.36. In a separate test using a Mk.36 Mosquito flown by an FIDS crew to intercept a Lakenheath-based B-29 flying at its maximum cruising speed, the report stated that the night-fighter took a full:[45]

> ten minutes to close the distance of less than 1 mile [1.6 km], even after the Mosquito had dived to gain speed in the initial stages of the interception. In view of this and the remarks in the earlier paragraphs comparing the Mosquito 36 with the Mosquito 38, it is considered that the Mosquito 38 is quite unsuited as a night-fighter, with a performance comparable to that of the B-29.[46]

CFE's report not only killed-off the Mosquito NF.38's career as a night-fighter, but also brought to an end the seven year programme to develop a home-grown replacement for AI.10.

AI.9 was formally cancelled in 1949, but not before Meteor VW413 underwent a further stretch of 15 inches (38 cm) to accommodate the Mk.9C variant. The aircraft flew this guise in December, which showed no changes to the machine's flying characteristics – a fact confirmed when it was flown by A&AEE pilots on 18 July 1950. At the end of July VW413[47] emerged yet again from Bitteswell's workshops with another nose extension, taking its overall length to 50 feet 6 inches (15.39 metres) in order to accommodate the scanner and equipment boxes of the American Westinghouse AN/APQ-43 radar. Flight testing by AWA once again showed the handling was not affected.[48,49]

CONCLUSIONS

For several reasons the Soviet ADD did not present a realistic threat to Great Britain at the end of the 1940s. First and foremost, the Soviet bomber force was yet to be equipped with an operational atomic bomb (A-bomb) and would not be so until well into the next decade. Second, whilst the Tu-4 was capable of carrying a moderately heavy conventional bomb load (19,840 lb/9,000 kg) over a reasonable distance (2,225 miles/3,580 km), there were insufficient of them to ensure a saturation of the UK defences, given that Britain would not have been the sole target. Total Tu-4 production amounted to somewhere just short of 1,300 aircraft.[50] Third, they would have to cross NATO aligned countries that provided a fighter barrier and an early warning system. This argument also ignores the possibility of retaliation from Bomber Command, or NATO, whose 'an attack on one is an attack on all' principle might trigger a nuclear counter-strike from the US, or in other words 'deterrence'. Given the size of the force, it is most probable the ADD's crews were insufficiently trained to carry out a bombing attack on the UK and return safely to their bases. In this respect one needs to consider the navigation difficulties faced by the Russian crews when flying outside the Soviet Union's borders.

Be that as it may, the RAF did not have a fighter in service at the end of the 1940s that could effectively intercept a B-29/Tu-4 bomber and destroy it, but the end of the piston engined night-fighter was clearly recognised by the Air Staff and the MoS. The cancellation of the AI Mk.9 programme, however, placed TRE and Fighter Command in the unenviable position of having no lock-follow radar with which to equip an interim night-fighter. The only practical solution open to all the parties (we Brits are very good at practical solutions) was to re-install refurbished examples of AI.10 and look once more towards the US for a modern AI radar.

Notes

1. By the war's end all the country's foreign investments had been sold and substantial debts had been accrued in fighting the war. In reality, since 1941 the country was living off US Lend-Lease which ceased immediately the conflict was over, leaving the Exchequer and the State in a parlous state.
2. Michael Armitage, *The Royal Air Force, An Illustrated History* (London: Brockhampton Press, 1996), p. 181.
3. The 1½ AMES Type 7 was still the majority GCI radar, supplemented by the centimetric Type 14 for GCI and 'extra low' (CHEL) working, the 50 cm MHz Type 11 GCI/CHL for use in areas subjected to *Luftwaffe* jamming, and the centimetric height finding Types 13 and 24. In addition to these were the fighter direction Type 16 sets and the Army's GL and SLC radars.
4. Air Marshal Sir James Robb, KBE, CB, DSO, DFC, AFC, replaced Air Marshal Sir Roderick Hill as AOC-in-C on 14 May 1945.
5. Jack Gough, *Watching the Skies, A History of Ground Radar for the Air Defence of the United Kingdom by the Royal Air Force from 1946 to 1975* (London: HMSO), 1993.
6. John Bushby, *Air Defence of Britain* (London: Ian Allan, 1973), p. 182.
7. The GPO's Air Defence Group, later the Post Office Air Defence Group, was responsible for the provision of all landline telecommunications links and associated equipment for UK air defence until the creation of British Telecom in 1982. The author was a member of the AD Group from 1973 to 1979.
8. Jack Gough, *op cit*, p. 46.
9. Jeff Jefford [2], *Post-War RAF Reserves to 1960* (RAF Historical Society, 2003), pp. 79–83.

10. It should be noted that from mid-1946 the RAF adopted the Arabic numbering scheme for its aircraft and radar equipment, hence the Spitfire XXI became the F.Mk.21 and AI Mk.X became AI Mk.10, or simply AI.10. This numbering scheme will be used from now on.

11. Later Air Marshal Sir Basil Embry, KBE, CB, DSO, DFC, AFC, and AOC-in-C Fighter Command from April 1949–April 1953.

12. Tony Freeman, *The Post-War Auxiliary Air Force* (RAF Historical Society, 2003), pp. 96–100.

13. Later Flight Lieutenant Ken Wright, who flew in Vampires with No. 151 Squadron, in Meteors with the Central Fighter Establishment, in Douglas Skynights with the US Marine Corps and finally in Meteors with No. 6 Joint Services Trials Unit at RAF Valley. He now lives in California.

14. Ken Wright in an e-mail to the Author dated 22 November 2005.

15. Sturtivant, Hamlin and Halley, *op cit*, pp. 230 and 240–2.

16. Later the Chief Engineer of the post-war TRE Aircraft Department at Pershore.

17. Barry Jones, *Trials & Testbed Meteors*, in Aeroplane Monthly, January 1996, p. 60.

18. The de Havilland Sea Hornet NF.21 entered service with the Royal Navy as a carrier-based night-fighter in January 1949.

19. Barry Jones, *op cit*, pp. 60 and 61.

20. Derek James, *Gloster Javelin*, in Aeroplane Monthly, January 2004, p. 61.

21. *Ibid*, p. 62.

22. Post-war MAP was replaced by the MoS who had responsibility for the procurement of all military equipment on behalf of the armed services.

23. Tony Buttler [1], *Gloster Javelin, Warpaint Series No. 17* (Milton Keynes: Hallpark Books Ltd), pp. 1 and 2.

24. *Ibid*, pp 2.

25. Edward Rickenbaker was America's premier World War One fighter ace and an advisor to President Roosevelt on aviation matters.

26. Yefim Gordon and Vladimir Rigmand [1], *Tupolev Tu-4 Soviet Superfortress* (Hinckley: Midland Counties Publishing, 2002), p. 3.

27. *Ibid*, pp. 9 and 10.

28. *Ibid*, pp. 15–26.

29. All identified Soviet bombers were allocated a NATO codename beginning with 'B'.

30. Jean Alexander, *Russian Aircraft Since 1940* (London: Putnam, 1975), p. 359.

31. *FIU ORB*, p. 56.

32. Jeremy Howard-Williams [3], *Head-Up, 50 Up*, in Flypast (Stamford: Key Publishing, December 1993), p. 48.

33. *Ibid*, p. 48.

34. *Ibid*, p. 48.

35. *Ibid*, p. 48.

36. *CFE Report No. 28*, p. 1.

37. *Ibid*, pp. 3, 11 and 12.

38. NATO's founding nations were the US, Canada, Great Britain, France, Belgium, The Netherlands, Denmark, Norway and Portugal.

39. John Bushby, *op cit*, pp. 185 and 186.

40. Bryan Philpott [1], *Meteor* (Cambridge: Patrick Stephens Ltd, 1986, an imprint of Hayes Publishing, Sparkford, Yeovil, Somerset), p. 103.

41. The E.1/44 fighter was an experimental, single-engined, jet powered day-fighter.

42. Bryan Philpott [1], *op cit*, pp. 103 and 104.

43. *CFE Report No. 129*, dated February 1948.

44. *Ibid*.

45. *Ibid*.

46. *Ibid*.

47. VW413 spent a further three and a half years flying at the RAE, Farnborough, before being delivered to No. 20 MU, Aston Down, where it was scrapped in March 1958.

48. The author is not certain if the full APQ-43 system was installed, since the equipment, which was a derivative of the AN/APS-57 and was later fitted to some marks of the Gloster Javelin as AI Mk.22, had two scanners (search and track) that could not have been fitted inside the Meteor's radome.

49. Barry Jones, *Gloster Meteor* (Crowood Press, 1998), pp. 116 and 117.

50. Gordon and Rigmant [1], *op cit*, p. 32.

CHAPTER FIFTEEN

An Interim Solution

1950–1956

In the Far East on the morning of 25 June 1950, soldiers from the Communist People's Democratic Republic of North Korea crossed the 38th parallel and began an all-out invasion of the southern half of the country. Taken completely by surprise, the US Government responded by calling a meeting of the United Nations (UN) Security Council, which promptly requested a ceasefire and the withdrawal of Communist troops. A second resolution granted military assistance to the Republic of South Korea and the creation of a UN force, led by America, to intervene in the south with the intention of driving the Communists back across the 38th parallel. By the end of September 1950 that objective had been accomplished, but the military situation was far from stable and would, in the following year, see the UN hard hit and back on the defensive when Chinese troops retaliated in response to incursions of their territory.[1]

The British Army and Air Force, whilst supplying assistance to the Americans in Korea in the form of a Commonwealth Brigade, were themselves hard pressed elsewhere in the Far East suppressing a Communist inspired civil war in Malaya. The so called 'Malayan Emergency' which began in 1948 and followed-on from Britain's exit from India and Palestine, was to drag on for twelve more years.[2]

As if these problems were not enough, Britain was also committed to supporting her NATO allies in the defence of western Europe, following the end of the Berlin Blockade and Air Lift, and the protection of the Canal Zone from attacks by an increasingly pro-nationalist Egyptian population. Taken all together, Britain was too overstretched militarily and economically to continue acting as one of the world's policemen.

With the show-down over the Berlin Blockade fresh in everyone's minds and various 'incidents' serving to raise the political temperature between East and West, the British and American governments continued to regard European defence as their primary responsibility. During the summer of 1950 NATO's allies began the process of establishing a command structure for Western Europe and adding substance to that structure in the form of the Mutual Aid Programme to the participating countries. With SHAPE up and running under the command of General Dwight D.Eisenhower in April 1951, NATO, and with it, Britain began a gradual expansion of her military capability in an attempt to counter the large numbers of Soviet forces based across the border in the East.[3] In May 1951 the US 3rd Air Force was created with its HQ at Ruislip on the western fringe of London, to control all permanently based USAF units in Britain that would eventually comprise one Fighter Interceptor Group,[4] a Fighter-Bomber Group[5] and a one Light-Bomber Group.[6] These

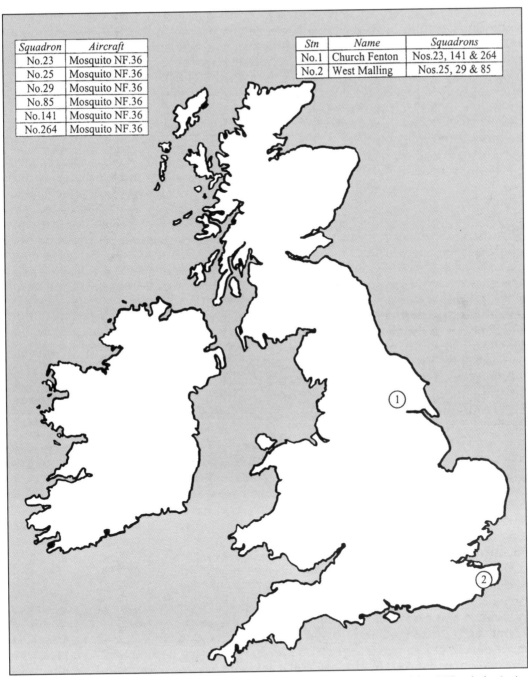

Squadron	Aircraft
No.23	Mosquito NF.36
No.25	Mosquito NF.36
No.29	Mosquito NF.36
No.85	Mosquito NF.36
No.141	Mosquito NF.36
No.264	Mosquito NF.36

Stn	Name	Squadrons
No.1	Church Fenton	Nos.23, 141 & 264
No.2	West Malling	Nos.25, 29 & 85

Figure 37.   Night-fighter Wings at June 1950. With the threat from the ADD regiments of the V-VS at the beginning of the Cold War oriented from the east, Fighter Command grouped its night assets within two wings at Church Fenton and West Malling. The former provided protection for the Bomber Command stations in Lincolnshire and East Anglia and northern England, whilst West Malling guarded the approaches to London, the south-east and East Anglia.

Groups were supplemented periodically by US-based squadrons deployed to Britain for exercises and in times of international stress – all of which would require protection from air attack. What then was the state of the UK's air defences to prevent such an attack and what was the true level of the ADD's bombing capability?

The outbreak of war in Korea and the continued rise and fall in East/West tension served to stimulate an improvement in UK defence preparedness, albeit at a significant cost to the Treasury and post-war recovery. In 1950 the responsibility for daylight air defence was broadly shared between the regular Air Force and the 'weekend warriors' of the Royal Auxiliary Air Force (RAuxAF).[7] By June 1950 all home-based day-fighter squadrons, with the exception of three operating the twin-engined de Haviland Hornet, were equipped with the Meteor F.Mk.4[8] or the Vampire FB.Mk.5, with the Auxiliaries also having begun their conversion to jet equipment in the form of the earlier marks of the Meteor and the Vampire F.1/F.3. The night-fighter force, however, remained firmly in the piston engined era, with just six home-based Mosquito squadrons available to defend the whole of UK airspace at night and provide detachments to the British Air Forces of Occupation (BAFO) in West Germany[9] (see Figure 37).

The following year brought brighter news in the form of progress in the delivery of the Meteor NF.11. On 24 July 1951 the NF.11's acceptability for MoS Controller of Supplies (Air) (CS[A]) release was discussed at a meeting at Baginton. The principal criticisms raised were the inability to see well enough to land in rain, excessive misting of the windscreen and hood and difficulty in operating the *Gee* Mk.3 navigation equipment. It was decided that before CS(A) release could be granted, a series of modifications had to be introduced. To overcome the visibility problem it was decided to incorporate a direct vision (DV) panel into the windscreen port quarter panel and provide an electrically heated air spray for de-misting. The difficulty with operating the *Gee* system was overcome when the whip aerial was moved to the cockpit roof. These changes could not be introduced until the sixty-first aircraft, with those already built requiring retrospective modification, which AWA estimated could not be completed before April 1952. This did not appear to affect the delivery of Meteor 11s to the RAF, since No. 29 Squadron, based at Tangmere, received its first aircraft, WD599, in August 1951. Conversion from the Mosquito NF.36 began that month, as did the task of working up to operational status, which involved the acclimatisation of the crews to the Meteor's higher speed and restricted range (compared to the Mosquito) and educating the fighter controllers in vectoring the new fighter at night. The accumulated knowledge was then passed to the other squadrons re-equipping with the Meteor. By August the Squadron had converted to the Meteor, when its Mosquitoes were returned to the MUs for the recovery of their AI.10 radar and the scrapping of the airframes.

The second squadron to convert to the Meteor NF.11, No. 141, formed part of the Coltishall night-fighter wing flying Mosquito NF.36s. No. 141 received its first Mk.11s in September 1951 and completed its conversion the following month and a return to operational status as part of the East Anglian defences that autumn. The third UK night air defence squadron, No. 85, based at West Malling, exchanged its Mosquitoes for Meteors in September and October. No. 264 Squadron, based at Linton-on-Ouse, received Meteor 11s in November 1951, as did No. 151 at Leuchars in March 1953 – the former in exchange for Mosquitoes and the latter for Vampire NF.10s. The final UK-based Meteor squadron, No. 125, was allocated Mk.11s when it reformed at Stradishall on 1 April 1955. With the re-equipment of the UK fighter squadrons more-or-less completed in 1951, the Air Staff turned its attention to the creation of a night-fighter force for 2TAF[10] in West Germany, where

Meteor 11s were allocated to two wings[11] in 1952. With the establishment of a jet night-fighter force in Germany the need for detachments from UK-based squadrons, other than for exercises and armament practise camps, was removed.

With ventral and wing tanks fitted, the NF.11 was capable of achieving 541 mph/Mach 0.79 (865 km/hr) at 30,000 feet (9,145 metres), a service ceiling of 40,000 feet (12,190 metres) and a maximum range of 950 miles (1,520 km). The NF.11 was, therefore, some 120 mph (190 km/hr) faster than the Mosquito NF.36 and flew 10,000 feet (3,050 metres) higher, carrying more or less the same equipment – pilot, N/R, AI.10 radar, radio equipment and four 20 mm cannon. More importantly, it was 50 per cent faster than the operational version of the Tu-4 which entered service with the ADD in April 1949, but would soon be rapidly outclassed by a clutch of Soviet bombers (the Tupolev Tu-16 *Badger*, Tu-20 *Bear* and Myasishchev Mya-4 *Bison*) some five years later. The latter were not considered to be too much of a problem, as by then it was anticipated the F.4/48 (Javelin) would be in service!

In radar terms, the AI.10 offered a small improvement in range over the Mosquito, which may be attributed to the use of metal targets, Lincolns, Washingtons and Transport Command Valettas, instead of the essentially wooden Mosquito. Overall this may have been responsible for an increase of between 2 and 6 miles (3.2–9.5 km) depending on the target size.[12] However,

> range in AI.10 was possibly more dependent on the skill of the operator and the [performance] of the individual set. One had to continually monitor the tuning of the receiver to match the transmitter, and to that end [one's] right hand was always on the knob which controlled that.[13]

With large numbers of Meteors being required to re-equip squadrons in the UK and Germany, there were insufficient aircraft coming off the production lines at AWA's Baginton factory to satisfy the RAF and NATO's demands for night-fighters. In a similar vein, the parent company was fully engaged in building Meteor day-fighters and gearing up for G.A.5/Javelin production. With delays in the Javelin programme raising the probability of a shortfall in the number of night-fighters available for UK air defence, the MoS and the Air Staff's Director of Operational Requirements (Air) (DOR[Air]) looked towards the de Havilland Company as a possible supplier of nocturnal fighters. This approach had the advantage of increasing the numbers of night-fighters in a fairly short timescale, or so they thought, at the expense of an anticipated short in-service period and an increase in logistical support.

In 1950, whilst pursuing the purchase of a night-fighter derivative of de Havilland's single-seat Venom fighter, of which more anon, the MoS was 'gifted' the possibility of acquiring a second fighter. A small quantity of this aircraft, the de Havilland DH.113 Vampire night-fighter, was made available to the RAF following a British government embargo on the sale to their original owner, the Egyptian Air Force.[14] Developed as a minimum cost, private venture fighter, the DH.113 flew for the first time on 28 August 1949 bearing the civil Class-B registration, G-5-2. Powered by the 3,350 lb (1,520 kg) thrust de Havilland Goblin Mk.3 turbo-jet and accommodating a pilot and navigator in side-by-side seats a lá Mosquito, with AI.10 radar and four 20 mm cannon, G-5-2 was pleasant to handle and slightly faster than Fighter Command's Vampire FB.Mk.5. Compared to the Meteor NF.11, the Vampire was slower in level flight and inferior in its rate-of-climb when fitted with two 100 gallon (455 litre) external tanks. G-5-2 was flown to A&AEE, Boscombe Down, for flight testing in March 1950, where it was joined by the first production aircraft,

WP232, and the second prototype, G-5-5. The trials confirmed the manufacturer's assessment, but raised questions concerning the aircraft's stall performance and the shape of the radome which spoilt the crew's view over the nose. Overall, however, the Vampire proved the more stable gun-platform in comparison with the Meteor and had a greater range.[15]

No. 25 Squadron based at West Malling was the first recipient of the de Havilland fighter, when WP238 was delivered on 2 July 1951, which more or less coincided with the arrival of the first Meteor NF.11 to No. 29 Squadron. With a limited number of airframes available to the Squadron by the year's end, the conversion and work-up extended into the winter of 1951/52. The second Vampire NF.10 unit, No. 23 Squadron at Coltishall, began its conversion with the arrival of WP256 on 11 September, but had to wait until February 1952 before there were enough aircraft to complete the conversion of 'A' Flight. 'B' Flight had to wait until June before it was able to convert and return its Mosquitoes to the MU – a process that had taken a full nine months since the delivery of the first aircraft. The third and final NF.10 squadron, No. 151, was re-formed at Leuchars on 15 September 1951 as part of the expansion of the UK's night air defences. Once again the scarcity of aircraft required some crews to be posted to No. 23 Squadron to train on their aircraft. It was not until 8 February 1952 that No. 151 received its first aircraft, WP252, with which to begin its re-equipment, a task that was not completed until the following summer.

In service the Vampire proved popular with the crews, was reliable and enjoyed good rates of serviceability. However, its lack of ejection seats and the difficulty in evacuating in an emergency due to the poor clearance offered by the upward hinging canopy, were examples of its less endearing characteristics. In practice interceptions against Bomber Command's B-29 Washingtons, the Vampire proved adequate for the task, but when attempting to intercept the recently introduced Canberra light-bomber or the US Boeing B-47B Stratojet, which flew both higher and faster, the NF.10 was completely out-classed.[16] On the completion of his tour flying Mosquitoes with No. 29 Squadron, Ken Wright was posted to No. 151 Squadron at Leuchars and gives his impressions of the Vampire from the navigator's viewpoint:

> The Vampire had side by side seating as in the Mosquito and now gave us the ability to fly up to 40,000 feet [12,190 metres], but it only had one engine. We flew it for over a year and never had an engine failure.
>
> With the higher operating altitudes now available to us, we encountered a little arcing problem in the beginning but that swiftly went away. One of the benefits of the AI.10 was that it could be used for ground map-reading on coastlines quite effectively. Just off the coast at Leuchars was an old S.B.A. abandoned tower in the sea which could easily be picked up on the radar. This coupled with the fact that one of the lattice lines on the *Gee* chart ran almost dead centre down the runway allowed the navigator to guide the pilot to a touchdown in bad weather. The radar response from the hangars gave the set operator another range point to help triangulate his position. Although not an official let down procedure, all navigators on the squadron became quite proficient in having this 'back-up' procedure should communications fail.
>
> Another use for the AI.10 was in anti-shipping attacks at night on MTBs. Operating at 5,000 feet [1,525 metres], the navigator could vector the pilot to a point where he could do a diving attack on the boats after they had been illuminated by flares. It was tried out at CFE but fortunately, the tactic never made it to squadron level.

The average range of radar contact with this equipment was 10–12 miles [16–19 km] when it moved into the jet age, an improvement over the Mosquito range of around 10 miles. There is no apparent reason for this small jump in performance, other than the sets were not subject to as much vibration. At altitude, I did pick up a target at 40 miles [64 km] range when on an exercise. The target turned out to be a United States Air Force (USAF)[17] Boeing B-47 Stratojet[18] and I had locked the beam on steady (the beam width at that range was quite considerable) which helped, plus the fact that the radar reflection of the B-47 was huge, but we couldn't get up to his altitude as he was flying at 40,000 plus [12,190 metres].

The time I spent on 151 Squadron and the practice interceptions done on Vampires and later Meteor NF.11s, almost 200 radar sorties, the set failures were less than 6 per cent. With the advent of the American APS-57 and the AI.17 coming along in the Javelins, the AI.10 had been in operation for over ten years. Quite an achievement. In the hands of a skilled operator, it had proved a deadly weapon in World War Two. By today's standards, with radars operating in 60 mile [95 km] plus ranges, development has come a long way.[19]

Whilst the NF.10 production saga might rightly be regarded as 'disappointing', in delivering what was after all an interim aircraft, it did (eventually) provide a useful, if short lived, reinforcement to the night defences.

Prior to the purchase of the Vampire NF.10, the de Havilland Company were hard at work developing its successor and trying to sell it to the RAF. Based around the single-seat Venom FB.Mk.1, the night-fighter variant was mated to a nacelle that was essentially that of the Vampire NF.10, but broadened to provide more room for the crew and accommodate the greater size of the 4,850 lb (2,200 kg) de Havilland Ghost 103 engine. Flown for the first time as G-5-3 on 22 August 1950, the aircraft was flown at HQ Fighter Command's request to CFE at West Raynham on 22 November for an evaluation by its All-Weather Wing. In trials that totalled some twenty-seven hours and thirty minutes in forty-four sorties, the Venom was shown to be 'more manoeuvrable, faster and had a better rate-of-climb than the Meteor NF.11'. Against B-29 targets at medium-speed / medium altitudes, the Venom was capable of affecting an interception within six minutes, in 90°-type (target crossing the fighter's path at right angles) and 180°-type (tail-chase) interceptions. At low level, 1,500 feet (460 metres), the average range was 2 nautical miles (3.5 km) against a Meteor or Mosquito. However, the report did criticise the aircraft's poor emergency exit for the crew, a shortage of space for additional equipment and the cramped nature of the radar and radio installation, which made servicing more difficult than on current aircraft (Mosquito and Meteor NF.11).[20]

G-5-3 was purchased by the MoS during January 1951 and given the military serial WP227, before being sent to Boscombe Down during April for preliminary handling tests by A&AEE. In their subsequent report to the MoS, the Boscombe Down pilots expressed their overall satisfaction with the aircraft's handling characteristics, but criticised its longitudinal stability, its poor roll-rate, elevator control and wing-drop near the stall. They also expressed the need for better rudder control to provide a more stable gun-platform and the provision of ejection seats for the crew. A number of these problems were partially cured by the removal of the 'acorn' fairings at the fin/tailplane junction, fitting an extended fillet on the fin and more responsive ailerons. However, by the time these modifications had been accomplished, production examples, designated Venom NF.Mk.2, were coming off the lines at Hatfield, with the first, WL804, flying on 4 March 1952.[21]

Further trials to improve the aircraft's high speed handling, combined with the redistribution of Venom work around de Havilland's factories,[22] served to delay the CS(A) release until May 1953. Consequently, it was not until 27 November 1953 that No. 23 Squadron at Coltishall received its first aircraft with which to begin its conversion from the Vampire NF.10. Structural problems with the wings, temporary groundings and altitude restrictions, served to impede the Squadron's re-equipment schedule and it was not until the spring of 1954 that it reached its full establishment of NF.2s. In the meanwhile the programme to re-equip further squadrons was halted until such time as the aforementioned modifications could be completed by the manufacturer.[23]

No. 23 Squadron did not have a happy time with the NF.2, as one of its pilots who had previously flow the Vampire NF.10, Noel Davies, recalls:

> The aircraft seemed similar to the Vampire, but the differences were critical: the range/ endurance was improved, so was the height and performance, but the radar was the same as the Vampire's with a range of around 4–10 miles [6.5–16 km]. Occasionally, the pilot would see the target, even at night, before the Nav/Rad[24] picked it up on his set!
>
> I only flew the Venom NF.2, the earliest form. Our aircraft had been in store for many months, awaiting modifications demanded by the Air Ministry, and all suffered from 'fatigue-type' problems. Although cramped, the cockpit accommodation was adequate for most, the side-by-side seating being preferred to the tandem version in the Meteor. The aircraft also had very narrow wheels to fit in the thinner wings, and were kept at very high pressures, which sounded rough on landing.
>
> During my last year with 23 Squadron, we lost ten men in ten months through flying accidents, including two commanding officers.[25] Within a couple of years three more of my squadron contemporaries had been lost. A sad time.[26]

As far as operations were concerned, the NF.2 proved capable of intercepting USAF Republic F-84F Thunderstreak fighter-bombers and North American B-45 Tornado light-bombers during exercise conditions. However, the Canberra remained immune to the NF.2's attentions, even though the Squadron frequently practised at 40,000 feet (12,190 metres) and up to 48,000 feet (14,630 metres) when the radar tracking network was up to it.[27]

No. 23 Squadron was destined to be the only unit equipped with the NF.2. Between November 1953 and November 1954, pilots at Boscombe Down tested a number of modified Venom airframes[28] to clear the different elements of the 'new' mark: upward hinging, transparent cockpit canopy, extended acorns at the rear of the tailplane, dorsal fin extensions a lá the Vampire T.Mk.11 and a flat topped fin and rudder taken from the Venom FB.Mk.4. Service clearance was granted sometime during the first half of 1954, with de Havilland's Chester factory delivering NF.2s to Marshall's Engineering at Waterbeach, Cambridgeshire, in June 1954 for conversion to the Venom NF.2A standard.[29]

Deliveries of the NF.2A to No. 23 Squadron began in August 1954, with the arrival of WR781 on 19 August. Aside from a number of problems that reduced their overall service-ability, No. 23 was able to participate in their first air defence exercise, *Battle Royal*, during September against Bomber Command Canberras and Lincolns. Sadly, the Squadron lost its second CO when Squadron Leader P.S. Englebach and his navigator, Flying Officer M.J. Wright, were killed on 15 February following a take-off accident at Coltishall. A change in Air Staff policy brought about a reversion to the wartime status of night-fighter squadrons, where the CO would now be of wing commander rank, with his two flight commanders being squadron leaders, in recognition of the importance attached to what were now described as 'all-weather' fighter operations. In this respect, No. 23's temporary CO, Squadron Leader C.R. Winter, DFC, was replaced by Wing Commander A.N. Davies, DSO, DFC, during July 1955.[30]

The second Venom NF.2A squadron, No. 253, was re-formed at Waterbeach during August 1954, having previously flown Spitfires and being disbanded since May 1947. What eventually became the Driffield Wing, was formed on 5 September, when No. 219 Squadron was re-established as a night-fighter squadron, alongside No. 33 on 15 October. These four squadrons were the only recipients of the NF.2A.

The Venom NF.2/2A's career in Fighter Command was destined to be short, due in part to the aircraft's increasing obsolescence and the reduction in fighter strength following the notorious Defence White Paper of 1957 – of which more later. No. 33 Squadron was disbanded on 3 June 1957, followed by No. 219 on 31 July and No. 253 on 31 August, with 253 Squadron's nameplate disappearing for ever from the RAF's order of battle. Only No. 23 Squadron survived to convert to the Venom NF.3 from October 1955.[31]

Whilst the performance of the Meteor NF.11 and the Venom NF.2/2A was adequate for the interception of the Tu-4, their radar-fit and armament was approaching obsolescence by comparison with developments in the US. With the cancellation of AI.9C in 1949, TRE was involved in the development of its successor, AI Mk.17, which was scheduled for installation in the Mk.1 version of the Javelin.[32] The AI.17 programme was a long-term development by TRE to produce a sophisticated radar system that was capable, not only of searching for and tracking a target, but of directing and firing its battery of de Havilland *Blue Jay*, later Firestreak, AAMs; This programme naturally had priority over the short-term needs of the Meteor and Venom. Ironically, a number of Meteor NF.11s were later employed on AI.17 trials in 1954 as part of the radar's development phase.[33] With nothing else in the 'armoury' the MoS and its aircraft designers had little option other than to seek the assistance of the US radar suppliers. If the equipment was to be supplied from America, Britain was able to request it be funded under the terms of the Mutual Defence Aid Programme (MDAP) which provided three possible radar options; AN/APQ-35, AN/APQ-43 or AN/APS-57.

The APQ-35 radar was built by the Westinghouse Corporation and used successfully in the USN's Douglas F3D Skynight night-fighter. In reality it comprised two radars, the AN/APS-21 search unit that was capable of detecting fighter-sized targets at ranges out to 20 miles (32 km) and the AN/APS-26 tracking system with a range of 2 miles (3.2 km). It was one of the first track-while-scan radars developed in the US.[34] The APQ-43 was another Westinghouse product that was not used on a US aircraft and like the APQ-35 employed the track-while-scan technique. Its twin dish aerial system was too large for the Meteor or Venom's radomes, although some accounts state that it was planned to be installed in one of the NF.11 prototypes, but it did see service later in the decade as AI Mk.22 in the Javelin. The choice went by default to the Westinghouse APS-57, which was adopted for RAF and Royal Navy service as AI Mk.21.[35]

AI.21 was a magnetron-based 3 cm X-Band radar operating on a frequency of 9.4 GHz. The system was based around the AN/APS-57 radar, with modifications restricted to those required to accommodate a British designed strobe unit. Full IFF and AI beacon facilities were available and the radar was able to operate in the ASV mode. The parabolic dish antenna[36] was capable of executing several scanning modes to project beams of varying width to suit the application required, at a peak power of 200 kW. The navigator's display comprised a control indicator unit, on which were located two CRT displays: a PPI presentation showing range/bearing and a C-Scope showing azimuth/elevation information. Targets were initially displayed on the PPI tube and selected on to the C-Scope by means of a strobe, with the maximum search range being stated as 25 miles (40 km), which, if the assessments by CFE are to be believed, was rarely, if ever met. The minimum range was a respectable 100 yards (90 metres), with beacon acquisition being calibrated to 200 miles (320 km). PRFs of 300, 550 and 2,450 Hz and pulse widths of 2.25,[37] 1.75 and 0.4 µsecs were available and selected automatically by the position of a Range Switch for each phase of the interception. As with AI.9C, AI.21 was capable of projecting aiming information into the pilot's gyro-gunsight. A facility known as 'Orientable Sector Scan' (OSS) enabled the navigator to select a narrow 15° or 30° beam once a target had been detected and concentrate the radar's energy upon it. This had the effect of increasing the accuracy of target tracking and the range, but restricted the interception as the range closed, particularly if the target was taking evasive action.[38]

Project studies by AWA in April 1951 in respect of updating the Meteor's powerplant, examined the feasibility of installing Armstrong Siddeley Sapphire engines, a pair of which had already been successfully tested in Meteor F.8, WA820,[39] in August 1950. The study results showed that whilst the Sapphires delivered twice the thrust of the NF.11's Derwents, they weighed twice as much and were over-sized for the airframe. Even with a 20 per cent increase in wing area, the engines increased the overall weight by 25 per cent, which resulted in a mere 12½ per cent increase in military load. For an aircraft that was still viewed as interim equipment pending the introduction of the Javelin, the development was hardly worth the effort. However, a modest increase in power, coupled with the installation of the APS-57 radar, all at a moderate cost, would find favour with the Air Staff and the MoS.[40]

In order to prove the Meteor/APS-57 concept, two Meteor NF.11s, WD670 and WD687, were converted to serve as prototypes for a new mark. The conversion required the Meteor's nose to be extended by a further 17 inches (43 cm) to accommodate the APS-57's equipment (scanner, power supply, modulator, transmitter/receiver and synchronizer unit), with the subsequent increase in weight being compensated by the installation of slightly more powerful 3,800 lb (1,725 kg) Derwent 9 engines. It was from these two aircraft that the Meteor NF.Mk.12 was developed. Testing at Boscombe Down with WD670[41] in June 1952, showed a tendency toward fin stalling and the locking of the rudder at high altitudes, which necessitated the introduction of area-increasing fillets on the fin at the junction with the tailplane. The additional square foot (0.09 sq metres) of side area cured the problem and restored the type's satisfactory handling characteristics following further trials at Boscombe Down in February 1953.[42] In performance terms, the NF.12 was broadly comparable with the earlier NF.11, with its increased weight being compensated by the slightly more powerful engines. With its flying trials successfully completed, the MoS signed a contract for the manufacture of 100 Meteor NF.12s, the first of which, WS590 flew from AWA's Baginton factory on 21 April 1953 with test pilot Eric Franklin at the controls.

A further refinement of the Meteor night-fighter, and the last Meteor fighter to enter service, the NF.Mk.14, was developed in parallel with the Mk.12. Retaining the forward section of the NF.12 and its AI.21 radar, the new mark introduced a two-piece, rearward sliding, fully transparent cockpit canopy and a revised windscreen of increased angle with direct vision panel. NF.11, WM261, was employed as the aerodynamic prototype, being fitted with spring tab ailerons and a two-axis auto-stabiliser to eliminate longitudinal instability at high altitude. The aircraft was flown to A&AEE for canopy jettisoning trials, with Bill Else accomplishing the first successful release on 8 October 1953. In performance terms, the Mk.14 was once again broadly comparable to the NF.11, but was regarded as the best of the NF-series, earning the accolade 'Queen of the Skies' by its crews,[43] or as Bill Gunston describes it, as 'almost adequate for the job'.[44] The MoS placed an order for 100 production NF.14s with Derwent 9 engines, the first of which, WS722, taking to the air from Baginton on 23 October 1953, again piloted by Bill Else.

Deliveries to Fighter Command began in March 1954, when No. 25 Squadron at West Malling received NF.12s and NF.14s in exchange for its Vampire NF.10s. From then on, as with many other night-fighter squadrons, No. 25 operated both types with each of its two flights being equipped with a single mark. No. 25's partner in the West Malling Wing, No. 85 Squadron, began its conversion from the Meteor NF.11 to the Mk.12/14 the following month, to be followed by two squadrons specially re-formed to operate the Meteor; No. 152 Squadron at Wattisham in July and No. 46 at Odiham in August, again mounted on the Mk.12 and Mk.14. The year was completed in October with the conversion of No. 264 at Linton-on-Ouse to both marks. Further deliveries enabled No. 153 Squadron to be re-formed at West Malling in February 1955 and convert two day-fighter squadrons to the night-fighter role: No. 72 Squadron at Church Fenton during February 1956 and No. 64 at Duxford in September (see Figure 38).[45]

The large number of surplus-to-requirements Meteor NF.11s that still had useful hours left on their airframes following their retirement in the mid-1950s, provided a useful pool for yet another version of the Meteor, the target towing Mk.20. Following a request from the Royal Navy for a jet-powered target tug and a similar requirement that emanated from the Royal Danish Air Force, AWA began the conversion of an NF.11, WD767, to the target towing role. Fitted with an M.L. Aviation Type-G, wind-driven winch on a pylon on the upper surface of the inner wing between the engine nacelle and the fuselage, that was capable of streaming radar and non-radar reflecting targets on the end of its 6,100 feet (1,860 metre) cable, a Saab near miss recording system and a winch operator in the rear cockpit, the resultant TT.Mk.20 was flown for the first time on 5 December 1956. Employed by the Navy's Fleet Requirements Unit at Hurn, No. 728 Naval Air Squadron at Hal Far, Malta, the Seletar and Changi Target Facilities Flights and No. 3 Civilian Anti-Aircraft Co-operation Unit at Exeter from March 1958, the TT.20 was tasked with towing targets for all three services until superseded by the Canberra TT.18 in 1970.[46]

Whilst AWA were undertaking their feasibility studies for an improved version of the Meteor, the design staff at Hatfield were working on the handling issues of the Venom NF.2/2A and likewise undertaking studies on an improved version. With deliveries of the long awaited Javelin expected in the near future, the MoS were anxious to avoid unnecessary expense, but did concede the need for a minimum cost upgrade to the airframe and its radar. The Company proposed the use of a reheated Ghost engine to improve the rate-of-climb and the increased speed at altitude, but in the end settled for an uprated version of the engine, the Ghost Mk.104 of 4,950 lb. Other improvements taken from the NF.2A development included a different shaped radome to enhance the crew's vision over the nose, an upward hinging, power operated, clear view canopy, plus AI.21 radar, a Maxaret braking system, power operated ailerons and the deletion of the outer tailplane elements. As with the Meteor NF.12 and 14, the opportunity was not taken to install ejection seats.

The silver painted prototype Venom NF.Mk.3, WV928, flew for the first time from de Havilland's Christchurch factory on 22 February 1953. Handling trials at Boscombe Down showed an improvement over the earlier Venoms, with its light controls being effective up to the limiting Mach number of M0.85. However, the increased weight of the new variant, up 706 lb (320 kg) on the NF.2, reduced the rate-of-climb and the time-to-height to 40,000 feet

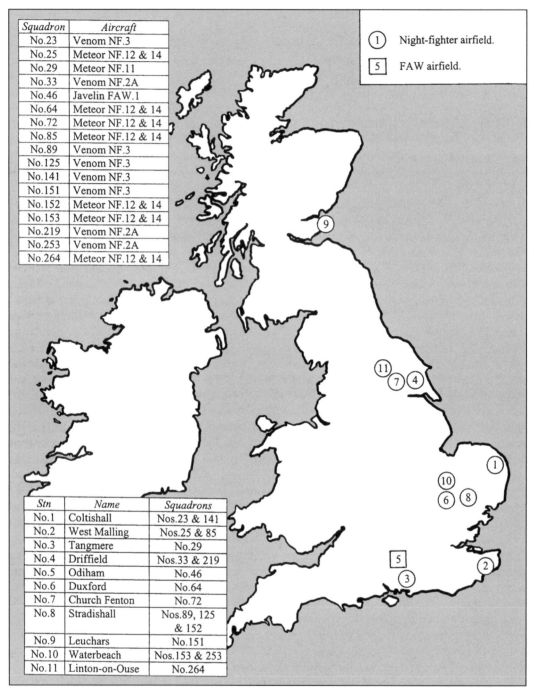

Squadron	Aircraft
No.23	Venom NF.3
No.25	Meteor NF.12 & 14
No.29	Meteor NF.11
No.33	Venom NF.2A
No.46	Javelin FAW.1
No.64	Meteor NF.12 & 14
No.72	Meteor NF.12 & 14
No.85	Meteor NF.12 & 14
No.89	Venom NF.3
No.125	Venom NF.3
No.141	Venom NF.3
No.151	Venom NF.3
No.152	Meteor NF.12 & 14
No.153	Meteor NF.12 & 14
No.219	Venom NF.2A
No.253	Venom NF.2A
No.264	Meteor NF.12 & 14

① Night-fighter airfield.

⑤ FAW airfield.

Stn	Name	Squadrons
No.1	Coltishall	Nos.23 & 141
No.2	West Malling	Nos.25 & 85
No.3	Tangmere	No.29
No.4	Driffield	Nos.33 & 219
No.5	Odiham	No.46
No.6	Duxford	No.64
No.7	Church Fenton	No.72
No.8	Stradishall	Nos.89, 125 & 152
No.9	Leuchars	No.151
No.10	Waterbeach	Nos.153 & 253
No.11	Linton-on-Ouse	No.264

Figure 38.   Night-fighter and FAW Squadrons at September 1956. February 1956 heralded the much awaited arrival of the Javelin, when No. 46 Squadron took delivery of its first FAW.1s. From thereon the night-fighter community would begin a gradual decline that would extend over the coming three years.

(12,190 metres) to sixteen minutes, a rate the Boscombe pilots considered to be barely adequate for the interception of high-speed bombers. The Establishment's final report considered it doubtful that the increased effectiveness of the radar would compensate for the decline in the aircraft's performance – hardly a ringing endorsement for the new fighter.[47]

With a number of the first production batch set aside for testing, the type received its CS(A) release in February 1955, with one aircraft, WX792, going to Manby for the writing of the pilot's notes and three, WX804, WX807 and W808, to CFE for operational testing. The first unit to convert to the NF.3, No. 141 Squadron based at Coltishall, received four aircraft on 5 July to begin its conversion from the Meteor NF.11. Its companion squadron in the Coltishall Wing, No. 23, took delivery of its NF.3s as replacements for the NF.2A during September. The Stradishall Wing began its conversion during July when the first deliveries of NF.3s was made to No. 125 Squadron, followed by No. 89 Squadron during September. No. 151 Squadron at Leuchars also received NF.3s during July, making a total of just five squadrons to operate alongside the remaining NF.2As.[48]

While the operational units were taking delivery of their first Meteors and Venoms, the Air Ministry authorised the delivery of two Meteor NF.12s to CFE's All-Weather Wing at West Raynham to conduct AI.21 testing under the watchful eye of its OC, Wing Commander Edward Crew.[49] During ninety-one hours of testing, twenty-five of which were at night, CFE's pilots and navigators showed that AI.21 was capable of detecting crossing targets[50] flying at 250 knots (440 km/hr) at medium heights, 20,000 feet (6,095 metres), at a maximum range of 14 miles (22.5 km) and head-on at 15¼ miles (24.5 km), with the latter being appreciably increased by the use of the radar's OSS facility. Ranges at low level, 1,500 feet (460 metres), in crossing flights over land were of the order of 5½ miles (8.75 km), increasing to 7¾ miles (12.5 km) over the sea. At high altitude, 35,000 feet (10,670 metres), an average crossing range of 13½ miles (21.5 km) was recorded. With respect to its performance, the report concluded the detection ranges were a considerable improvement over those obtained by AI.10 and the OSS facility improved the maximum detection range, albeit at the expense of close quarter interception when the target was taking evasive action. They considered it would take an experienced AI.10 operator just five hours to convert to the new radar. Mutual interference from another AI.21 aircraft operating in the vicinity, and much like that experienced with AI.9, remained a problem, even when the aircraft were separated by 30,000 feet (9,145 metres) and 50 miles (80 km). From the servicing viewpoint, with the exception of the mechanical elements of the scanning unit, the overall performance of the equipment was 'good'.[51]

A second trial to assess the radar's tactical capability, was conducted by CFE between April and October 1954 using a Meteor Mk.12 and Mk.14. In this far reaching trial, where some 270 hours were flown, 140 of which were at night, against Lincoln, Canberra and Meteor targets during exercises, CFE's report provided some comfort with respect to the current Soviet threat (the Tu-4), but foreshadowed serious shortcomings that would require addressing in the not too distant future. Approximately 100 interceptions were flown against Canberras flying at 40,000 feet and speeds of M0.7 in crossing and head on attacks. During these interceptions a 'key position' was established, whereby the target was tracked under GCI control and brought to a range somewhere between 4 and 6 miles (6.5–9.5 km), depending on altitude, with a heading between the fighter and his target of 100°, or 10° off the fighter's nose (see Figure 39). From this position few problems were experienced in turning inside the target's line of flight and completing the interception. However, if these conditions were breached, or should the fighter fail to achieve the necessary high rate-of-

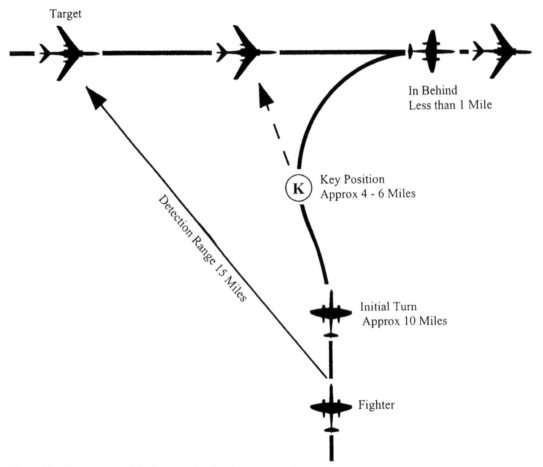

Figure 39.   The concept of the key position in night interception.

turn required in this form of attack (a manoeuvre that was difficult to accomplish in the Meteor at high altitude, but not impossible given the expertise of the crew), the interception would invariably fail.[52]

The report also identified an improvement in the radar's performance (compared to AI.10) in its resistance to *Window* jamming and praised the beacon facility and ASV mode as 'useful aids to navigation'. The operation of the guns, however, caused a vibration of the AI displays, and on one occasion a complete breakdown of the system, whilst the GGS was 'only of limited value'. Serviceability was once more effected by failures in the scanner mechanism and a lack of test equipment. Overall, AI.21 might be regarded as an improvement on AI.10.[53]

With the choice of radar settled and its testing more or less completed by 1955, it is perhaps worth examining another facet of the night-fighter, its armament. The standard armament of fighter aircraft in the 1950s, be they operating by day or by night, was a battery of four 20 mm Hispano cannon, that were not too far removed from those first installed in the Beaufighter in 1940. As in those days, the jet night-fighter was required to intercept the target and close to gun-range, about 200 yards (180 metres), before opening fire. Whilst undertaking this manoeuvre they were exposed, although not dangerously so, to the defensive armament of the bomber, six optically aimed 23 mm cannon in the case of the Tu-4. Nevertheless, large aircraft of the size of the Tu-4 and its successors required a greater deal of hitting power than the 20 mm cannon could provide. Post-war examinations of German cannon, particularly that of the 30 mm MK 108, and their development by the Royal Ordnance Factories, eventually led to the 30 mm Aden revolver cannon that saw extensive service in such aircraft as the Hunter, Swift and the Javelin, from 1953 onwards. However, as far as the author is able to judge, no work was ever undertaken by AWA or de Havilland to fit 30 mm cannon to the Meteor or the Venom. What was really needed was some form of air-to-air guided missile.

Although by 1951 work was underway with Fairey Aviation on the development of the beam riding *Blue Sky* AAM, later renamed 'Fireflash', and the IR-homing de Havilland Propellers' *Blue Jay*, later 'Firestreak', the lifespan of the Meteor and Venom was such that they would not be considered to carry either weapon. Ironically, a number of Meteor NF.11s were employed as test-beds[54] for the Fireflash and its beam projecting radar with the RAF's No. 12 Joint Service Trials Unit (JSTU) on the Woomera Range in Australia and with CFE at RAE, Aberporth, overlooking Cardigan Bay. Similarly, two Venom NF.2s, WL813 and WL820, were employed on Firestreak development between December 1952 and January 1957, with WL813 launching the first unguided round on 16 April 1953.[55]

Much of a 1950s night-fighter squadron's time was taken up with training the aircrews to an operational level of efficiency, with each flight operating by day and the other by night, and alternating once a week. During this period, a squadron's aircraft establishment varied at around sixteen, plus 'spares' and one or two instrument rating Vampire T.11s on the Venom units and Meteor T.7s on the Meteor squadrons. A pair of day-fighter Meteor F.8s were often on strength for target towing or use as squadron 'hacks'. Much of the operational training was taken up with PIs, with one aircraft taking the role of the 'target', whilst the other operated under GCI control and attempted to intercept him. Fighter Command aircraft also participated in bomber exercises, or *Bomex*, in which Bomber Command aircraft under-took simulated attacks on targets in the UK and the night-fighters attempted to intervene. In this respect, the Canberra, which was capable of attaining 48,000 feet (14,630 metres) or more and speeds up to 550 mph (880 km/hr), represented by far the greatest threat and was the most difficult aircraft to intercept. In Roger Lindsay's view, 'in the hands of a skilled crew the Venom night fighter's superb high altitude performance usually enabled it to make successful interceptions up to about 43,000 feet (13,105 metres), above which the Canberras were virtually immune.[56] In this situation the Meteor stood no chance at all.

Peter Verney describes his introduction to night-fighting, which began with a tour at No. 1 Air Navigation School (ANS) at Hullavington where he completed the basic course in air navigation between December 1950 and August 1951. However, such was the need for AI operators to support the rapidly expanding night-fighter force, the 'survivors' of the course were posted to No. 228 OCU for conversion to Nav/Rads:

The course was in two parts, the first, the flying classroom stage on the Brigand and Wellington, which carried two pupils and an instructor who could demonstrate and then supervise the pupils in varying types of interception, up to and including violent evasive action. Ground instruction included a detailed description of the current radar [AI.10], where we were even taught which fuse to change (out of about 100) to cure various faults …. Those who successfully completed this stage were awarded the much coveted 'wing'.[57]

On leaving No. 228 OCU as a qualified Nav/Rad, Peter was awarded his 'N' brevet and promoted to Sergeant on 23 October 1951, prior to teaming-up with his pilot, Joe Halkiew, and a posting with No. 39 Squadron flying Mosquito NF.36s and later Meteor NF.13s in the Canal Zone. In October 1954 he joined No. 152 Squadron at Wattisham flying the Meteor NF.12/14 where he participated in several exercises:

This Squadron had only recently been reformed on the Meteor NF.12 and 14, and still had a squadron leader as CO.[58] There was much emphasis on high level PIs over the North Sea, as well as the usual cine-gun exercises, but very little gunnery until the annual visit to the APC [Armament Practice Camp] at RAF Acklington in Northumberland …. The Javelin was being designed to replace the Meteor, which was frankly obsolescent and could not cope with the Canberra, let alone the V-bombers. The Javelin was designated as an 'All-Weather Fighter' and to this end we were expected to fly in almost any weather conditions, which led to a few anxious moments and diversions.

   The end of my tour was extended to include one of the large UK air defence exercises, which included two more dicey incidents. Our last night interception was against a Valiant, and, as usual, we had been scrambled far too late. While in a near stalled condition, struggling for height, I obtained a close range contact and had to order an immediate turn to try to get into position behind the target. We flew through his slipstream during this turn with the result that the aircraft practically turned over. I regained contact, but we had no speed and lost height so were doomed to an unequal tail chase which had to be abandoned. The very next day we were scrambled to 40,000 feet [12,190 metres] in poor weather and while still climbing at about 38,000 feet [11,580 metres] the hood opened. The resultant explosive decompression was interesting to say the least and we had no option but to return to Wattisham and my career in night fighters was over.[59]

In a similar vein, Venom squadron pilot John Pugh who joined No. 141 Squadron following officer training at Cranwell and jet flying at Driffield in November 1955, comments on the NF.3's capability:

Overall, the Venom 3 was a nice aircraft to fly, but was no great improvement on the Meteor NF.11; marginally quicker perhaps, but lacking solidarity – and the reassurance of two engines! Other points included the lack of ejection seats, fuel measured in gallons, a pretty useless AI Mk.21 radar, carrying out interceptions at night without lights, unreliable artificial horizons, an 'up-rated' engine that still coughed every time you crossed the coast at night, and the fact that 'bone domes' were optional – we tended not to wear them on air-to-air combat flying as they kept banging into the hood as we moved our heads around.

   My personal memories of the aircraft may be clouded by the engine failure we suffered at low level on 16 January 1956. We force landed in a field, but unfortunately hit an old wartime air-raid shelter at about 60 knots [105 km/hr]. My navigator [Flying Officer] Stan Perry, suffered two broken legs after the AI fell on top of them, and I was knocked unconscious as the [Gee] box threw me on to the gunsight (I was wearing a helmet!). Three farmers pulled us to safety after the aircraft caught fire – and were subsequently awarded MBEs and a BEM for their bravery. The cause of the accident was a drive belt failure to the fuel pump, which didn't please the Board of Inquiry who were trying to nail me for running out of fuel![60]

In fairness to the Venom, No. 141 Squadron was experiencing a period of poor serviceability (down to three aircraft by the end of December 1995), brought about by a lack of tradesmen and an inadequate supply of spares, both of which may possibly have contributed to John Pugh's accident and the loss of Flying Officers Ian Jarvis and Don Parsons in WX795 on 2 January.[61]

CONCLUSIONS

The night-fighter expansion programme, based around the interim fighters, the Meteor and Vampire/Venom, that were intended to provide a viable night air defence force pending the introduction of the Javelin, reached its zenith in 1956. In what might be described as the first phase of that programme, the RAF introduced the Meteor NF.11, Vampire NF.10 and the Venom NF2/2A, which were, in terms of their equipment (AI.10 radar, the *Gee* navigation aid and armament comprising four 20 mm cannon), the jet equivalent of the Mosquito NF.30/36, with only their performance distinguishing them from the wartime fighter. These aircraft were in turn matched against such types as the B-29 and the Lancaster/Lincoln, that remained firmly rooted in the piston-engine era and over which they maintained a number of advantages in respect of their performance, aircrews (many of whom had wartime experience), and the organisation of the ground radar tracking and control organisations that supported them. These advantages were in turn reinforced by an embryo Soviet bomber force that was striving to acquire the necessary skills in handling and maintaining very large aircraft and their sophisticated electronic navigation, bombing and ECM systems and create the support and training infrastructure necessary to keep them in the air. Altogether, a significant undertaking for a country that was in the late 1940s/early 1950s relatively inferior to the western powers in terms of its technology and organisational skills. With these elements in mind, it may be concluded that the first phase of jet night-fighters in concert with the ground organisation, were capable of providing a proper night defence for Great Britain because there was no realistic opposition.

This situation would not, however, last forever. The introduction of the jet-powered British Canberra in June 1951 and the deployment of the American B-47B the same year, both of which were nuclear capable, foreshadowed their development in the Soviet Union (of which more in Chapter 16). Against these aircraft, the first phase of night-fighters were inferior in terms of their performance, radar and armament – matters that could only be addressed in the longer term by the introduction of the Javelin. In the meanwhile a relatively quick and cost-effective 'fix' was required. In the event, the 'second phase' aircraft, the Meteor NF.12/14 and the Venom NF.3, which were introduced in the mid-1950s, did not provide the 'edge' the defence required. In *Bomex* and air defence exercises and during other forms of operational training, the Meteor and Venom were completely outclassed by the Canberra, the latter perhaps less so, in respect of their high altitude performance and armament. Their AI.21 radar, however, was a significant improvement in terms of its range and resistance to *Window* jamming, but lacked a true lock-follow capability (the OSS function went some way towards providing this), was susceptible to mutual interference that did not enhance the GGS capability. In conclusion, the second phase fighters in the mid-1950s could not provide a proper defence against nuclear capable, high performance, jet bombers – a situation that was set to worsen as the decade went by. Fortunately for Britain, the long awaited Javelin was about to enter service.

Notes

1. Mark Hichens, *op cit*, pp. 279 and 280.
2. *Ibid*, p. 293.
3. Large numbers of troops and aircraft were maintained in East Germany, Poland, Czechoslovakia, Hungary, Bulgaria, Romania and the Soviet homeland.
4. The 81st Fighter Interceptor Group at Bentwaters, Suffolk.
5. The 406th Fighter-Bomber Group at Manston, Kent.
6. The 47th Bomb Group (Light) at Sculthorpe, Norfolk.
7. The AAF was retitled the RAuxAF by HM King George VI in 1948.
8. No. 245 Squadron was beginning its conversion to the Meteor F.8.
9. Bill Taylor, *Royal Air Force Germany Since 1945* (Hinckley: Midland Publishing, 2003), p. 33.
10. BAFO had its title changed to the 2nd Tactical Air Force (2TAF) on 1 September 1951.
11. Nos 68 and 87 Squadrons at Wahn and Nos 96 and 256 Squadrons at Ahlhorn.
12. Peter Verney in an e-mail to the author dated 28 February 2006.
13. Peter Verney in an e-mail to the author dated 2 March 2006.
14. The embargo was imposed by the British Government in 1950 on the sale of weaponry to Egypt and Israel, who were then on the verge of open warfare.
15. David Watkins [1], *De Havilland Vampire, The Complete Story* (Thrupp: Sutton Publishing, 1996), pp. 91–3 and 95.
16. Roger Lindsay [1], *De Havilland Vampire NF.Mk.10 in RAF Service* (Luton: Alan W. Hall [Publications] Ltd), pp. 3 and 4.
17. The USAF was created as an independent member of the US Armed Forces in July 1947.
18. The Boeing B-47 Stratojet entered service with the USAF's Strategic Air Command in October 1951 and was capable of cruising at 557 mph (890 km/hr) at 38,500 feet (11,735 metres).
19. Ken Wright in a letter to the author dated 2 November 2005.
20. CFE Report dated February 1951.
21. David Watkins [2], *Venom, De Havilland Venom and Sea Venom, The Complete History* (Thrupp: Sutton Publishing, 2003), pp. 126 and 127 and Roger Lindsay [2], *de Havilland Venom* (Privately Published, 1974), p. 6.
22. Eight NF.2 airframes, WL804 to WL810 and WL812 were built and Hatfield, after which production was switched to Chester where deliveries began in December 1952.
23. Roger Lindsay [2], *op cit*, p. 6.
24. Post-war the title of 'navigator/radio' (N/R) was retained in Fighter Command, as opposed to Bomber Command who introduced the title 'navigator/radar' for use in the V-Force. However, the term 'radio observer' (R/O) was reintroduced during the mid-1950s to cover those who had not completed the full navigator's course. Source, Peter Verney in an e-mail to the author dated 27 January 2006.
25. It is known that Flying Officer A. Towel was killed on 29 December 1953, the CO, Squadron Leader A.J. Jacomb-Hood, DFC, and his N/R, Flying Officer A.E. Osbourne, on 21 January 1954 and Sergeants P.B. Jackson and H. Drabble (N/R) in March 1954.
26. Quoted in David Watkins [2], *op cit*, p. 132.
27. *Ibid*, pp. 132 and 133.
28. WL810, dorsal fairings, WL811, transparent cockpit canopy, WL809, spinning trials, WL806, first representative airframe, WL814 service release trials and WL807 spinning trials and Ghost 104 development.
29. David Watkins [2], *op cit*, pp. 134–7.
30. *Ibid*, p. 137 and Roger Lindsay [2], *op cit*, p. 6.
31. Roger Lindsay [2], *op cit*, p. 7.
32. AI.17 was later installed in the Javelin FAW.Mks1, 4, 7 and 9.
33. Between January and September 1954, Ken Wright undertook air-to-air trials of AI.17 and describes his luck with the equipment as 'pretty abysmal' – letter to the author, undated.
34. Information derived from www.vectorsite/net/avsky.html.
35. The author has sought to establish the background to the APS-57, but with little success.
36. Post-war the American term 'antenna' was adopted in place of the British 'aerial'.

37. It is interesting to note that both the 2.25 μsec/2,450 Hz PRF and 1.75 μsecs/550 Hz PRF combinations yield an average power of 196 Watts, whilst the 2.25 μsecs/300 Hz PRF combination gives 135 Watts. Lew Paterson suggests that since such radars tended to be mean power limited, perhaps the 2.25 figure is a misprint for 3.25 μsecs – which would give the same average power as the other modes.
38. AP2913D, Volume 1, Second Edition, June 1955.
39. On 31 August 1951, WA820 established four time-to-height records.
40. Bryan Philpott [1], *op cit*, p. 121.
41. It should be noted that WD670 and WD687 retained their lower powered Derwent 8 engines.
42. Barry Jones, *Gloster Meteor* (Crowood Press, 1998), p. 122.
43. Tony Buttler [2], *Gloster Meteor*, in Warpaint Series No. 22 (Milton Keynes: Hall Park Books), p. 35.
44. Bill Gunston [1], *Night Fighters* (Cambridge: Patrick Stephens Ltd, 1976), p. 138.
45. *Ibid*, pp. 122 and 123.
46. Tony Buttler [2], *op cit*, pp. 51 and 52.
47. David Watkins [2], *op cit*, pp. 145 and 146.
48. Roger Lindsay [2], *op cit*, pp. 7 and 8.
49. Wing Commander Crew, DSO*, DFC, flew Blenheims and Beaufighters with No. 604 Squadron during the Winter Blitz, commanded a flight in No. 85 Squadron in 1943 and later commanded No. 96 Squadron. He ended the war with thirteen enemy aircraft and twenty-one V-1 flying bombs to his credit and retired from the RAF in 1973 as an Air Vice Marshal with the CB.
50. Targets flying at 90° to the Meteor's line of flight.
51. *CFE Report No. 247*, dated May 1954.
52. *Ibid*.
53. *Ibid*.
54. NF.11s WD743, WD744, WD745, WM372, WM373 and WM374.
55. Barry Jones, *op cit*, p. 161 and David Watkins, *op cit*, pp. 203–5.
56. *Ibid*, p. 7.
57. Peter Verney in a letter to the author dated 3 December 1995.
58. No. 152 Squadron was re-formed at Wattisham on 30 June 1954.
59. Peter Verney in a letter to the author dated 3 December 1995.
60. Quoted in David Watkins [2], *op cit*, p. 150.
61. *Ibid*, pp. 148 and 149.

CHAPTER SIXTEEN

Decline and Fade

1956–1959

The year of 1956 could reasonably be seen as Britain's year of 'Imperial Over-Stretch'. By means of some very effective operations by the C-in-C Malaya, General Sir Gerald Templar, the Malayan Emergency was more or less brought to an end in 1954. Nevertheless, troops and aircraft would be required for another six years before the Communists finally ceased their guerrilla operations. In 1955, Greek Cypriot forces under Colonel Grivas began a campaign of organised guerrilla warfare against British forces in Cyprus. His movement known as EOKA, sought to gain independence from Britain and align the country with Greece by violent means. Once again British troops were called on to fight EOKA and keep the peace between the Greek and Turkish communities. This war was to last for another five years until the island gained its independence in 1960.[1]

Britain's oil interests, in what was then Persia (modern day Iran), were taken over by the Persian government, followed in 1952 by the overthrow of King Farouk of Egypt by a group of dissident army officers under the leadership of Colonel Gamal Nasser. With British troops and RAF forces withdrawn from the Canal Zone, Nasser set about the task of antagonising Israel and spouting anti-British propaganda. In 1955 he signed an arms deal with the Soviet Union, which upset the delicate military balance in the region and extended the Cold War to the Middle East. With Britain withdrawing the last of her troops in 1956, Nasser sought funding from Britain and the US to finance the Aswan Dam project, whom he then played-off against Russia. This caused the US to withdraw their support, which Nasser countered by threatening to national the Suez Canal to pay for the Aswan project. With the Prime Minister, Sir Anthony Eden, unwilling to accept the situation and with the active connivance of France and Israel, Britain[2] and her allies (excluding the US) invaded Egypt during October 1956. However, faced by threats from the US to implement economic sanctions against Britain and with world and UN opinion against her, Eden was forced into a humiliating climb-down and withdrawal from Egypt.[3]

Whilst Britain and the world's attention was diverted towards the Middle East, Soviet air and ground forces invaded Hungary during October. The invasion was in response to the Hungarian people's desire for independence from Russian control, withdrawal of their country from the Warsaw Pact, the Soviet equivalent of NATO, and an end to Stalinist brutality and ill treatment. This was a threat the ex-Stalinist commissar and President of the Soviet Union, Nikita Khrushchev, could not ignore and in went the troops. By the time the revolt was put down 25,000 Hungarians had lost their lives and 140,000 became refugees. With Britain over extended in the Middle and Far East, and fears within NATO of nuclear conflict with Russia, there were insufficient forces available, nor the political will, for anyone to come to the aid of the Hungarian people. But just how far had the Russians gone by 1956 in creating a strategic air force and how great was the threat of nuclear bombardment to Great Britain?[4]

To answer these questions we need to return to the development of the Tu-4 in the late 1940s. The testing of the Tu-4 was carried out in parallel with the development of the Soviet Union's first atomic weapon, the plutonium-based RDS-1 (also known as *izdeliye 501*), for which some thirty ballistic tests were undertaken at a site near Totskoye, in the Orenburg region of central Russia. Following the successful detonation of the RDS-1 atomic device in September 1949, five series-production bombs (*izdeliye 501M*) were built, but were not deployed by the V-VS before the scientists' attention switched to the development of a uranium device, which (potentially) offered a greater yield. The resultant uranium/plutonium, mixed-core bomb, was designated RDS-3 (*izdeliye 503*).[5]

Following the successful detonation of the first RDS-1 device, the Tu-4 was designated as the carrier aircraft for the air-drop weapon and for which a new variant was prepared. Designated Tu-4A, the new model was provided with an electrically heated and environmentally controlled bomb-bay to protect the weapon in transit, a redesigned suspension system to hold the bomb, shielding in the cabin to protect the crew from radiation in flight and an automatic weapon control system. Two prototypes were built to the A-model standard and were available for nuclear weapons testing by the summer of 1951.[6]

On 18 October 1951, a Tu-4A, operating under the authority of the C-in-C of the V-VS and a State Commission headed by Major General G.O. Komarov, and commanded by Lieutenant Colonel Konstantin Oorzhuntsev,[7] successfully dropped an improved RDS-3 bomb from a height of 32,800 feet (10,000 metres) over the test range at Totskoye. The weapon detonated with a yield of 42 kilotons.[8] The RDS-3 and its Tu-4A carrier aircraft were committed to production in 1951 (at about the same time as the MoS was thinking about introducing the AI.21 versions of the Meteor and Venom), with an aircraft from the first batch being used to drop an RDS-3 in 1952. In the summer of 1953 four Tu-4As were deployed to Kazakhstan to undertake further live weapons drops over the Semipalatinsk Nuclear Test Range and take measurements of the radiation levels and the drift of the fall-out cloud. These tests probably used-up the first production run of three RDS-3 bombs.[9]

In September 1953, another Tu-4A successfully dropped an RDS-5 weapon from a height of 29,500 feet (9,000 metres) over the Semipalatinsk Range. The yield of this weapon is not known, but was almost definitely more powerful than the RDS-3. Testing of the Tu-4A/RDS-5 combination continued throughout the remainder of the year to clear the weapon for service use and check the proper operation of its neutron initialisation system.[10] Given the amount of testing the RDS-5 required and the production of the Tu-4A amounted to just ten aircraft, it would be reasonable to conclude that the majority of the A-bombs available to the V-VS in 1953, never more than a dozen, were expended in testing. By comparison, the US' nuclear arsenal amounted to some 1,350 bombs![11]

The vast majority of the 1,200–1,300 Tu-4s built between 1948 and 1953[12] were completed as conventional bombers, where they were used to replace the wartime Soviet-built Il'yushin Il-4 and Petlyakov Pe-8 light and heavy-bombers and examples of the American B-25 Mitchell, B-17 Fortress and B-24 Liberator. The first Tu-4s were allocated to Heavy Bomber Air Regiments (*dahl'nebombardiro-vochnyy aviapolk* – DBAP) based on airfields in the western military districts of the Soviet Union, specifically at Nezhin, Poltava and Priluki in the Ukraine, Lodeinoye Pol'e near Leningrad, Karelia, Narva in Estonia and in Belorussia. In time of war and during exercise periods these regiments were forward deployed to airfields in the Soviet Union's satellite states, where they were just a few hundred miles from NATO's bases in Europe and Britain. The V-VS's 13th Long-Range Bomber Division (*dahl'nebom-bardiro-voch-naya aviadiveeziya* – DBAD) based at Poltava, was chosen as the first formation to be equipped with the new bomber, with the 185th Guards DBAP converting to the Tupolev bomber during May 1949.[13]

The Tu-4s operated in a similar manner to their erstwhile enemies in SAC, with groups of bombers being scrambled to patrol along NATO's borders in readiness for a conventional strike on tactical and strategic targets in Europe. For this role they were usually loaded with combinations of 1,500 kg (3,305 lb) and 3,000 kg (6,615 lb) high explosive bombs and 250 or 500 kg (550 or 1,100 lb) chemical bombs[14] that were aimed by optical or radar means. However capable the aircraft, the ADD's aircrews had to master a machine whose complexity and maintenance requirements were far in excess of those previously employed by the V-VS. These required more thorough and consequently, longer periods of training for all members of the crew: pilots, navigator, bomb-aimer, radio operators, gunners and the like. From the maintenance viewpoint, groundcrews had to master new technologies, particularly in the fields of electronics, pressurisation and complex powerplants, and carry out regular routines in severe weather conditions.[15]

Only ten Tu-4s were converted to the A-standard,[16] these being allocated to ADD crews to compile and test the command's nuclear operating procedures and conduct experimental weapons drops over the Semipalatinsk Ranges, as has already been mentioned. Whilst testing of the RDS-3 and RDS-5 nuclear weapons continued into 1954, it does not appear the Tu-4A was required to stand on quick reaction alerts (QRA) in defence of the Soviet Homeland. This function was reserved for the three jet-bombers that would follow the *Bull*.[17]

The origin of the first of a trio of capable and imaginative first generation of Soviet bombers, the Tu-16, began as the V-VS was in the process of accepting the first deliveries of the Tu-4 in April 1948. At about that time, the V-VS' commanders began the formulation of a requirement for a jet-powered, strategic bomber, capable of carrying a payload of 3,000 kg (6,600 lb) over a distance of 6,000 km (3,750 miles), at a maximum speed at 1,000 metres (3,300 feet) of 900 km/hr (560 mph). This requirement was based on the need to be able to strike at the heart of American and allied centres of power, but more particularly, that of western naval power, the defeat of which the Soviet Union saw as key to victory in any future war.[18]

The prototype, of what would become the Tu-16, the Type-88/1, was designed and built by Andrei Tupolev's OKB-156 and flown for the first time from the V-VS' Flight Test Centre at Zhukovskiy, near Moscow, on 27 April 1952. Factory testing quickly showed that although the aircraft had a good performance, it was overweight by 5 or 6 tons (5.08–6.1 tonnes). During testing the aircraft was damaged in a heavy landing and the opportunity was taken to lighten the second prototype, the Type-88/2, and continue the trials. Type 88/2 undertook its first flight, again with pilot Rybko in command, on 6 April 1953. The results of the following trials were sufficiently promising for Tupolev to order the Type-88/2 into production before state approval was granted – a very brave decision which saved a great deal of time in getting the type into service![19]

Production of what was now described as the Tupolev Tu-16, began at the Kasan Production Plant during 1953, when two aircraft of the baseline Tu-16 conventional bomber were completed before the year was out. Between then and the end of 1958, three factories

(Kasan, Kuibyshev and Voronezh) built a total of 294 Tu-16s, with many having an in-flight refuelling (IFR) capability built-in or added at a later date. The Tu-16, or *Badger-A* in NATO parlance, entered squadron service with the V-VS in February/March 1954, and came to public notice the following May when a formation of nine aircraft participated in the May Day celebrations over Moscow's Red Square. By the mid-1950s the bomber was being built in large numbers, sufficient to equip several ADD long-range bomber regiments and units of the Soviet Navy (*Aviatsiya Voenno-morskoi flot* – AVMF).[20]

In November 1953 work began on the conversion of the Tu-16 to the nuclear capable Tu-16A standard. Two Tu-16s were modified with the addition of an electrically heated bomb-bay that was insulated and thermally regulated to ensure the safe working of the weapon – the RDS-3 or RDS-5. Special protection to shield the crew from the effects of the heat and shock of the nuclear flash was also provided, alongside sytems for monitoring and releasing the weapon by means of a bomb-sight that was linked to the bomb-aiming radar and autopilot. Active ECM equipment and long-range radio navigation aids were also provided. Bearing no external difference to the standard Tu-16, except for an overall white paint scheme, and weighing just 410 lb (185 kg) more, the Tu-16A retained its NATO *Badger-A* designation. Deliveries to the ADD and the AVMF from the Kasan factory began in 1954, with 453 Tu-16As being built when production ended sometime in 1958. One aircraft, serialed 4200503 and designated as a Tu-16V, was used as a carrier of the Soviet Union's hydrogen bomb, but was not deployed operationally.[21]

The Tu-16/*Badger-A* was a swept wing, medium range, strategic bomber with a crew of five or six, powered by two Mikulin AM-3M turbo-jet engines of 19,290 lb (8,750 kg) thrust and capable of carrying a maximum bomb-load of 19,840 lb (9,000 kg). Its performance was broadly comparable to that of the B-47 (see Table 8).[22] Like those of the B-47, the Tu-16 regiments were forward deployed on the airfields of the Soviet Union's satellites, bringing them within striking range of NATO's European allies and the British Isles. Its classification as a medium bomber precluded nuclear strikes from the Soviet Union's western airfields over the North Pole (over the pole) against the continental US and Canada, as its range, even with IFR, was insufficient for the return flight. This requirement was, therefore, put on hold pending the availability of larger and more powerful aircraft.

Whilst the Tupolev OKB persevered to get the Tu-16 built and through its State Acceptance Trials, the V-VS was able to deploy the Tu-4 as a stopgap strategic bomber. Russia's experiences in Korea had shown that, when faced with jet-power MiG-15s, the American B-29s were vulnerable to fighter attack and could no longer be operated by day without incurring serious losses. Like Germany in the Second World War, the USAF's bomber groups were forced to operate at night and suffer the consequent reduction in bombing accuracy. This point was not lost on Stalin, who cancelled further development of the Tu-4 in favour of a true long-range, jet-powered, strategic bomber. Stalin's action put a stop to the development of the Tu-85, itself a derivative of the Tu-4, which did possess the range to bring most of America's industrial centres within its grasp. Tupolev's experience in the development of the Tu-16 had shown that Soviet engine technology was insufficient to meet Stalin's requirement for a bomber capable of carrying a 5 tonne (4.9 tons) fission bomb, over an unrefuelled range of 16,000 km (10,000 miles). However, he was convinced he could build an inter-continental range bomber if it was powered by propeller-turbine (turbo-prop) engines. Using the Tu-4's fuselage matched to powerful turbo-props and large swept wings, the latter reminiscent of Boeing's B-47 and B-52, Tupolev believed he could build a bomber with a longer range than that of a pure jet – an assertion that was later proved to be correct.[23]

However, with the V-VS and the Council of Ministers maintaining their demands for a jet-powered, nuclear bomber, Vladimir Myasishchev, the head of OKB-23 based at Fili in the western suburbs of Moscow, proposed an aircraft powered by four of the Tu-16's AM-3 engines, having a maximum speed of 590 mph (945 km/hr) and a range of 8,000 miles (12,800 km).[24,25] Both designs were authorised for development by Stalin in 1951, who effectively started a two-horse race to produce the Soviet Union's first nuclear armed, strategic bomber with inter-continental range.[26]

Tupolev's Tu-95 was the first aircraft to fly, when it undertook a fifty minute flight from the Zhukovskiy Flight Test Centre on 12 November 1952, closely followed by the M-4 on 20 January 1953. Whilst the M-4 met many of its design requirements, that for range was lower than expected, only achieving 6,125 miles (9,800 km), as opposed to Stalin's specified minimum of 10,000 miles (16,000 km). Flight testing of the Tu-95 suffered a setback on 11 May 1953 following the failure of the gearbox on one of the Kuznetsov 2TV-2F turbo-props, which started a fire that destroyed the aircraft. The 'official' fault was traced to the failure of an intermediate gear in the turbo-prop's contra-propeller gearbox, brought on by fatigue and poor design. A second prototype was made ready and flew for the first time on 16 February 1955 and later confirmed the type's maximum range as 8,600 miles (13,900 km), which, although lower than the official specification, was a remarkable feat and sufficient to enable the bomber to reach targets in North America over the pole.[27]

Although by comparison with the Tu-95, the M-4's range performance was undoubtedly disappointing, the decision was taken in 1954 to commit the type to production in its own right and as an insurance in the event of the failure of the Tu-95 programme. Engine fires continued to trouble the Tu-95's testing, which delayed the type's commitment to production until January 1955 and its service entry to the following year. The Kuibyshev factory produced thirty-one Tu-95s in 1957, before switching to the Tu-95M equipped with improved NK-12M engines. Despite the delays and a change of model on the production lines, the Tu-95 entered V-VS service in 1956. Thirty-one M-4s were also produced between January 1955 and June 1956, but these aircraft had a poor accident record and were replaced on the Fili lines by the M-3 and M-6. Like the Tu-95, the M-4 entered service in 1956 and like the Tupolev aircraft, was configured for conventional and nuclear bombing.[28] The Tu-95 was allocated the reporting name *Bear-A* in NATO's coding system and the M-4 that of *Bison-A*.[29] By 1960, two regiments of the 22nd Heavy Bomber Aviation Division at Engels Air Force Base were equipped with the M-4 and -3M versions of the *Bison*, in addition to two further divisions, the 79th with two regiments at Semipalatinsk in Kazakhstan and the 106th with three regiments at Uzine in the Ukraine, with the Tu-95M version of the *Bear*.[30]

Like the aircraft, their nuclear weapon-loads had also been improved. The bomber's initial armament comprised the 42 kiloton RDS-4 fission bomb, but this was superseded by the 2.9 megaton, thermonuclear RDS-37, which remained the standard free-fall weapon throughout the remainder of the 1950s and on into the early sixties. Considered by some in the V-VS to be 'excessive', the RDS-37 was frequently replaced by the lower yield RP-30 and RP-32 weapons of 200 kilotons.[31]

In respect of western bombers, the M-4's performance was broadly comparable with that of the B-52B Stratofortress which entered USAF service in June 1955, excepting its ceiling capability, which was significantly lower, and a range which was slightly down on the Boeing.[32] The Tu-95 was broadly comparable in terms of range with the early models of the B-52, but slower and with a lower ceiling.

Although not telling the complete story, since there is no assessment of aircrew expertise, nor of the quality of their armament and ECM systems, Table 8 provides a broad comparison between the performance of the Soviet Union's strategic bombers and Britain's night-fighters in 1956 – the last year in which the interim jet night-fighters would predominate in Fighter Command's order of battle. The table also ignores the fighter's state of serviceability and the performance of their radar, which, in reality, were not always on the 'top line'.[33]

As can be seen, every one of the Soviet bombers had an operational ceiling in excess of 40,000 feet (12,190 metres), with that of the M-4 reaching 50,000 feet (15,240 metres), and, with the exception of the Tu-95, speeds approaching 600 mph. With speeds that matched, or at best were a little slower than that of their adversaries, and ceilings that were comparable, the defending fighters would have been hard pressed to intercept their targets and bring them to gun range. Further, having to close to 200–300 yards in a tail pursuit, would in turn have exposed the fighters to return fire from the bomber's radar-directed 23 mm cannon. However, with the accuracy of the NR-23 guns and their directing radar unknown, they may have been less exposed at night. In terms of range, the Soviet bombers were capable of reaching any target in Europe, the British Isles and, in the case of the Tu-95, the continental US without IFR. When IFR capability was fitted to all three aircraft in the late 1950s/early 1960s, they were all capable of striking targets in the continental US and NATO shipping in the Atlantic from the Russian homeland. Fortunately for Britain her air defences were strengthened in 1955 by the arrival of the Type 80 surveillance radar and the following year 1956 with delivery of the first batch of Javelin all-weather interceptors.

In 1954 Britain was supplied with a few examples of the American FPS-3, L-band (1,200 MHz) radar, one of which was installed in the Northern Sector (see Chapter 14, Figure 30,) at Boulmer on the Northumberland coast. This radar introduced the possibility of a reversion to the direct method of fighter control used during the war, where the FPS-3's high discrimination and long range, 200 miles (320 km), provided the controller with early warning and short range interception information, combined on to one PPI display. Following a practical trial during the 1954 *Exercise Beware*, under the control of Squadron Leader Eric Holmes, who also initiated the idea, the Boulmer FPS-3 showed that if a senior controller was given the authority to scramble the sector's fighters and allocate them to his controllers, who sat in front of duplicates of his display, the defence's reaction time was considerably decreased and the flow of accurate information was much improved.[34]

TABLE 8. A COMPARISON OF SOVIET BOMBER PERFORMANCE WITH BRITISH NIGHT-FIGHTERS IN 1956[35]

Type	Maximum speed	Ceiling	Range	Armament
Tu-16A Badger-A	585 mph at 32,000 ft	42,650 ft	3,600 miles with 6,600 lb bomb-load	6 × twin 23 mm radar directed cannon in three turrets
Tu-95A Bear-A	500 mph at 41,000 ft	44,290 ft	9,000 miles with 11,000 lb bomb-load	3 × twin 23 mm radar directed cannon in three turrets
M-4 Bison-A	595 mph at 39,370 ft	51,180 ft	6,100 miles with 11,200 lb bomb-load	4 × twin 23 mm radar directed cannon in four turrets
Meteor NF.11/12/14	535 mph at 30,000 ft	40,000 ft	725 miles	4 × 20 mm cannon
Venom NF.2/2A	630 mph	44,000 ft	950 miles	4 × 20 mm cannon
Venom NF.3	630 mph	45,000 ft	1,000 miles	4 × 20 mm cannon

The result of Holmes' trial was greeted with some enthusiasm by higher authority, who began to show some interest for what might be described as 'autonomous control'. In the early months of 1955 Boulmer received one of the new Type 80 long-range surveillance radars, which promised better results than the FPS-3. With its large aerial rotating at six revolutions per minute and operating in the 10 cm band (3 GHz), the Type 80 had a range in excess of 200 miles (320 km), and gap-less cover with pin-point accuracy. The new radar was trialled at Boulmer using Holmes' ideas in *Exercise Formulate*, that was conducted during the early months of 1955. The *Formulate* results were monitored by a representative of Fighter Command's Operational Research Branch, Mr Jimmy Tedd, whose reports and recommendations on the success of the concept were sufficient to bring about a further reorganisation of fighter control. Using a system of MRS, Tedd abandoned the concept of the SOC with its multiplicity of manual plotting, filter centres and interconnected telephone landlines and introduced autonomous control. An MRS replaced the SOC in each of five sectors (Caledonian, Northern, Eastern, Metropolitan and Southern)[36] and took over the responsibility for the control of all interception operations within its boundary. The new system showed an immediate improvement in performance by pushing the interception point further away from coast – something that was to be applauded as the air defence network was now having to contend with the destruction of nuclear armed bombers as far away from Britain as was possible.[37]

The prototype G.A.5/Javelin, WD804, flew for the first time from Gloster's Moreton Valence airfield, in the hands of the Company's chief test pilot, Squadron Leader Bill Waterton,[38] on 26 November 1951. WD804 was followed in relatively slow succession by four further prototypes, WD808[39] on 20 August 1952, WT827 on 7 January 1953, WT830 on 14 January 1954 and the last, WT836, on 20 July. A sixth machine, XD158, was ordered as a replacement for the prototype, which was lost following the failure of both elevators and the subsequent wrecking of the machine during a forced landing at Boscombe Down on the 29 June 1952.[40,41]

The MoS placed an order for 200 Javelins on 7 June 1952, with the first designated FAW.Mk.1 and carrying the serial XA544, being flown for the first time on 22 July 1954 by Gloster's new chief test pilot, Wing Commander Dicky Martin.[42] Fitted with 8,300 lb Armstrong Siddeley Sapphire Sa.6 engines, AI Mk.17 S-Band radar equipped with lock-follow and a blind-fire capability and four 30 mm Aden cannon, the FAW.1 received a restricted Controller of Aircraft (CA) release on 30 November 1955.[43,44] Whilst still under MoS ownership, two FAW.1s, XA554 and XA559, participated in Fighter Command's October 1955 annual air defence *Exercise Beware*. Flying from Coltishall with RAF crews,[45] alongside the station's resident Venom NF.3s of No. 23 and No. 141 Squadrons, the new aircraft 'claimed' the destruction of at least eight Canberras. Thus, for the first time, with the Javelin able to reach 52,000 feet (15,850 metres)[46] and speeds up to 540 mph (865 km/hr) at 40,000 feet (12,190 metres), the Canberra, and for that matter the Soviet 'three' (*Bear*, *Badger* and *Bison*), were no longer immune to interception.[47]

Deliveries to the RAF proper began on 3 January 1956, when the first of three FAW.1s, XA565, was delivered to CFE's All-Weather Wing at West Raynham, tasked with defining the type's operational procedures. No. 46 Squadron, based at Odiham, Hants, was chosen as the first front-line squadron to convert to the new fighter and immediately demonstrated its superiority to the Meteor NF.12s and 14s it replaced. Dwarfing the Meteor, the Javelin was quicker to turn-round due to the greater accessibility of the equipment bays, a single point, pressure refuelling connector and cartridge starters for the engines. Possessing a performance, which for its day was described as 'respectable', the Mk.1 was airborne at 120–125 knots (210–220 km/hr) after a run of just 550 yards (500 metres) and could reach 45,000 feet (13,715 metres) in 9.8 minutes, or approximately half the time taken by the Meteor. Operating at 48,000 feet (14,630 metres), the Javelin remained highly manoeuvrable, with a good range, 750 miles (1,200 km), and endurance, a radar set capable of locking-on to its target, a decent set of guns with fifteen seconds-worth of ammunition, a well equipped navigation suite and Martin Baker Mk.3J ejection seats for both crew members.[48]

By the month's (February) end, the Squadron's Meteors had been flown away to the MUs and the unit was fully Javelin equipped. Such was the pace of Javelin development, that only two squadrons were to see service with the FAW.1, No. 46, as already mentioned, and No. 87 Squadron operating from Brüggen in West Germany, which was equipped mostly with former aircraft drawn from No. 46 Squadron when that unit received Javelin FAW.2s in May 1957.[49] Similar to the Mk.1, the FAW.2 introduced the American APQ-43 radar (AI Mk.22) to RAF service, but was built in small numbers.

The importance of the Javelin as an interceptor was emphasised in the 1956 Statement on Defence, which confirmed that:

> For some time to come the manned fighter must continue to provide the backbone of our defence. The firepower and lethality of fighter aircraft will be markedly increased by equipping them with air-to-air guided missiles. The first generation of [Fireflash] missiles will become available in the course of 1956–57. They will be brought to service with a special mark of Swift [the Mk.7] and will be used to gain experience of this type of weapon. Although manned fighter aircraft and their weapons will improve [with the Firestreak], the surface-to-air guided missile may well play a predominant part in air defence.[50]

These were indeed prophetic words, bearing in mind the iminent introduction of Firestreak on the Javelin and the Government's decision in 1957 to abandon the manned fighter altogether in favour of surface-to-air missiles (SAM).

Due to the ever increasing number of variants being developed and introduced to the service in small numbers and with aircraft being required to re-equip squadrons in the UK and Germany, deliveries of the Javelin to Fighter Command in 1956 had little impact on the night-fighter force. With the advent of the Mks 4,[51] 5[52] and 6[53] that were built in reasonable numbers by Gloster and AWA, the Javelin re-equipment programme began to make some progress during 1957. By the time of the annual air defence exercise in September, the rundown of the Venom squadrons was well underway. The Coltishall-based Nos 23 and 141 Squadrons had exchanged their Venom NF.3s for Javelin FAW.4s by May 1957, No. 151 at Turnhouse for the FAW.5 in June and No. 89 at Stradishall for the FAW.6 in November, whilst Nos 125, 219 and 253 Squadrons were disbanded by September. The aircrews no doubt were relieved they would no longer be required to fly night patrols over the North Sea with one engine and no ejection seats.

In April 1957, the Secretary of State for Defence, Mr Duncan Sandys, placed a White Paper before the House of Commons which proposed a realignment of British defence policy. In order to check the ever rising cost of defence, the Minister proposed the country's defence could best be served by the creation of a QRA deterrent force, the V-Force, armed with nuclear weapons, whose bases would be protected, not by aircraft, but by SAM weapons systems. Sandys had been appointed to his defence post in January by Prime Minister

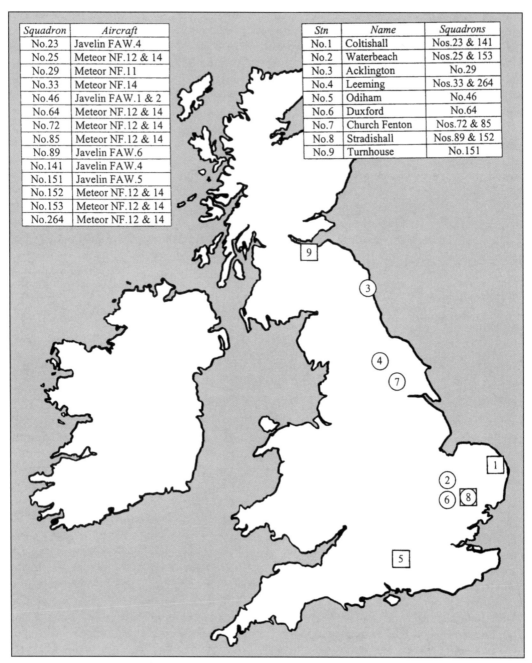

Squadron	Aircraft
No.23	Javelin FAW.4
No.25	Meteor NF.12 & 14
No.29	Meteor NF.11
No.33	Meteor NF.14
No.46	Javelin FAW.1 & 2
No.64	Meteor NF.12 & 14
No.72	Meteor NF.12 & 14
No.85	Meteor NF.12 & 14
No.89	Javelin FAW.6
No.141	Javelin FAW.4
No.151	Javelin FAW.5
No.152	Meteor NF.12 & 14
No.153	Meteor NF.12 & 14
No.264	Meteor NF.12 & 14

Stn	Name	Squadrons
No.1	Coltishall	Nos.23 & 141
No.2	Waterbeach	Nos.25 & 153
No.3	Acklington	No.29
No.4	Leeming	Nos.33 & 264
No.5	Odiham	No.46
No.6	Duxford	No.64
No.7	Church Fenton	Nos.72 & 85
No.8	Stradishall	Nos.89 & 152
No.9	Turnhouse	No.151

Figure 40. Night-fighter and FAW Squadrons at September 1957. By September 1957, five of the fourteen squadrons available to the night defence force were FAW equipped.

Harold Macmillan, with a remit to formulate a defence policy that would bring about a reorganisation of the Armed Forces to ensure a 'substantial reduction in expenditure and manpower'.[54] This policy, which relied heavily on the development of missile technology, allowed the Conservative Government, or so they thought, to reduce the RAF's strength, whilst maintaining a credible deterrence, and make significant savings in the defence budget. However, since this policy could not be realised until well into the 1960s, fighter aircraft would still be needed to protect the V-Force bases, albeit in smaller numbers, until the SAMs were deployed. Therefore, with the exception of the work being carried out on the English Electric Company's P.1 interceptor, later named 'Lightning', the development of all future fighter aircraft would cease.

The White Paper had the immediate effect of causing the disbandment of a number of squadrons and reducing the development of the Javelin in preference to the Mach 2 Lightning, which, in its P.1B form flew for the first time on the very day the Secretary of State made his infamous statement in the House of Commons. In addition to those units already mentioned, Fighter Command lost further squadrons in 1957/58. No. 264 was renumbered No. 33 Squadron on 30 September 1957, following the latter's disbandment at Driffield on 31 July. The Javelin-equipped No. 141 Squadron based at Coltishall, was renumbered No. 41 Squadron on 16 January 1958, followed by the disbandment of the Meteor equipped of No. 152 Squadron at Stradishall during July, and No. 153, also Meteor equipped and based at Waterbeach. However, the latter was renumbered No. 25 Squadron that same month. The resultant reduction of four squadrons is reflected in the order of battle at September 1958 (see Figure 40), by which time the Meteor force was in terminal decline.

The year of 1958 saw the introduction of the latest version of the Javelin, the FAW.Mk.7, which was based on the earlier Mk.5 (AI.17) and sported up-rated 11,000-lb thrust Sapphire Sa.7 engines, a reprofiled and extended rear fuselage to improve drag and the addition of four wing pylons to carry Firestreak AAMs or four 100 gallon (455 litre) drop tanks.[55] The Mk.7 was built in greater numbers than any other mark, with 142 being delivered from the Gloster and AWA production lines. The first production machine flew on 9 November 1956, but an extended period of testing did not see the type complete its CA Release until late in 1957. No. 33 was the first to deploy the Mk.7, when it converted from the Meteor during July 1958 at Leeming.[56]

From there on, the replacement of the Meteor quickened pace. No. 25 Squadron (which had been re-established by renaming No. 153 Squadron at Waterbeach in July 1958) followed suit during December 1958 when it converted to the Javelin Mk.7. The last Meteor unit and the last UK-based night-fighter squadron to convert to the FAW role, No. 72, at Leconfield, received its first batch of Javelin FAW.4s in April 1959, which brought about the demise of the Meteor by the end of June.

Despite the Meteor being removed from UK air defence in June 1959, it continued to give good service in the Far East, when the Singapore-based No. 60 Squadron was re-formed at Leeming in May to convert to the Meteor NF.14. On its return to Tengah, No. 60 provided the sole night/all-weather defence force in the Far East until August 1961, when it too succumbed to the delights of the Javelin. In Germany, the seriously obsolete Meteor NF.11s of No. 11 Squadron at Geilenkirchen were displaced by Javelin FAW.4s in February 1960, followed by No. 5 during the spring which re-equipped with FAW.5s at Laarbruch. These three units, therefore, brought to an end the operational service of the RAF's radar-equipped night-fighter squadrons.[57]

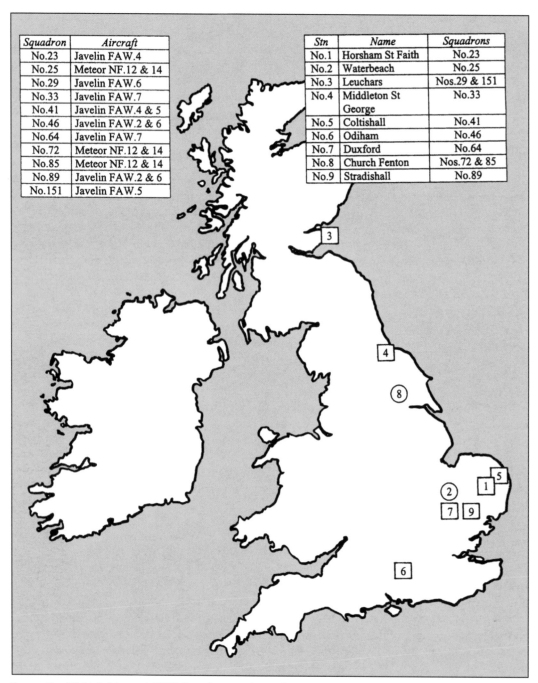

Squadron	Aircraft
No.23	Javelin FAW.4
No.25	Meteor NF.12 & 14
No.29	Javelin FAW.6
No.33	Javelin FAW.7
No.41	Javelin FAW.4 & 5
No.46	Javelin FAW.2 & 6
No.64	Javelin FAW.7
No.72	Meteor NF.12 & 14
No.85	Meteor NF.12 & 14
No.89	Javelin FAW.2 & 6
No.151	Javelin FAW.5

Stn	Name	Squadrons
No.1	Horsham St Faith	No.23
No.2	Waterbeach	No.25
No.3	Leuchars	Nos.29 & 151
No.4	Middleton St George	No.33
No.5	Coltishall	No.41
No.6	Odiham	No.46
No.7	Duxford	No.64
No.8	Church Fenton	Nos.72 & 85
No.9	Stradishall	No.89

Figure 41. Night-fighter and FAW Squadrons at September 1958. A year later and only three night-fighter squadrons, all equipped with the Meteor NF.12/14, remain in service. Note how the FAW force has extended further north to protect Scotland and the north of England, but with the main element grouped in East Anglia to protect the V-bomber bases in that region and Lincolnshire.

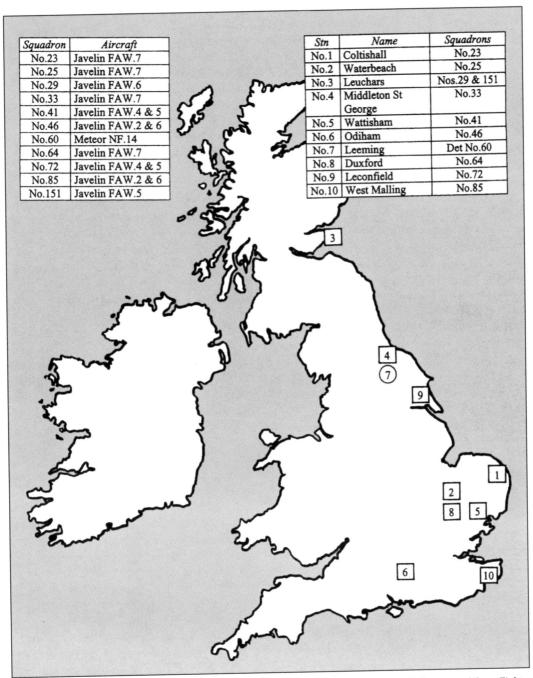

Squadron	Aircraft
No.23	Javelin FAW.7
No.25	Javelin FAW.7
No.29	Javelin FAW.6
No.33	Javelin FAW.7
No.41	Javelin FAW.4 & 5
No.46	Javelin FAW.2 & 6
No.60	Meteor NF.14
No.64	Javelin FAW.7
No.72	Javelin FAW.4 & 5
No.85	Javelin FAW.2 & 6
No.151	Javelin FAW.5

Stn	Name	Squadrons
No.1	Coltishall	No.23
No.2	Waterbeach	No.25
No.3	Leuchars	Nos.29 & 151
No.4	Middleton St George	No.33
No.5	Wattisham	No.41
No.6	Odiham	No.46
No.7	Leeming	Det No.60
No.8	Duxford	No.64
No.9	Leconfield	No.72
No.10	West Malling	No.85

Figure 42.   FAW Squadrons at 1 July 1959. On this date the classic night-fighter aircraft disappeared from Fighter Command's inventory in the UK. The sole active Meteor NF.14 unit, a detachment of No. 60 Squadron, is working-up at No. 228 OCU at Leeming prior to its return to the Far East.

CONCLUSION

By the middle of 1957 Fighter Command's first generation of jet night-fighter had been replaced with the improved Meteor NF.12/14 and the Venom NF.3 offering the updated AI.21 radar and little else. By contrast the Soviet Government, and more particularly its Council of Ministers, had recognised the need to create a strategic nuclear bomber force capable of countering that of the USAF's SAC. Designed to strike any of the West's principal seats of government with thermo-nuclear weapons (H-bombs), the resultant Tu-16 *Badger*, M-4 *Bison* and Tu-95 *Bear* strategic bombers had performances equal to, or better than Britain's second generation of jet night-fighters until the advent of the Javelin in 1957. The delays and over-extended development of the Javelin-series necessitated the retention of the Meteor and Venom into obsolescence and beyond. Fortunately for Great Britain, it was only by great good fortune that the V-VS was able to deploy few, if any bombers equipped for nuclear strike operations until well into 1957. The introduction of MRS system and the concept of autonomous fighter control in 1955 and the deployment of the Javelin the following year, marked the beginning of an improvement in the UK's air defences. By these two means, the country's air defence border was extended further from its shores and the interception performance was significantly improved.

The Conservative Government's need to reduce defence expenditure by a combination of nuclear deterrence and a dependence on SAM technology and its exposure to parliamentary scrutiny in the infamous 1957 White Paper, foreshadowed a gradual reduction in conventional fighter aircraft. Nevertheless, the process of gradual improvement in the Javelin force and the removal of the elderly night-fighters was allowed to continue into 1958 and beyond. The availability in reasonable numbers of the Javelin FAW.4/5 brought about the demise of the Venom and the introduction of the missile equipped Mk.7, ensured the removal of the Meteor from UK air defence by mid-1959 and in Germany by 1960.

With respect to the *Badger*, *Bear* and *Bison*, the Javelin missile-armed Javelin FAW.7 and the later Mk.8 and Mk.9 represented a credible deterrent in performance terms, if not in numbers. Ironically, by then (1960), the Soviet Union was in the process of transferring its nuclear delivery capability from bombers to intercontinental ballistic missiles (ICBM). On 7 May 1960 the Council of Ministers created the Strategic Rocket Forces as a separate branch of the Soviet armed forces and in late-1960 fielded its first ICBM, the NATO named *SS-7 Saddler*, with a range of more than 6,250 miles (10,000 km) and carrying a 5 megaton warhead.[58] From then on, the V-VS gradually transitioned the ADD's strategic bombers to the maritime role in support of submarine-launched ballistic missiles, as air-to-surface missile (ASM) carriers and long-range reconnaissance platforms.

The introduction of the ICBM and the redeployment of the Soviet Union's strategic bombers precipitated the decline in Fighter Command's importance as Britain's first line of defence – a task that was passed to the Royal Navy's Polaris submarine force in 1965. Again ironically, by the time the Javelin force reached its peak in 1960/61, its replacement in the form of the all-weather, Mach 2 Lightning, was already in being[59] and by 1968 the Javelin had completely disappeared from the RAF's front line inventory.[60] Fighter Command's gradual fall from grace culminated in its disappearance the same year, on 30 April, when it was reduced to the status of a subordinated group within the new Strike Command.

Throughout its thirty year life Fighter Command formed the country's primary means of national defence, with its radar equipped, two-seat, night-fighters being responsible for protecting the *Dark Sky* for twenty of those years. During that period the Command fielded six types of night-fighter, in eighteen variants, with nine types of radar sets, and defined the principles and practice of night interception. Without its pioneering work, Europe, with the single exception of Germany, America and the former Soviet Bloc countries, would not have established night-fighting as an indispensable arm of their national defence – an achievement which the civilian scientists and engineers of AMRE/TRE and the night-fighter air and ground crews of Fighter Command can be justifiably proud.

Notes

1. Mark Hichens, *op cit*, pp. 293–5 and 301–3.
2. The US proposed to sell its holdings of Sterling.
3. Mark Hichens, *op cit*, pp. 297–9.
4. *Ibid*, p. 311.
5. Steven J. Zaloga, *The Kremlin's Nuclear Sword, The Rise and Fall of Russia's Strategic Forces* (Scraton: Smithsonian, 2002), pp. 10–12.
6. Gordon and Rigmant [1], *op cit*, pp. 33 and 34.
7. Colonel Oorzhuntsev was subsequently awarded the Order of Lenin by the Presidium of the USSR Supreme Soviet for his work on 18 October 1951.
8. Where 1 kiloton equates to the equivalent detonation of 1,000 tons of TNT.
9. Steven Zaloga, *op cit*, p. 12.
10. Gordon and Rigmant [1], *op cit*, pp. 34–6.
11. Steven Zaloga, *op cit*, p. 21.
12. *Ibid*, p. 32.
13. *Ibid*, p. 69.
14. These weapons were filled with a combination of jellified mustard gas and lewsite that had a persistance of seventy-two hours, *Ibid*, p. 69.
15. *Ibid*, pp. 69–72.
16. Yefim Gordon and Vladimir Rigmant [2], *Tupolev Tu-16 Badger* (Hinckley: Midland Publishing, 2004), p. 13.
17. Gordon & Rigmant [1], *op cit*, pp. 75 and 76.
18. Gordon & Rigmant [2], *op cit*, p. 3.
19. *Ibid*, pp. 14–17.
20. *Ibid*, pp. 18–22 and 103.
21. *Ibid*, pp. 20–2 and 25–7.
22. *Ibid*, p. 23, and William Green and Gordon Swanborough, *The Observer's Soviet Aircraft Directory* (London: Frederick Warne, 1975), pp. 212–14.
23. Yefim Gordon, Myasishchev M-4 and 3M, *The First Soviet Strategic Bomber* (Hinckley: Midland Publishing, 2003), pp. 4–6.
24. Plant No. 23 has long since been absorbed within the city's limits.
25. Yefim Gordon and Vladimar Rigmant [3], *Tupolev Tu-95/-142 Bear, Russia's Intercontinental-Range Heavy Bomber* (Hinckley: Midland Publishing, 1997), p. 11.
26. Steven Zaloga, *op cit*, p. 23.
27. *Ibid*, pp. 25–30.
28. *Ibid*, p. 31.
29. Steven Zaloga, *op cit*, pp. 24–9.
30. *Ibid*, p. 29.
31. *Ibid*, pp. 31 and 32.
32. *Ibid*, pp. 83–4
33. This would include such aspects as the adjustment and tuning of the AI and the maintenance of the aircraft's engines and systems, its armament, radio, navigation equipment and the aircrew's operational training and competence.
34. John Bushby, *op cit*, pp. 187 and 188.
35. Data taken from Green and Swanborough, Gordon and Rigmant [2], Gordon and Rigmant [3] and Yefim Gordon and Roger Lindsay [2], *op cit*, Edward Shacklady, *The Gloster Meteor* (London: Macdonald & Co Ltd, 1962, and A.J. Jackson, *De Havilland Aircraft* (London: Putnam, 1962).

36. The old Western Sector shown in Figure 30 was abandoned in 1953.
37. John Bushby, *op cit*, pp. 188 and 189.
38. Squadron Leader W.A. Waterton, AFC.
39. WD808 was lost on 11 June 1953 along with its pilot, Peter Lawrence.
40. For his bravery in saving the aircraft and its precious recording equipment, Bill Waterton was subsequently awarded the George Medal.
41. Tony Buttler [1], *op cit*, pp. 9–11.
42. Wing Commander R.F. Martin, DFC, AFC.
43. The restrictions were applicable to stalling and looping, pending the outcome of spinning trials to define the type's recovery technique.
44. Tony Buttler [1], *op cit*, p. 12.
45. This statement is not totally accurate, as one of the Javelins was flown by Dicky Martin, a retired RAF officer.
46. Although capable of reaching 52,000 feet, the FAW.1 rarely operated above 48,000 feet (14,630 metres) – Roger Lindsay [3], *Service History of the Gloster Javelin Marks 1 to 6* (Privately published in 1975), p. 32.
47. *Ibid*, p. 6.
48. *Ibid*, p. 7.
49. No. 46 Squadron's first Javelin FAW.2 was received in May 1957, but no more followed until July 1957 and even then, deliveries were protracted.
50. Quoted in Bryan Philpott, *English Electric/BAC Lightning* (Wellingborough: Patrick Stephens Ltd, 1984), pp. 36 and 37.
51. Based around the FAW.1, with Sa.6 engines, AI.17 radar and an all-flying tailplane.
52. FAW.1 with AI.17, the all-flying tailplane, modified wings to carry 125 gallons (570 litres) of additional fuel and provision for four de Havilland Firestreak AAMs.
53. FAW.2 with all-flying tailplane, modified wing, Sa.6 engines and AI.22 radar.
54. Bryan Philpott [2], *Lightning* (Wellingborough: Patrick Stephens Ltd, 1984), p. 36.
55. Although the Javelin FAW.7 was equipped to carry drop tanks, they were rarely, if ever, carried. The later Mk.9 variant was 'plumbed' for two 230 gallon (1,050 litre) tanks, which were deployed operationally.
56. Tony Buttler [1], *op cit*, p. 16.
57. Following their retirement from front-line service a number of Meteor NF.14s were issued to Nos 1 and 2 Air Navigation Schools in 1961 to provide fast-jet experience for trainee navigators. These were finally retired in 1965 with the introduction of the Hawker Siddeley Dominie T.Mk.1 at Stradishall.
58. David Baker, *The Rocket, The History and Development of Rocket & Missile Technology* (London: New Cavendish Books, 1978), p. 192.
59. The Lightning F.1-equipped No. 74 Squadron was formed at Coltishall in June 1960.
60. No. 60 Squadron at Tengah, Singapore, disbanded on 30 April 1968.

APPENDIX 1

A Rough Guide to Radar

Radar technology is a complex subject and it is not, therefore, the author's intention to provide anything other than a guide for those who do not have a background in radio theory or electronics. In this section the author proposes to describe the basic principles in as simple a way as is possible, such that the aviation enthusiast might understand the technology described in the main body of the text.

The Radio Spectrum

Radar systems share the radio spectrum with a number other systems: broadcasting, communications and navigation systems, as shown below. However, as far as we are concerned, in relation to wartime and post-war radar systems, the radar bands are concentrated in the very high frequency (VHF), the ultra high frequency (UHF) and super high frequency (SHF) bands (see Table 9).

The Basic Principle of Pulsed Radar

All radar sets be they air, sea or grounded-based, operate on the principle that if a pulse of electro-magnetic (radio) energy is transmitted through space, or the atmosphere, and strikes an object, preferably metallic, a very small part of that energy will be reflected back to a receiver at the source. The time taken for the pulse of energy to reach an object (target) and be reflected back to a receiver, can be measured accurately. Given the speed (velocity) at which electro-magnetic waves travel (propagate) through the atmosphere is 300,000,000 metres/sec (or 3×10^8 m/sec), it is possible by using the well known formula:

$$\text{Speed} = \frac{\text{Distance}}{\text{Time}} \quad \text{or} \quad \text{Velocity (m/sec)} = \frac{\text{Range (m)}}{\text{Time (secs)}}$$

TABLE 9. THE RADIO SPECTRUM AT 1960.[1]

Frequency band	Frequency range	Wavelength range	Usage
Very low frequencies (VLF)	10–30 KHz	30,000–10,000 m	Long distance comms
Low frequency (LF)	30–300 KHz	10,000–1,000 m	Marine navigation aids
Medium frequencies (MF)	300–3,000 KHz	1,000–10 m	Broadcasting and marine
VHF	30–300 MHz	10–1 m	Radar, TV, broadcasting and comms
UHF	300 MHz–3 GHz	1 m–10 cm	Radar, TV and microwave comms
SHF	3–30 GHz	10–1 cm	Radar, radio relay and navigation
Extremely high frequencies (EHF)	30–300 GHz	1–0.1 cm	Experimental

the range may be calaculated:

$$\text{Range} = \text{Velocity} \times \text{Time}$$

However, the distance the pulse travels to the target is twice the real distance (outward journey plus the return journey), therefore, the formula has to be modified to reflect the true range, as follows:

$$\text{Range} = \text{Velocity} \times \tfrac{1}{2}\,\text{Time}$$

It should be noted that whilst the transmitter pulse is going out, the receiver is switched off. This is done to avoid the transmitter pulse being fed back directly to the sensitive circuitry in the receiver and appearing as a very close target.

Frequency and Wavelength

Frequency and wavelength are directly proportional and governed by the formula:

$$\text{velocity (v)} = \text{frequency (f) in Hz} \times \text{its wavelength (l) in metres, where } v = 3 \times 10^8$$

In reality this says, 'the higher the frequency, the shorter the wavelength'. For example, radar sets operating on 1½ metre wavelengths operate on a frequency of 200 MHz, whilst those on 10 cm have a frequency of 3,000 MHz, or 3 GHz. During the Second World War, as now, frequencies were grouped into frequency bands for convenience and to aid identification. The relevant radar bands used throughout the Second World War to the 1960s and their approximate frequencies are shown below:

Wartime bands	Frequency	Wavelength
P	220–390 MHz	1.36–0.77 m
L	390–1,550 MHz	77–19 cm
S	1.5–5.2 GHz	19–5.8 cm
X	5.2–10.9 GHz	5.8–2.7 cm
K	10.9–36 GHz	2.7–0.83 cm

The Relationship Between Pulse Repetition Frequency and Pulse Width

The radar sets described in this book use pulsed transmissions and are, therefore, defined as 'pulsed radars'. The rate at which the number of pulses were transmitted each second was defined during wartime as the 'pulse repetition rate' (PRR) and was measured in pulses per second (pps). However, in modern terms the repetition rate is defined as the 'pulse repetition frequency' (PRF) and is measured in Hz. A radar set having a PRR of say 750 pps, equates to a PRF of 750 Hz. In reality, the two terms mean the same thing. In deference to current terminology, PRF will be used throughout this book.

The value of the PRF depends on the use to which the radar is to be put, but the rate must be such that there is sufficient time for the pulse to travel to the target and the echo be returned before the next pulse is sent out. Failure to observe this principle will result in false range indications being displayed. In general terms, 'the longer the range of the radar, the lower the PRF', or conversely, 'the shorter the range the higher the PRF'. Since we are

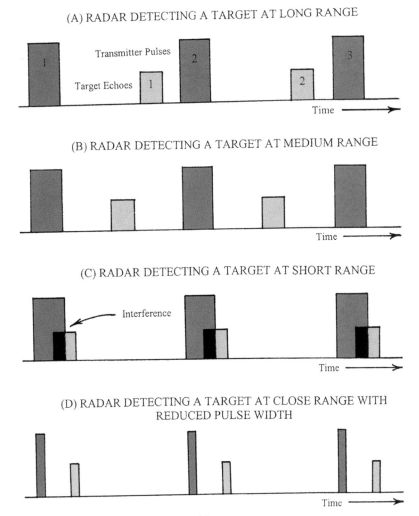

(A) RADAR DETECTING A TARGET AT LONG RANGE

Transmitter Pulses

Target Echoes

Time →

(B) RADAR DETECTING A TARGET AT MEDIUM RANGE

Time →

(C) RADAR DETECTING A TARGET AT SHORT RANGE

Interference

Time →

(D) RADAR DETECTING A TARGET AT CLOSE RANGE WITH
REDUCED PULSE WIDTH

Time →

Figure 43.   Interference problem with long pulse widths.

dealing with short-range fighter radars, the latter will apply. For 1½ metre radars, a PRF of 750 Hz was quite common.

The duration of a pulse is termed the 'pulse width' and is measured in seconds, or more usually, µseconds. 2.8 µsecs was a typical value for 1½ metre AI radars. There is a realtionship between PRF and pulse width that must be maintained if ranging ambiguities are to be avoided. If the pulse width is long, it will work satisfactorily at long ranges. However, as the range closes the echo will begin to be returned as the next transmitter pulse is going out (see Figure 43), when the receiver is switched-off!

If the range was permitted to close further, the target will disappear off the radar screen(s). To overcome the problem the pulse width has to be substantially reduced (as shown above) in order that the transmitter pulse does not mask the incoming echo. It was the difficulties with circuit design to produce such short pulse widths, that plagued the early radar engineers and caused so much trouble in producing radars with a good minimum range.

Factors Effecting the Operating Range

The strength of the echo pulse, and hence the range of a radar set, is dependent upon a number of factors: the transmitter's output power, the efficiency of the aerial system, the size and construction of the target (large metal aircraft reflect more energy than small wooden ones) and the distance of the target from the carrier aircraft. These, therefore, need to be understood in greater detail.

As with all electrical apparatus, the output power of the transmitter and the cabling that carries it to the aerial system, is not finite. The output is limited by the amount of power the aircraft's power supplies can generate (in the case of the Blenheim, Beaufighter and Mosquito, not a great deal by comparison with today's radars), the transmitter's frequency (of which more later) and the power level the aerial system will handle without breakdown (this was very poor until the introduction of polyethylene insulation to the manufacture of co-axial cable).

As mentioned earlier, the size and construction of the target aircraft will greatly influence the strength of the return echo. Large metal aircraft with angular surfaces (a factor applicable to most medium and heavy-bomber and transport aircraft of the Second World War period) are good reflectors of radar energy, whereas those built from wood (the Mosquito and many training aircraft) are nowhere near as good. The target's size also directly effects the amount of energy returned to the receiver. This relationship may be described by a simple formula:

Target Area (A_T) in sq. metres is proportional (∞) to Distance to Target (d)^2 in metres

or

$$A_T \infty\, d^2 \text{ metres}$$

Therefore, if the range to a target was say, 1 km, the target area illuminated would be 1 sq. metre. If the range was increased to 3 km, the target area illuminated would rise to 9 sq. metres, but the amount of transmitted energy striking the target would diminish to 1/9th of that at 1 km. The same is also true of the reflected energy, such that the echo

Figure 44. Target area (A_T).

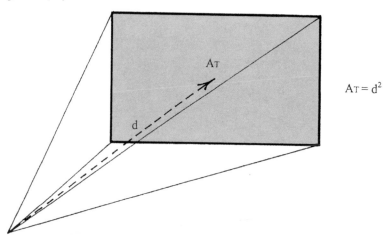

A_T

$A_T = d^2$

d

returned to the receiver on a 3 km target would diminish to $1/9\text{th} \times 1/9\text{th} = 1/81\text{st}$ of that transmitted and so on. Increasing the range to 4 km reduces the signal even further; i.e. $1/16\text{th} \times 1/16\text{th} = 1/256\text{th}$ of that transmitted and so on. Therefore 'the distance to a target is proportional to the signal strength returned'.

The reason why metal surfaces reflect greater amounts of energy than wooden surfaces is due to metal being the better conductor of electricity. The greater the conductivity of the material, the greater will be the electrical field created within it and the greater will be the energy returned in the echo.

The final aspect relating to the operating range is the selection of the transmitter frequency. Radars operating on metric wavelengths employ relatively large dipole aerial systems that produce a main balloon-shaped lobe (see Figure 45).

It will be seen that this comprises two parts, a main lobe that contains the majority of the radar energy and a number of weaker side lobes that leak energy due to the aerial being less than 100 per cent efficient. When energy within the main lobe strikes the ground directly below the carrier aircraft, a strong echo is returned to the receiver and displayed on the CRTs of the indicating unit. These echoes, termed the 'ground return', are very much stronger than the echoes from the target aircraft. Since 'the range of the ground returns is equal to the height of the carrier aircraft above the ground', any target flying at a height greater than that of the carrier aircraft will not be detected.

Consider a night-fighter flying at 20,000 feet (6,100 metres) – a not unreasonable height for the likes of a Beaufighter or Mosquito - whence, the maximum range of its radar would be 3.8 miles (6 km). Any aircraft flying below that altitude would be detected. Conversely, if it was flying at 10,000 feet (3,050 metres), a bomber flying at 17,000 feet (5,180 metres) would remain undetected.

Figure 45. Polar diagram of height versus range for 1½ metre radar.

MAIN LOBE

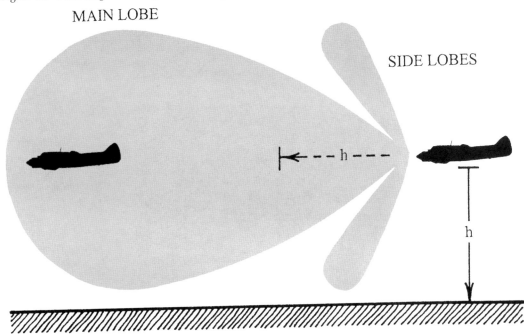

SIDE LOBES

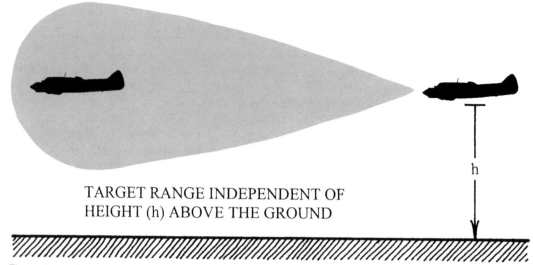

Figure 46.   Polar diagram of height versus range for centimetric radar.

One of the principal failings of the metric radars was their inability to track targets flying at low level, say 5,000 feet (1,525 metres), or lower, where the maximum range diminished to less than a mile (1.6 km). With the introduction of centimetric (S-Band) radar, which operated on a wavelength of 10 cm, and the use of parabolic dish aerials, the beams were considerably sharpened (see Figure 46).

Centimetric radars generally had longer ranges than their metric predecessors and were more effective at low altitudes. Some also had an ASV capability – AI Mk.VIII and Mk.21 being particular examples.

1½ Metre Aerial Systems

All 1½ metre radar sets employed dipole aerial arrays, based around the 75 ohm (Ω) impedance, half-wave dipole (see Figure 47).

Figure 47.   Half-wave dipole aerial.

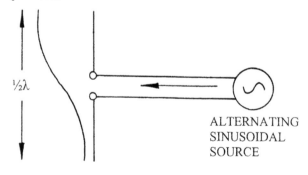

WHERE λ = WAVELENGTH IN METRES

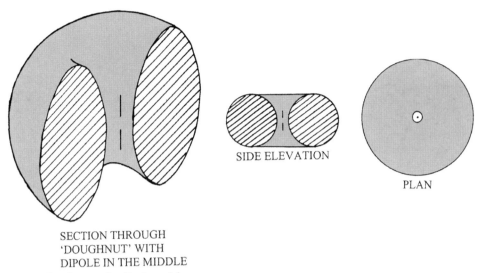

SIDE ELEVATION

PLAN

SECTION THROUGH
'DOUGHNUT' WITH
DIPOLE IN THE MIDDLE

Figure 48.   Polar diagram for dipole aerial.

When an electrical sinusoidal signal is sent down an open (balanced) transmission line, it will produce a standing wave on a dipole that has half the wavelength and twice the amplitude of the original. It also exhibits an impedance[2] (Z) of $75\,\Omega$ at its centre. This 'characteristic' impedance is the same whether the dipole is employed as a transmitting or receiving aerial. The shape of the aerial's radiation pattern (usually referred to as its 'polar diagram') is similar to that of a doughnut, with the dipole at its centre (see Figure 48)

Ordinary open wire transmission feeders of the type employed during the early-years of radar development, whilst providing the best match for dipole aerial, and hence the maximum power transfer from transmitter to aerial, were prone to interference and proved difficult to maintain in the field. Shielded (unbalanced) feeders of the co-axial type, however, provide good shielding (screening) characteristics and may be adapted to dipole aerials (see Figure 49)

It will be seen that the output of the transmitter, or input to the receiver, is connected to the central conductor of the co-axial cable and the outer screen is connected to earth, or the aircraft's structure in the case of airborne radar. By this means signals to or from the transmitter/receiver are delivered to the dipole and screened to the airframe at one and the same time. Polyethylene co-axial feeders were used on all $1\frac{1}{2}$ metre AI sets.

Figure 49.   Co-axial connections to transmitter/receiver.

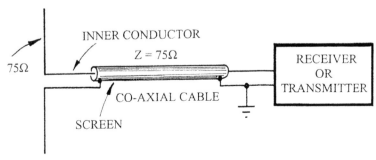

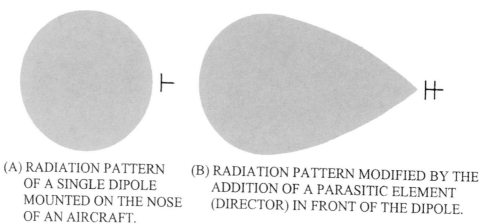

(A) RADIATION PATTERN
OF A SINGLE DIPOLE
MOUNTED ON THE NOSE
OF AN AIRCRAFT.

(B) RADIATION PATTERN MODIFIED BY THE
ADDITION OF A PARASITIC ELEMENT
(DIRECTOR) IN FRONT OF THE DIPOLE.

Figure 50.   Beam sharpening with parasitic element.

With the dipole transmitting aerial mounted on the nose of the aircraft in the vertical plane (usually the case with 1 ½ metre AI, but there were exceptions, see AI Mk.I–Mk.III), the rear portion of the doughnut was lost in the aircraft's fuselage and structure. Therefore, in reality a single dipole produces a circular polar diagram when viewed from the side of an aircraft (see Figure 50). This type of 'floodlighting' radiation has poor directional qualities when its comes to defining a target's position to the left or right of the carrier aircraft, or above or below it. Some means, therefore, had to be found to improve the directionality, whilst retaining the maximum power transfer from transmitter to aerial. This was accomplished using 'parasitic' elements.

In the 1930s a Japanese radio specialist, Professor Yagi, discovered that if a parasitic element was placed in front of, or behind the fed dipole, the aerial's radiation pattern would be sharpened and its directivity improved. The more parasitic elements that were employed, the sharper became the radiation pattern (see Figure 50).

Figure 51 shows a transmitting array mounted on a common supporting arm, with the driven dipole spaced quarter-wavelength from the parasitic element. It is worth noting at this point that parasitic elements placed in front of the fed dipole are termed 'directors' and

Figure 51.   Yagi transmitting aerial with director.

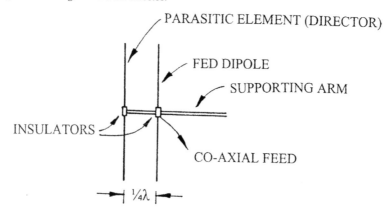

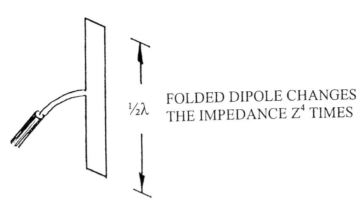

Figure 52. Folded dipole.

are slightly shorter than the dipole and those placed behind the fed dipole are termed 'reflectors' and are slightly longer than the dipole. Generally, directors are used on transmitter arrays and reflectors on receiving arrays.

Unfortunately, the impedance of the aerial changes from its normal $75\,\Omega$ when parasitic elements are employed. In order to restore the characteristic impedance to the original $75\,\Omega$, a folded dipole is employed. The dipole shown in Figure 52 has an impedance that is the square of the ordinary dipole. Thus, conveniently, the reduction in the impedance of the ordinary dipole with its parasitic element is replaced exactly by that of the folded dipole. For this reason, all transmitting aerials on $1\,\tfrac{1}{2}$ metre radars employed folded dipoles with single directors (see Figure 52 and also Figure 5 in Chapter 4).

NOTE : From the aforementioned, it will be seen that the length of dipoles for all $1\,\tfrac{1}{2}$ metre AI aerials was one half of the operating wavelength, i.e. approximately 0.75 metres, or 30 inches.

Centimetric Aerials

The very high frequencies employed in centimetric radar aerial systems precluded the use of dielectric-filled co-axial cables to feed the radiating elements, since such cables become increasingly 'lossy' at SHF and EHF frequencies and are not capable of carrying the high power generated by the magnetron-based transmitter devices. Transmission lines for use at these frequencies were therefore designed and developed using hollow, thin-walled, rectangular tubes fabricated from high-conductivity metals such as brass, copper or aluminium, which overcame the losses. This type of low-loss transmission line is described as a 'waveguide'. The most common rectangular waveguides are designed with a width-to-height ratio of 2:1. Thin-wall circular metal tubes are also used as waveguides, manufactured from similar high-conductivity metals. The interior walls of these tubes are occasionally plated with silver or gold to further reduce transmission losses and to protect against corrosion. The dimensions of both rectangular and circular waveguide designs are specified according to the range of wavelengths each can support, whilst still possessing the desired low-loss transmission characteristics. The propagation of microwaves in waveguides is a complex subject involving electromagnetic field theory, but, as a rule of thumb, 'the normal operating range of wavelengths for a rectangular or circular waveguide lies between 110 and 150 per cent of the broad dimension of a rectangular waveguide and

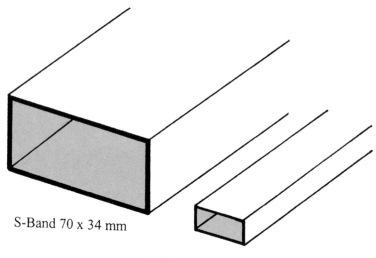

S-Band 70 x 34 mm

X-Band 25 x 12½ mm

Figure 53.   Waveguides for S-band and X-band systems.

around 150 per cent of the diameter of a circular waveguide'. From this, wavelengths in the 10 cm region (S-Band) require a rectangular waveguide measuring, approximately 70 × 34 mm (2.85 × 1.34 inches), whilst a 3 cm (X-Band) system will require waveguides of 25 × 12½ mm (1 × ½ inch) (see Figure 53).[3]

As Yagi aerial systems will not easily support microwave propagation and radar engineering became more of a 'plumbing job', parabolic reflector, or dishes, became the standard form of centimetric radar aerial (antenna).[4] As with waveguides there is a relationship between the curvature of a dish antenna and the frequencies it is carrying, which in many ways are analogous to the reflection of light off curved surfaces. The advantage of this type of antenna is its ability to provide very high signal gains and to operate over a broad range of frequencies. It also has the property of projecting or collecting radio waves from a 'focus' (see Figure 54) and converting them from spherical waves into plane (flat) waves.

There are two ways of collecting/transmitting radio energy from a parabolic dish antenna: by direct means and indirect means. Wartime radars employed the former with a con-

Figure 54.   The principle of the parabolic reflector and focus.

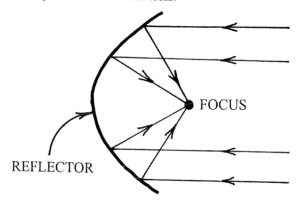

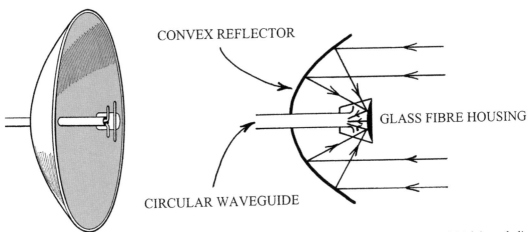

CONVEX REFLECTOR

GLASS FIBRE HOUSING

CIRCULAR WAVEGUIDE

Figure 55. Parabolic dish antenna focus methods; (left) parabolic dish antenna with dipole and (right) parabolic dish antenna with sub-reflector.

ventional dipole at the focus to collect or radiate the radio energy (see Figure 55) as used in AI Mk.VII and Mk.VIII, with later radars employing the use of a sub-reflector.

Two scanning techniques were employed on centimetric airborne radar systems during and after the war, spiral and helical. Spiral scanning (see Figure 56) was employed on the early British AI systems and was achieved by spinning a simple dipole at the focus of a

Figure 56. Spiral scanning.

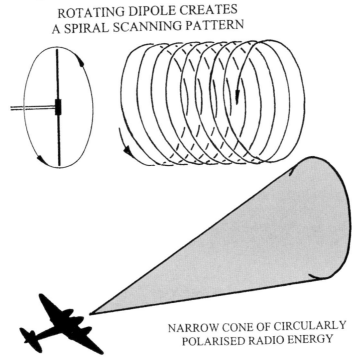

ROTATING DIPOLE CREATES
A SPIRAL SCANNING PATTERN

NARROW CONE OF CIRCULARLY
POLARISED RADIO ENERGY

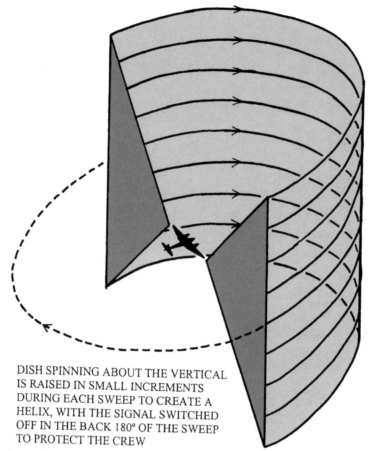

DISH SPINNING ABOUT THE VERTICAL
IS RAISED IN SMALL INCREMENTS
DURING EACH SWEEP TO CREATE A
HELIX, WITH THE SIGNAL SWITCHED
OFF IN THE BACK 180° OF THE SWEEP
TO PROTECT THE CREW

Figure 57.   Helical scanning.

parabolic dish. This produced a rotating spiralling motion that formed a narrow cone of radio energy and a circular pattern on the observer's display (see also Chapter 9).

Helical scanning was introduced by the Americans on their SCR-720 equipment and the British AI Mk.IX (see also Chapter 12). This technique required a parabolic dish to be spun around the vertical plane in a circular (360°) fashion, with the dish being tilted up slightly after each revolution to describe a helix in the sky. When it was at the top of its scan, the dish was automatically dropped to the lowest point and the process began all over again. In order not to expose the crew to high doses of RF radiation, the transmitter was switched-off when the dish was facing the crew, i.e. for the rearward 180° as shown in Figure 57.

Notes

1. Taken from Henry Jacobowitz, *Electronics Made Simple* (London: W.H. Allen, 1965), p. 201.
2. Impedance in ac circuits may be regarded as resistance that varies with frequency.
3. Ned Cartwright in an e-mail to the author dated 21 April 2006.
4. The American term 'antenna' superseded the British term 'aerial' during the war and it is commonplace today to refer to aerials as antenna.

Principal Characteristics of AI Radars, 1939–1959

Metric Radars

AI	Mk.I	Mk.II	Mk.III	Mk.IV	Mk.V	Mk.VI
Wavelength	1½ metres	1½ metres	1½ metres	1½ metres	1½ metres	1½ metres
Frequency	Approx 193 MHz	approx 193 MHz	193 MHz	193 MHz	193 MHz	193 MHz
PRF	Not known	Not known	750 Hz	750 Hz	670 Hz	670 Hz
Pulse width	Not known	Not known	Not known	2.8 µsecs	2.8 µsecs	2.0 µsecs
Peak pulse power	Not known	Not known	10 kW	10 kW	10 kW	10 kW
Aerial system	Yagi	Yagi	Yagi	Yagi	Yagi	Yagi
Aerial polarisation	Horizontal	Horizontal	Horizontal	Vertical	Vertical	Vertical
Maximum range	2–3 miles	2–3 miles	2–3 miles	3½ miles	3½ miles	2½–3 miles
Minimum range	Possibly 900 ft	Very poor	Not known	400 ft	400–500 ft	500 ft
D/F sharpness	Not known	Not known	Not known	5° @ dead ahead	5° @ dead ahead	5° @ dead ahead
Observers displays	Range/azimuth and range/elevation	Range/azimuth and range/elevation	Range/azimuth and range/elevation	Range/azimuth and range/elevation	Range and direction	Range and direction
Pilot's indicator	Not fitted	Not fitted	Not fitted	Not fitted	Yes	Yes
IFF	Not fitted	Not fitted	Not fitted	Yes	Yes	Not fitted
Beacon	Not fitted	Not fitted	Not fitted	Yes	Yes	Not fitted
Beam approach	Not fitted	Not fitted	Not fitted	Yes	Yes	Not fitted
Power supply (watts and frequency)	500W @ 80v and 1600 Hz	500W @ 80v and 1600 Hz	500W @ 80v and 1600 Hz	500W @ 80v and 1600 Hz	500W @ 80v and 1600 Hz	500W @ 80v and 1600 Hz
Weight	Not known	Not known	Not known	118 lb	135 lb	134 lb
Aircraft	Blenheim Mk.If and Mk.IVf	Blenheim Mk.If	Blenheim Mk.If and Mk.IVf	Beaufighter Mk.If and Mk.VIf; Mosquito NF.II	Mosquito NF.II	Hurricane IIc

Centimetric Radars

AI	Mk.VII	Mk.VIII	Mk.IXB	Mk.X/SCR-720	Mk.21
Wavelength	9.1 cm	9.1 cm	9.1 cm	9.1 cm	3.2 cm (X-band)
Frequency	3.3 GHz (S-band)	3.3 GHz (S-band)	3.3 GHz (S-band)	3.3 GHz (S-band)	9.4 GHz
PRF (AI)	2,500 Hz	2,500 Hz	667 Hz	1,500 Hz	2,450 & 550 Hz
PRF (Beacon)	Not available	930 Hz	Not known	375 Hz	550 and 300 Hz
Pulse width (AI)	1 µsec	1 µsec	1 µsec	¾ µsec	0.4 & 1.75 µsec
Pulse width (Beacon)	Not available	3 µsecs	Not known	2¼ µsec	1.75 and 2.25 µsec
Peak pulse power	5 kW	approx 25 kW	200 kW	70 kW	200 kW
Aerial system	Parabolic dish	Parabolic dish	Parabolic dish	Parabolic dish	Parabolic dish
Scanning method	Spiral	Spiral	Helical	Helical	Helical
Aerial polarisation	Vertical	Vertical	Vertical	Vertical	Vertical
Maximum range	3 miles	5½ miles	10 miles[1]	6 miles	20 miles
Minimum range	400–500 ft	400–500 ft	800 ft[2]	300 ft	300 ft
D/F sharpness	1.3° @ dead ahead	1.3° @ dead ahead	Not known	5°	3°
Observers displays	Spiral spot tube	Spiral spot tube	Spot tube and range/azimuth	C-scope & B-scope[3]	PPI and C-scope
Pilot's indicator	Not fitted	Not fitted	Yes	via GGS	via GGS
Auto lock-follow	Not fitted	Not fitted	Yes	Not fitted	Not fitted
IFF	Not fitted	Yes	via Lucero	via SCR-729	Yes
Beacon	Not fitted	Yes	via Lucero	Yes	Yes
Beam approach	Not fitted	Yes	via Lucero	via SCR-729	Not known
Power supply (watts and frequency)	1,200W @ 80v and 1600 Hz	1,200W @ 80v and 1600 Hz	1,400W @ 80v[4]	1,200W @ 80v and 1600 Hz	Not known
Weight	Not known	212 lb[5]	Not known	500 lb	Not known
Aircraft	Beaufighter Mk.If and Mk.VIf	Beaufighter Mk.VIf; Mosquito NF.XII and XIX	Experimental installation in Mosquito NF.XII and XVII	Mosquito NF.XVII, XIX, 30 and 36; Meteor NF.11 and 13; Vampire NF.10; Venom NF.2	Meteor NF.12 and 14; Venom NF.3

Notes

1. Recorded on CFE trials in April 1945.
2. Recorded on CFE trials in April 1945.
3. B-scope provides range/azimuth information and C-scope azimuth/elevation.
4. The generator frequency is not stated, but is most probably 1,600 Hz.
5. Six main units only.

British Night-Fighters 1939–1959

Bristol Blenheim Mk.If

Blenheim If 'YX-N' of No. 54 OTU, Church Fenton, September 1941.

Originally designed to meet a Fighter Command requirement for a long-range escort fighter, the Bristol Aeroplane Company modified Blenheim Mk.I, L1424, in 1938 to act as the fighter prototype. The 'fighter' was a simple adaptation of the standard Mk.I light-bomber by the installation of a basic gun-pack containing four 0.303-inch (7.69 mm) Browning machine-guns and 2,000 rounds of ammunition (500 rounds per gun). This was attached directly to the bomb beams in the bomb-bay and weighed less than the maximum bomb load, whilst its relatively low profile produced little drag and had no appreciable effect on the aircraft's performance. A total of 1,375 of these gun-packs were manufactured in the workshops of the Southern Railway Company at Ashford for use by Fighter and Coastal Commands. The aircraft retained its fuselage mounted, forward firing Browning for use by the pilot and the dorsal turret mounting a single Vickers 'K' machine-gun. When AI Mk.III radar was fitted from the summer of 1940, the turret was generally removed to reduce weight and improve performance.

The Blenheim fighter entered service with Fighter Command during December 1938 and remained as Britain's principal night-fighter until late in 1940. From February 1940 onwards the dedicated night-fighter squadrons were provided with small quantities of AI Mk.II radars, before standardising on AI Mk.III the following summer. The Blenheim If was withdrawn from night-fighter operations in 1941 and allocated in substantial numbers to the night OTUs, where it remained until 1943.

TECHNICAL DATA[1]

Description: Twin-engined, long-range, day and night-fighter with a crew of two/three, of all-metal, stressed skin, construction.

Manufacturers:	Bristol Aeroplane Co Ltd, Filton, Bristol.
Power plant:	Two 840-hp, 9-cylinder, supercharged, Bristol Mercury VIII radial engines.
Dimensions:	Span, 56 ft 4 ins (17.17 metres). Length, 39 ft 9 ins (12.11 metres). Height, 9 ft 10 ins (9.83 metres). Wing area, 469 sq. ft (43.57 sq. metres).
Weights:	Empty, 8,100 lb (3,675 kg). Loaded, 12,500 lb (5,670 kg).
Performance:[2]	Maximum speed, 279 mph (446 km/hr) at 15,000 ft (4,570 metres).
(fully laden)	Cruising speed, 200 mph (320 km/hr). Initial rate-of-climb, 1,785 ft/min (545 m/min). Time to 15,000 ft (4,570 metres), 8.8 mins. Service ceiling, 30,000 ft (9,145 metres). Endurance, 5.65 hours. Range, 1,000 miles (1,600 km).
Armament:	One fixed forward firing 0.303 ins (7.69 mm) Browning machine-gun, one 0.303-inch Vickers Type-'K' machine-gun in dorsal turret and four 0.303-inch Browning machine-guns, with 500 rounds per gun (rpg), in ventral pack.
Radar:	AI Mk.III.

UNIT ALLOCATION (see also Appendix 6)

Night-fighter:	Nos 23, 25, 29, 64, 68, 600, 601 and 604 Squadrons.
Night-fighter training:	Nos 51, 54 and 60 OTUs.

Bristol Blenheim Mk.IVf

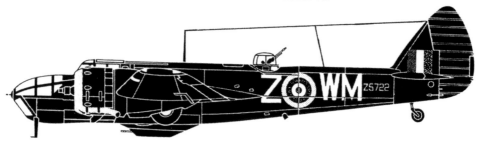

Blemheim IVf Z5722/WM-Z of No. 68 Squadron, Catterick, May 1941.

Relatively few Blenheim IVfs were issued to operational night-fighter squadrons. Based on the Mk.IV bomber airframe, the Mk.IVf employed gun-packs similar to those used on the Blenheim If – the IVf's packs were deeper and cut-back to enable the rounds better to clear the underside of the fuselage. The Mk.IV featured an extended nose to accommodate the navigator/bomb aimer's position and more powerful Mercury XV engines of 995-hp, which provided a useful increase in range. The first IVfs were allocated accidentally to 'C' Flight, No. 25 Squadron at Northolt in the late summer of 1939, when Mk.Ifs had been specified by AMRE's Airborne Group. A small batch of IVfs were allocated to No. 68 Squadron as interim equipment when it reformed at Catterick in January 1941, but the majority of the deliveries went to Coastal Command. Mk.IVfs and straight Blenheim IVs served with the OTUs for pilot and R/O training. Its performance was broadly comparable to that of the Mk.If.

Bristol Beaufighter Mk.If

Beaufighter If R2101/NG-R of No. 604 Squadron, Middle Wallop, Winter 1940/41.

The Beaufighter Mk.I was the basic design for Coastal (c) and Fighter Command (f) variants of Bristol's design for a twin-engined, long range, day and night fighter. Based around a pair of 1,400-hp Bristol Hercules III radial engines, the Beaufighter was capable of a maximum speed of 323 mph (517 km/hr) at 15,000 ft (4,572 metres) and had a service ceiling of 30,000 ft (9,145 metres). The maximum loaded weight was 21,120 lb (9,580 kg). Armament comprised four 20 mm Hispano cannon and six 0.303-inch (7.69 mm) Browning machine-guns. The cannon on the first 400 aircraft were fitted with sixty-round drums that were changed by the R/O. The 401st machine introduced the Chatellerault, recoil operated Mk.I feed in September 1941, which fed the rounds from larger magazines and required no assistance from the R/O. The Beaufighter prototype, R2052, flew for the first time on 3 July 1939 and entered service with Fighter Command in September 1940. With its crew of two and AI Mk.IV radar, the Beaufighter If represented a significant improvement in performance and capability over the Blenheim. The introduction of the centimetric AI Mks.VII and VIII radar in April 1942, brought about the conversion of a significant number of Beaufighter's to accommodate a Bristol designed fibre-glass radome for the system's mechanically scanning parabolic 'dish' aerial.

The If was built at Bristol's Filton and Weston-super-Mare factories and at Fairey's factory at Stockport, Cheshire and remained in front-line service with Fighter Command until June 1943. It was used extensively by the night-fighter OTUs until the war's end.

TECHNICAL DATA[3]

Description:	Twin-engined, night-fighter with a crew of two, of all-metal, stressed skin, construction.
Manufacturers:	Bristol Aeroplane Co. Ltd, Filton, Bristol, and Weston-super-Mare. Fairey Aviation Co. Ltd, Stockport, Cheshire.
Power plant:	Two 14-cylinder, super-charged, 1,400-hp Bristol Hercules III, X, or XI radial engines.
Dimensions:	Span, 57 ft 10 ins (17.62 metres). Length, 41 ft 4 ins (13.0 metres). Height, 15 ft 10 ins (4.83 metres) Wing area, 503 sq. ft (46.73 sq. metres).
Weights:	Empty, 13,800 lb (6,260 kg). Loaded, 21,120 lb (9,580 kg).
Performance: (*fully laden*)	Maximum speed, 323 mph (517 km/hr) at 15,000 ft (4,572 metres), or 330 mph (530 km/hr) at 16,000 ft (4,875 metres) with Hercules XI. Maximum cruising speed, 272 mph (435 km/hr). Initial rate-of-climb, 1,850 ft/min (564 m/min). Climb to 20,000 ft (6,095 metres), 14 mins. Service ceiling, 30,000 ft (9,145 metres). Endurance, 5.65 hours. Range, approximately 1,100 miles (1,760 km).

Armament: Four 20-mm Hispano cannons in forward lower fuselage and six
                     0.303-inch (7.69 mm) Browning machine-guns with 1,000 rpg.
Radar: AI Mk.IV, Mk.VII and Mk.VIII.

UNIT ALLOCATION (see also Appendix 6)

UK night-fighter: Nos 25, 29, 89, 141, 153, 219, 255, 256, 307, 600 and 604 Squadrons.
Night-fighter training: Nos 51 and 60 OTUs.

Bristol Beaufighter Mk.IIf

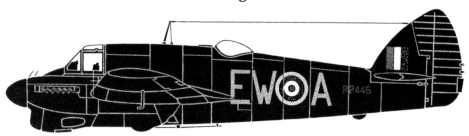

Beaufighter IIf R2445/EW-A of No. 307 (Polish) Squadron, late 1941.

The marriage of the single-stage R-R Merlin 20-series engine to the Mk.I airframe, produced the basis for the Beaufighter Mk.II. Designed as an alternative to the Hercules-powered Mk.I, the Mk.II was built in smaller numbers. The initial engine, the Merlin XX of 1,250-hp, whose nacelle, taken from the Lancaster bomber, was adapted by means of an intermediate 'pick-up' bay to fit that of the Beaufighter. The first Mk.II, R2058, powered by Merlin Xs, up-rated to the Mk.XX standard, flew from R-R's airfield at Hucknall on 14 June 1940, followed shortly after by the second, R2061. A third prototype, R2062, was destroyed in an air raid on the Filton works.

The longer engine nacelles of the Mk.II increased the take-off swing and, therefore, required even more careful handling, whilst the less than fully-feathering Rotol airscrews and Schwartz wooden propellers, added further to the pilot's problems. Both marks of Beaufighter suffered from a low-frequency longitudinal instability in the climb that was more pronounced on the Mk.II, which also suffered from a slight tail heaviness. To overcome these problems, Mk.II, R2257, was fitted with a 12° dihedral tailplane which improved the situation, but rendered the aircraft too stable for night-fighting. Production Mk.IIs and later Mk.VIs were built with both types of tailplane to suit fighter and coastal operations.

By comparison, the Mk.II was slightly lighter than the Mk.I (see below) and consequently had a higher ceiling, but a lower rate-of-climb. Its maximum speed was also marginally faster, at 337 mph (540 km/hr) at 22,000 feet. (6,705 metres). Its loaded weight was 21,077 lb (9,560 kg). In all other respects the two marks were broadly similar in performance, with the Mk.II requiring the more careful handling on the ground and in the climb. Like the Mk.If, the IIf was fitted with AI Mk.IV radar and an identical armament.

The Mk.IIf was introduced to Fighter Command during April 1941 and remained in front line service as a night-fighter until May 1943, when it was replaced by the Beaufighter Mk.VIf and the night-fighter variants of the Mosquito. Small numbers were introduced to the night-fighter training organisation from February 1942.

TECHNICAL DATA[4]

Description: Twin-engined, night-fighter with a crew of two, of all-metal, stressed skin, construction.

Manufacturers: Bristol Aeroplane Co. Ltd, Filton, Bristol.

Power plant: Two 12-cylinder, super-charged 1,250-hp R-R Merlin XX in-line engines.

Dimensions: Span, 57 ft 10 ins (17.62 metres). Length, 42 ft 9 ins (13.03 metres). Height, 15 ft 10 ins (4.83 metres). Wing area, 503 sq. ft (46.73 sq. metres).

Weights: Empty, 13,800 lb (6,260 kg). Loaded, 21,077 lb (9,560 kg).

Performance: Maximum speed, 337 mph (540 km/hr) at 22,000 ft (6,705 metres). Maximum cruising speed, 272 mph (435 km/hr). Time to 15,000 ft (4,570 metres), 8.8 mins. Service ceiling, 32,600 ft (9,935 metres). Range, approximately 1,000 miles (1,600 km).

Armament: Four 20 mm Hispano cannons in forward lower fuselage and six 0.303 inch (7.69 mm) Browning machine-guns with 1,000 rpg.

Radar: AI Mk.IV.

UNIT ALLOCATION (see also Appendix 6)

UK night-fighter: Nos 96, 125, 255, 307, 406, 409, 410, 456, 488, 515 and 600 Squadrons.

Night-fighter training: No. 54 OTU.

Bristol Beaufighter Mk.VIf

Beaufighter VIf RO-L of No. 219 Squadron, early 1943.

The original design proposal for the Beaufighter always envisaged the use of the two-stage supercharged Hercules VI of 1,670-hp. MAP had hoped to introduce the Hercules VI to the Filton lines on the completion of Beaufighter II production, for which two Weston built machines, X7542 and X7543, were converted to Mk.VI power to assess the new type's endurance. The resultant Beaufighter Mk.VI, differed from the Mk.I only in having Hercules VIs in place of the Mk.Is Hercules III or XI. These and other tests showed the Hercules VI-powered MK.IVf's performance to be slightly better than that of the Mk.If, returning a maximum speed of 333 mph (535 km/hr) at 15,600 feet (4,755 metres). Engine ratings of 1,670-hp in 'M' gear at 7,500 ft (2,285 metres) and 1,500-hp in 'S' gear at 17,000 ft (5,180 metres) were also recorded. The maximum weight was raised to 21,600 lbs (9,800 kg).

Incorporating all the modifications implemented in the course of the Mk.I's development, including the 12° tailplane, but not on all machines, the VIf was originally fitted with the equipment and aerial system for AI Mk.IV. However, with the introduction of the centimetric AI Mks.VII and VIII radar, the Beaufighter's nose was adapted to accommodate a fibre-glass radome for the mechanically scanning parabolic 'dish' aerial. The Bristol developed radome-nose was previously used on the Mk.If, when a number of squadrons converted to the new radar from in the spring of 1942.

The Beaufighter VIf was introduced to Fighter Command in May 1942 and remained in UK front line service until August 1944 and overseas (Italy) to February 1945.

TECHNICAL DATA[5]

Description:	Twin-engined, night-fighter with a crew of two, of all-metal, stressed skin, construction.
Manufacturers:	Bristol Aeroplane Co. Ltd, Filton, Bristol, and Weston-super-Mare.
Power plant:	Two 14-cylinder, super-charged, 1,600-hp Bristol Hercules VI or XI radial engines.
Dimensions:	Span, 57 ft 10 ins (17.62 metres). Length, 41 ft 4 ins (13.0 metres). Height, 15 ft 10 ins (4.83 metres). Wing area, 503 sq. ft (46.73 sq. metres).
Weights:	Empty, 14,900 lb (6,350 kg). Loaded, 21,322 lb (9,670 kg).
Performance:	Maximum speed, 333 mph (530 km/hr) at 15,600 ft (42,495 metres). Maximum cruising speed, 272 mph (435 km/hr). Time to 15,000 ft (4,570 metres), 7.8 mins. Service ceiling, 30,000 ft (9,145 metres). Range, 1,540 miles (2,465 km)
Armament:	Four 20 mm Hispano cannons in forward lower fuselage and six 0.303-inch (7.69 mm) Browning machine-guns with 1,000 rpg.
Radar:	AI Mk.IV, Mk.VII, or Mk.VIII.

	UNIT ALLOCATIONS (see also Appendix 6)
UK night-fighter:	Nos 29, 68, 96, 125, 141, 153, 219, 255, 256, 307, 406, 409, 456, 488 and 600 Squadrons.
Night-fighter training:	No. 54 OTU.

Douglas Havoc Mk.I and Mk.II

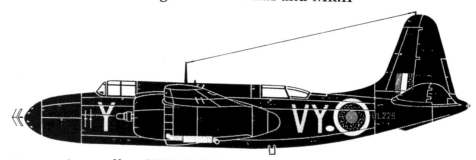

Havoc I BL228/VY-Y of No. 85 Squadron, Spring 1941.

Known by its manufacturing designation, the Douglas DB-7 light-bomber was ordered by the French government in modest numbers to equip the bomber squadrons of the *Armee de l'Air*. Incorporating modifications requested by the *Armee de l'Air*, the French Purchasing Commission placed an order for 100 DB-7s on 15 February 1939. With the demise of the

French state in June 1940, the British government intervened with the US government and Douglas Aircraft, to divert the remaining DB-7s of the French contract to the RAF. Once this arrangement had been settled, the government placed orders for a further 781 Boston bombers. The direct British purchased bombers were designated DB-7B by Douglas and Boston Mk.III by the RAF. Those from the French orders with single-speed supercharged SC3-Gs were designated Boston Mk.I and Mk.II for those with two-speed SC4-G engines.

Difficulties were encountered when the first twenty DB-7A aircraft from the French contract were received in the UK and readied for service with the RAF. The Boston's workshop manuals were written in French, as were the aircraft's instruments, whose dials were calibrated in metric units. The throttles operated in the French manner (backwards for an increase in power and forwards to reduce power) and had to be reversed for RAF operation. As a result, the early Boston Mk.Is were relegated to training duties.

Following an evaluation by the RAF, when it was established that the Boston I and II had insufficient range for bombing operations from UK bases, the decision was taken to modify the Boston II for night intruder and fighter duties at the Burtonwood Aircraft Repair Depot. With British instrumentation, armament and AI Mk.IV or V radar installed, the aircraft were designated 'Havoc Mk.I (Night Fighter)' and 'Havoc Mk.I (Intruder)'. Fitted with a solid nose designed by the Martin Baker Company, that held eight 0.303-inch (7.69 mm) machine-guns, and with no provision for rear protection, or the ability to carry bombs, the first Havoc Is (Night Fighter) were delivered to Fighter Command in the early months of 1941.

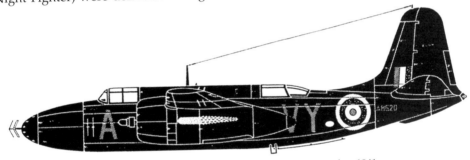

Havoc II AH520/VY-A of No. 85 Squadron, September 1941.

A second variant based on the French DB-7A with 1,600-hp R-2600-A5B 14-cylinder radials, was designated Havoc Mk.II. These were similar to the Mk.I in all respects, excepting the eight-gun pack was exchanged for one having twelve 0.303 Browning guns. The conversion of these aircraft was also the responsibility of the Burtonwood facility. As with all British night-fighters of the period, the Havocs were finished in night-black. A very small number of Boston IIIs, believed to be three, are known to have been fitted with the Helmore light in a similar manner to those of the Havoc I and II. These were designated 'Boston III (Turbinlite)'.

Only one squadron, No. 85, was destined to be completely re-equipped with Havoc Is in February 1941 and Mk.IIs in the following July, which operated both types until they were withdrawn in September 1942. The Havoc (Night Fighter) does not appear to have seen service with the OTUs.

TECHNICAL DATA (HAVOC I and II)[6]

Description: Twin-engined, night-fighter with a crew of two, of all-metal, stressed skin, construction.

Manufacturers:	Douglas Aircraft Company, El Segundo, California, USA.
Power plant:	(Havoc I) Two 14-cylinder, super-charged, 1,100-hp Pratt & Whitney R-1830-S3C4-G Double Wasp radial engines.
	(Havoc II) Two 14-cylinder, super-charged, 1,600-hp Wright R-2600-A5B radial engines.
Dimensions:	Span, 61 ft 3 ins (18.66 metres). Length, 46 ft 11 ins (14.32 metres). Height, 15 ft 10 ins (4.83 metres). Wing area, 464 sq. ft (43.11 sq. metres).
Weights:	(Havoc I) Empty, 11,400 lb (5,170 kg). Gross, 19,040 lb (8,637 kg).
	(Havoc II) Empty, 13,674 lb (6,203 kg). Gross, 19,322 lb (8,764 kg).
Performance:	(Havoc I) Maximum speed, 295 mph (472 km/hr) at 13,000 ft (3,965 metres). Time to 12,000 ft (5,443 metres), 8 mins. Service ceiling, 25,800 ft (7,865 metres). Combat range, 996 miles (1,595 km).
	(Havoc II) Maximum Speed, 323 mph (517 km/hr) at 12,800 ft (3,900 metres). Cruising speed, 275 mph (440 km/hr). Service ceiling, 27,680 ft (8,437 metres). Combat range, 490 miles (785 km).
Armament:	(Havoc I) Eight 0.303-inch (7.69 mm) Browning machine-guns.
	(Havoc II) Twelve 0.303-inch (7.69 mm) Browning machine-guns.
Radar:	AI Mk.IV or Mk.V.

UNIT ALLOCATION (HAVOC I and II)

Havoc I (night-fighter):	Nos 85, 531, 533 and 537 Squadrons.
Havoc II (night-fighter):	No. 85 Squadron.

Havoc (Turbinlite) Mk.I and Mk.II

Havoc II (Turbinlite) AH470/F of No. 1459 Flight/No. 538 Squadron, March 1942.

Built in response to a requirement for an aircraft to operate as the hunter element of a hunter-killer duo, with a 2,700 million candle-power Helmore airborne searchlight, the Turbinlite was based on the Havoc and Boston airframes. The Havoc I (Turbinlite), of which at least twenty-one were converted, was based on the Havoc I airframe, with all armament removed and a flatter nose profile to accommodate the Helmore light. Fitted with AI Mk.IV or V radar that had twin arrowhead transmitting aerials either side of the nose plate, whose edges were finished with a Townsend ring to smooth the airflow over the blunt nose, and a bank of lead-acid accumulators (wet batteries) in the bomb-bay, the Havoc I (Turbinlite) entered Fighter Command service in May 1941. A further thirty-nine Havoc IIs were converted to Turbinlite configuration.

The complexity of the interception technique and the Turbinlite's poor success rate, coupled with the availability of sufficient 'standard' night-fighters, brought about the type's retirement during the early months of 1943.

TECHNICAL DATA (HAVOC [TURBINLITE] I and II)

The data for the Havoc (Turbinlite) I and II is identical to those for the Havoc I and II, excepting for their length being slightly shorter at 45 ft (13.72 metres).

UNIT ALLOCATION (see also Appendix 6)

Havoc I (Turbinlite): Nos 531, 532, 533, 534, 535, 537, 538 and 539 Squadrons, and Nos 1422, 1452, 1453, 1454, 1455, 1456, 1458, 1459 and 1460 Flights.

Havoc II (Turbinlite): Nos 533, 534, 535, 536 and 538 Squadrons and Nos 1422, 1451, 1454, 1455, 1456, 1457, 1459 and 1460 Flights.

de Havilland Mosquito NF.II

Mosquito NF.II DD737/RS-B of No. 157 Squadron, 1942.

Like the Beaufighter Mk.If, the Mosquito NF.II was to form the basis of a series of night-fighters that would serve in Fighter Command from 1942 to 1955. Based around the proto-type fighter, W4052, the NF.II was powered by the single-stage R-R Merlin XXI or 23-series engines, driving three-bladed de Havilland constant speed propellers. Armament com-prised four 20 mm Hispano cannon in the lower fuselage and four 0.303-inch (7.69 mm) Browning machine guns in the extreme nose. Accommodation was provided for a pilot and R/O sitting side-by-side under a fully glazed canopy, with a flat, bullet proof windscreen. AI Mks.IV or V were fitted as standard. With a full operational load (guns, AI and 410 gallons [1,863 litres] of fuel) the Mk.II weighed 18,100 lb (8,210 kg) and achieved a maximum speed of 340 mph (545 km/hr) at 20,000 feet (610 metres).

Deliveries of the Mosquito NF.II began in March 1942, with a further 465 being delivered from de Havilland's factories at Hatfield and Leavesden, where they replaced Beaufighters in Fighter Command. A version specifically adapted to night-intrusion by No. 23 Squadron, the Mk.II (Special), had the AI gear removed and increased fuel capacity installed. The NF.II served with Fighter Command until April 1945, after which large numbers of time-expired airframes were transferred to the night-fighter OTUs, where they served until the war's end.

TECHNICAL DATA[7]

Description: Twin-engined, night-fighter with a crew of two, of mixed wooden monocoque and stressed skin construction.

Manufacturers: De Havilland Aircraft Company Ltd, Hatfield and Leavesden.

Power plant: Two 12-cylinder, super-charged 1,460-hp R-R Merlin XXI or 23 in-line engines.

Dimensions: Span, 54 ft 2 ins (16.5 metres). Length, 40 ft 6 ins (12.34 metres). Height, 12 ft 6 ins (3.8 metres). Wing area, 454 sq. ft (42.17 sq. metres).

Weights: Tare, 13,431 lb (6,090 kg). Full operational load, 18,649 lb (8,460 kg).

Fuel capacity: Maximum, 547 gallons (2,485 litres). Operationally loaded, 410 gallons (1,865 litres).

Performance: Maximum speed, 341 mph (545 km/hr) at 20,000 ft (6,095 metres). Maximum cruising speed, 252 mph (405 km/hr) at 20,000 ft. Initial rate-of-climb, 3,000 ft/min (915 metres/min). Service ceiling, 36,000 ft (10,970 metres). Maximum range, 1,705 miles (2,730 km).

Armament: Four 20-mm Hispano cannons in forward lower fuselage, with 200 rpg, and four 0.303-inch (7.69 mm) Browning machine-guns with 500 rpg.

Radar: AI Mk.IV or AI Mk.V.

UNIT ALLOCATION (see also Appendix 6)

UK night-fighter: Nos 25, 85, 141, 151, 157, 264, 307, 410 and 456 Squadrons.

Bomber Support: Nos 515 and 605 Squadrons.

Night-fighter training: Nos 54 and 60 OTUs.

de Havilland Mosquito NF.XII and XIII

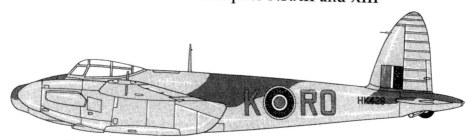

Mosquito NF.XIII HK428/RO-K of No. 29 Squadron, May 1943–April 1944.

The introduction of centimetric radar in the form of AI Mk.VII and Mk.VIII necessitated the development of the Mosquito airframe to accommodate it. Accordingly, NF.II, DD715, was taken off the Hatfield line in July 1942 and fitted with a perspex 'thimble' radome to accommodate the scanning mechanism and equipment boxes of AI Mk.VIII. Fitted with its new nose the aircraft was flown for the first time during August and transferred to TFU at Defford for flight trials in September. On the successful completion of the trials a further ninety-seven Mk.IIs were flown from the Leavesden factory and delivered to Marshall's Flying School on the outskirts of Cambridge for conversion to the NF.Mk.XII standard. The first of these aircraft, HJ945 and HJ946 arrived at Cambridge on 2 January 1943. Being the production prototype, HJ945 was delivered to Defford on 13 February for further trials. Powered by the same engines as the Mk.II (Merlin 21 or 23s rated at 1,460-hp) and with its four Browning machine guns removed to accommodate the radome, the new mark was in all other respects similar to the NF.II. The all up weight of DD715 and production aircraft was recorded as 18,441 lb (8,365 kg).

The first Mk.XIIs were delivered to Fighter Command in March 1943, with production quantities enabling ten night-fighter squadrons to be formed on the new aircraft. The Mk.XII served to the war's end, with the last UK-based squadrons standing them down during January 1945. The NF.XII also served in the night-fighter training role with the OTUs.

The production version of the Mk.XII, the NF.Mk.XIII, employed the stronger wing of the fighter-bomber Mk.VI variant, with provision for external drop tanks, and the thimble, or in later production aircraft, the bull-nose taken from the NF.Mk.XIX. The aircraft was also capable of carrying a pair of bombs to the rear of the cannon ammunition boxes, although these were rarely fitted in the night-fighter role. The first production aircraft, HK363, flew from the Leavesden factory during August 1942, before passing to Boscombe Down for service trials. Late production aircraft employed the more powerful Merlin 25-series engine rated at 1,635-hp. Fighter Command received the first examples of the new mount during October 1943, followed by further deliveries in December. Altogether eleven squadrons operated the Mk.XIII at home and abroad, with the final unit standing its aircraft down in September 1945. Like its stable-mate, the NF.XII, the Mk.XIII also saw service with the OTUs.

TECHNICAL DATA (NF.XII)

Description:	Twin-engined, night-fighter with a crew of two, of mixed wooden monocoque and stressed skin construction.
Manufacturers:	De Havilland Aircraft Company Ltd, Hatfield and Leavesden.
Power plant:	Two 12-cylinder, super-charged 1,460-hp R-R Merlin 21 or 23 in-line engines.
Dimensions:	Span, 54 ft 2 ins (16.5 metres). Length, 40 ft 5 ins (12.32 metres). Height, 12 ft 6 ins (3.8 metres). Wing area, 454 sq. ft (42.17 sq. metres).
Weights:	Tare, 13,696 lb (6,210 kg). Full operational load, 19,700 lb (8,935 kg).
Fuel capacity:	Maximum, 547 gallons (2,485 litres). Operationally loaded, 410 gallons (1,865 litres).
Performance:	Maximum speed, 341 mph (545 km/hr) at 20,000 ft (6,095 metres). Maximum cruising speed, 252 mph (405 km/hr) at 20,000 ft. Initial rate-of-climb, 3,000 ft/min (915 metres/min). Service ceiling, 36,000 ft (10,970 metres). Maximum range, 1,705 miles (2,730 km).
Armament:	Four 20-mm Hispano cannons in forward lower fuselage, with 200 rpg.
Radar:	AI Mk.VIII.

UNIT ALLOCATION (NF.XII)

UK night-fighter:	Nos 29, 85, 151, 256, 307, 406, 488 and 604 Squadrons.
Night-fighter training:	No. 54 OTU.

TECHNICAL DATA (NF.XIII)[8]

Description:	Twin-engined, night-fighter with a crew of two, of mixed wooden monocoque and stressed skin construction.
Manufacturers:	De Havilland Aircraft Company Ltd, Hatfield and Leavesden.
Power plant:	Two 12-cylinder, super-charged 1,460-hp R-R Merlin 21, 23 or 1,635-hp 25 in-line engines.
Dimensions:	Span, 54 ft 2 ins (16.5 metres). Length, 40 ft 5 ins (12.32 metres). Height, 12 ft 6 ins (3.8 metres). Wing area, 454 sq. ft (42.17 sq. metres).

Weights:	Tare, 15,300 lb (6,940 kg). All-up 20,000 lb (9,070 kg). Overload with two 50-gallon (230 litres), 20,278 lbs (9,200 kg).
Fuel capacity:	Maximum, 716 gallons (3,255 litres). Operationally loaded, 453 gallons (2,060 litres).
Performance:	Maximum speed, 394 mph (630 km/hr) at 13,800 ft (4,205 metres). Maximum cruising speed, 255 mph (410 km/hr) at 20,000 ft. Initial rate-of-climb, 3,000 ft/min (915 metres/min). Service ceiling, 30,000 ft (9,145 metres). Maximum range, 1,860 miles (2,730 km).
Armament:	Four 20-mm Hispano cannon in forward lower fuselage with 200 rpg.
Radar:	AI Mk.VIII.

UNIT ALLOCATION (NF.XIII)

UK Night-fighter:	Nos 29, 85, 96, 151, 264, 409, 410, 488 and No. 604 Squadrons.
Night-fighter training:	No. 54 OTU.

de Havilland Mosquito NF.XVII

Mosquito NF.XVII DZ659/ZQ-M of the Fighter Interception Unit.

As with the NF.XII, the introduction of the American SCR-720B/AI Mk.X during the autumn of 1943 required a variant of the Mosquito to carry the new radar. The Mk.XVII was based on the Mk.II airframe, but employed more powerful 1,635-hp Merlin 25 engines and a 'bull-nose' radome to house the AI Mk.X's helical scanner. The first NF.Mk.XVII, a converted Mk.XII HK195, flew during March 1943 and was followed by ninety-eight production aircraft built in three batches by the Leavesden factory.

The Mk.XVII was built in small numbers and consequently only saw service with six night-fighter squadrons, beginning in November 1943. By March 1945 the Mk.XVII had disappeared from Fighter Command's inventory, with a few being transferred to the night-fighter training organisation.

TECHNICAL DATA[9]

Description:	Twin-engined, night-fighter with a crew of two, of mixed wooden monocoque and stressed skin construction.
Manufacturers:	De Havilland Aircraft Company Ltd, Hatfield and Leavesden.
Power plant:	Two 12-cylinder, super-charged 1,635-hp R-R Merlin 25 in-line engines.
Dimensions:	Span, 54 ft 2 ins (16.5 metres). Length, 40 ft 6 ins (12.34 metres). Height, 12 ft 6 ins (3.8 metres). Wing area, 454 sq. ft (42.17 sq. metres).
Weights:	Tare, 13,224 lb (6,000 kg). All-up 19,200 lb (8,710 kg).

Fuel capacity: Maximum, 547 gallons (2,485 litres). Operationally loaded, 403 gallons (1,830 litres).

Performance: Maximum speed, 370 mph (630 km/hr) at 13,000 ft (3,960 metres). Maximum cruising speed, 255 mph (410 km/hr) at 20,000 ft. Initial rate-of-climb, 3,000 ft/min (915 metres/min). Service ceiling, 30,000 ft (9,145 metres). Maximum range, 1,705 miles (2,730 km).

Armament: Four 20-mm Hispano cannons in forward lower fuselage with 200 rpg.

Radar: AI Mk.X (SCR-720B).

UNIT ALLOCATION (see also Appendix 6)

UK Night-fighter: Nos 25, 68, 85, 125, 219 and 456 Squadrons.

Bomber support: No. 85 Squadron.

Night-fighter training: No. 54 OTU.

de Havilland Mosquito NF.XIX

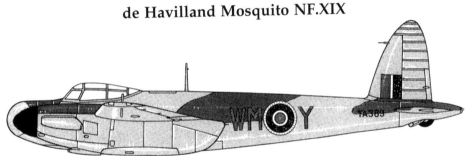

Mosquito NF.XIX TA389/WM-Y of No. 68 Squadron circa September 1944.

The poor supply situation that afflicted the introduction of AI Mk.X sets from the US in the summer and autumn of 1943 and faults associated with its installation in the Mosquito, caused the Air Staff to consider a reversion to British AI Mk.VIII radar as an interim measure. To this end, a new model that was to be capable of taking either radar was ordered into production. Similar to the Mk.XVII airframe with the bull-nose radome capable of accommodating the spiral scanner of AI Mk.VIII or the helical scanner of AI Mk.X, the NF.Mk.XIX flew during April 1944. Powered like the Mk.XVII by Merlin 25 engines fitted with paddle-blade propellers, the prototype, MM624, proved capable of a maximum speed of 378 mph (605 km/hr) at 13,200 feet (4,025 metres) with a full operational load (including two 50-gallon [225 litres] drop tanks). The Mk.XIX was built in moderate numbers by the parent company at Hatfield, who delivered 230 aircraft in seven batches.

The first Mk.XIXs were issued to Bomber Command's No. 100 Group in May 1944, with sufficient aircraft being delivered to equip eight squadrons by the war's end. Only three squadrons served in the UK air defence role and these were all stood down by August 1945. However, in light of AAF policy to restore a number of reserve squadrons to the night-fighter role, No. 500 (County of Kent) Squadron was briefly equipped with the Mosquito NF.19 for a short period in 1947.

TECHNICAL DATA[10]

Description: Twin-engined, night-fighter with a crew of two, of mixed wooden monocoque and stressed skin construction.

Manufacturers:	De Havilland Aircraft Company Ltd, Hatfield and Leavesden.
Power plant:	Two 12-cylinder, super-charged 1,635-hp R-R Merlin 25 in-line engines.
Dimensions:	Span, 54 ft 2 ins (16.5 metres). Length, 40 ft 6 ins (12.34 metres). Height, 12 ft 6 ins (3.8 metres). Wing area, 454 sq. ft (42.17 sq. metres).
Weights:	Tare, 14,471 lb (6,565 kg). Operationally loaded, 20,420 lb (9,260 kg).
Fuel capacity:	Maximum, 716 gallons (3,255 litres). Operationally loaded, 453 gallons (2,060 litres).
Performance:	Maximum speed, 378 mph (605 km/hr) at 13,200 feet (4,025 metres). Maximum cruising speed, 255 mph (410 km/hr) at 20,000 ft. Initial rate-of-climb, 3,000 ft/min (915 metres/min). Service ceiling, 30,000 ft (9,145 metres). Maximum range, 1,705 miles (2,730 km).
Armament:	Four 20-mm Hispano cannons in forward lower fuselage with 200 rpg.
Radar:	AI Mk.VII or AI Mk.X (SCR-720B).

UNIT ALLOCATION (see also Appendix 6)

UK Night-fighter:	Nos 68 and 500 Squadrons.
Bomber Support:	Nos 157 and 169 Squadrons.

de Havilland Mosquito NF.30

Mosquito NF.30 NT362/HB-S of No. 239 Squadron, 1945.

In March 1944 a new Mosquito variant was devised by marrying the airframe of the Mk.XIX to the two-stage Merlin engine installation of the PR.VIII/IX, to produce an aircraft with a greater performance at higher altitude, more fuel and longer range. AI Mk.X was installed in the prototype, HK364, which flew for the first time during March 1944. Designated NF.30 in RAF service, the new mark was powered by two Merlin 72 engines of 1,680-hp, giving it a maximum speed of 424 mph (680 km/hr) at 26,500 feet (8,075 metres), an operational ceiling of 30,000 feet (9,145 metres) and a range of 1,180 miles (1,890 km). Being a relatively straight forward development of the Mk.XIX airframe, production models began leaving de Havilland's Leavesden lines in April 1944.

The first Mk.30s were delivered to units in England in May 1944, with the last being withdrawn from front line service in 1949.

TECHNICAL DATA[11]

Description:	Twin-engined, night-fighter with a crew of two, of mixed wooden monocoque and stressed skin construction.
Manufacturers:	De Havilland Aircraft Company Ltd, Leavesden.

Power plant: Two 12-cylinder, two-stage super-charged 1,680-hp R-R Merlin 72, or
                    1,710-hp Merlin 76 in-line engines.

Dimensions: Span, 54 ft 2 ins (16.5 metres). Length, 41 ft 4 ins (12.57 metres). Height,
                    12 ft 6 ins (3.8 metres). Wing area, 454 sq. ft (42.17 sq. metres).

Weights: Tare, 15,241 lb (6,915 kg). Operationally loaded with two 100-gallon
                    (455 litre) wing tanks, 23,496 lb (10,660 kg).

Fuel capacity: Maximum, 716 gallons (3,255 litres). Operationally loaded, 453 gallons
                    (2,060 litres).

Performance: Maximum speed, 424 mph (680 km/hr) at 26,500 feet (8,080 metres).
                    Maximum cruising speed, 220 mph (410 km/hr) at 30,000 ft (9,145
                    metres). Initial rate-of-climb, 2,250 ft/min (685 metres/min). Service
                    ceiling, 35,000 ft (10,670 metres). Maximum range, 1,300 miles (2,080 km).

Armament: Four 20 mm Hispano cannons in forward lower fuselage with 200 rpg.

Radar: AI Mk.X (SCR-720B).

SQUADRON ALLOCATION (see also Appendix 6)

UK Night-fighter: Nos 264, 410, 456, 488, 500, 502, 504, 605, 609 and 616 Squadrons.
Bomber Support: Nos 239, 307 and 406 Squadrons.
Night-fighter training: No. 54 OTU and No. 228 OCU.

de Havilland Mosquito NF.36

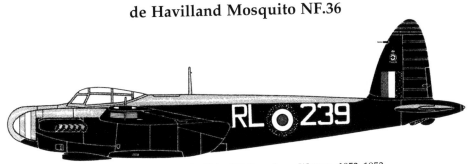

Mosquito NF.36 RL239 of No. 199 Squadron, Watton, 1952–1953.

The final version of the Mosquito night-fighter delivered to the RAF, the NF.Mk.36, was to form the backbone of the service's night defence force in the post-war era. Differing from the earlier NF.30 only in the engine installation, with Merlin 113s of 1,690-hp replacing the Mk.30's 73-series, the first Mk.36, RK955, made its maiden flight during May 1945. Production at the Leavesden factory began in May 1945 with RK955 and ended with the delivery of the 162nd aircraft, RL247, in January 1947.

Deliveries to Fighter Command began during January 1946, where the type formed the backbone of UK night air defence until 1950, when it was gradually replaced by jet-powered fighters. By October 1952 the last of the Command's piston-engined night-fighters had been removed from the inventory.

TECHNICAL DATA

The Mosquito NF.36 was broadly comparable to the NF.30 in all respects.

UNIT ALLOCATION (see also Appendix 6)

UK Night-fighter: Nos 23, 25, 29, 85, 141 and 264 squadrons.

de Havilland Vampire NF.10

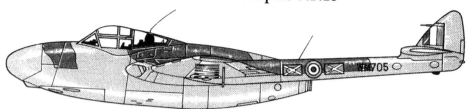

Vampire NF.10 WM705 of No. 151 Squadron, Leuchars, 1952–1953.

Developed to an in-house de Havilland specification for a two-seat, night and all-weather fighter, the de Havilland fighter was designed as a replacement for the Mosquito then in service with a number of foreign air forces. Based around the Vampire FB.Mk.5, the Company's designers added a new wooden monocoque fuselage that accommodated the two-man crew, AI Mk.10 radar and four 20 mm cannon beneath the cockpit floor. Like the Mosquito before it, the Vampire night-fighter's crew were seated side-by-side, with the N/R slightly to the rear on the starboard side, under a heavily framed canopy that opened upwards for entry and egress. Ejection seats were not fitted and the fuel capacity was the same as that for the FB.5. Due to the increased weight of the radar and its equipment boxes, the aircraft was powered by the more powerful 3,350 lb (1,520 kg) thrust de Havilland Goblin 3 engine.

Work on the project began in 1947 under the designation DH.113, for which the Company approved the construction of two prototypes with the Class 'B' registrations G-5-2 and G-5-5. The first flight of G-5-2 took place at Hatfield on 28 August 1949 with de Havilland test pilot Geoffrey Pike at the controls. During flight testing the new fighter was found to be pleasant to fly and manoeuvrable despite the additional 270 lb (120 kg) of radar equipment. However, stability problems were encountered, but these were overcome by revising the fin shape, extending the tailplane fairings and fitting acorn fairing at the junction of the fin and the tailplane.

The prototype was exhibited at the Farnborough Air Show on 6 September 1949 and secured an order from the Egyptian Air Force for twelve aircraft. Unfortunately for de Havilland, tension between Egypt and its Israeli neighbour forced the British Government to intervene and embargo the sale. Faced with a delay in the delivery of de Havilland Venom NF.2s and more advanced versions of the Meteor to the RAF, the Air Ministry intervened and agreed to takeover the order and introduce the Vampire as an interim jet night-fighter. To prepare what was now described as the Vampire NF.Mk.10 for RAF service, G-5-2 was flown to A&AEE Boscombe Down in March 1950 for flight trials. The first production aircraft, WP232, joined the test programme in March 1951 and flew alongside the second prototype, G-5-5, to clear the armament, radar, radio, IFF and *GEE* Mk.3 navigation equipment. Apart from a tendency to spin at the stall and complaints with the radome shape that spoiled the forward view, Boscombe Down's pilots confirmed the manufacturer's findings and proved the Mk.10 to be a more stable gun-platform than its Gloster stable-mate and have more range.

The Vampire's night-career was destined to be brief, with the first examples being delivered to Fighter Command in July 1951 and the last being withdrawn from squadron service in January 1954. Some aircraft were, however, transferred to Flying Training Command to serve in the fast-jet, navigator training role.

TECHNICAL DATA[12]

Description:	Single-engined, jet-powered night-fighter with a crew of two, of mixed wooden monocoque and stressed metal skin construction.
Manufacturers:	De Havilland Aircraft Company Ltd, Hatfield and Chester.
Power plant:	One 3,350 lb (1,520 kg) static thrust (st) de Havilland Goblin 3 turbojet.
Dimensions:	Span, 38 ft (11.58 metres). Length, 34 ft 7 ins (34.58 metres). Height, 6 ft 7 ins (6.58 metres). Wing area, 261 sq. ft (24.25 sq. m).
Weights:	Tare, 6,984 lb (3,170 kg). All up, 11,350 lb (5,150 kg).
Fuel capacity:	Internal, 330 gallons (1,500 litres). Maximum with two 100-gallon (455 litre) wing tanks, 530 gallons (2,410 litres).[13]
Performance:	Maximum speed, 550 mph (880 km/hr) at 20,000 ft (6,095 metres). Initial rate-of-climb, 4,500 ft/min (1,370 metres/min). Service ceiling, 40,000 ft (12,190 metres). Maximum range with two 100-gallon external tanks, 1,220 miles (1,950 km).
Armament:	Four 20-mm Hispano cannons in forward lower fuselage with 200 rpg.
Radar:	AI Mk.X (SCR-720B).

UNIT ALLOCATION (see also Appendix 6)

UK Night-fighter:	Nos 23, 25 and 151 Squadrons.
Training:	Nos 1, 2 ANS and CATCS.

de Havilland Venom NF.2/2A

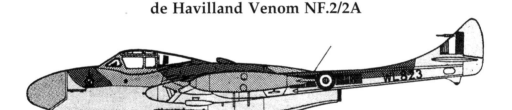

Venom NF.2 WL823 of No. 23 Squadron, Coltishall, 1955.

Having already proposed a design for a Vampire NF.10 replacement to the Air Ministry early in 1948, and had it more or less rejected, but believing there was nevertheless a need for an interim jet night-fighter, the de Havilland Company proceded with the design of a two-seat Venom night-fighter based on the Vampire NF.10.

Taking the Vampire NF.10 fuselage nacelle, complete with its AI.10 radar and 20 mm cannon installation, as the starting point, the de Havilland engineers installed a de Havilland Ghost Mk.103 engine of 4,850 lb st (2,200 kg st), fitted the slightly swept wings of the Venom single-seaters and incorporated the twin booms, triangular fins and rudders of the NF.10. The interior of the fuselage nacelle was enlarged to accommodate the pilot, N/R and additional R/T equipment, with the nose also seeing an extension to house the scanner and electronic boxes of the AI.10 radar under an upward hinging glass-fibre radome and sideways opening access doors. The heavily framed and upward hinging cockpit canopy, fuel tankerage, undercarriage and dive brakes of the NF.10 were also retained.

The Venom night-fighter prototype, G-5-3, was flown for the first time by de Havilland test pilot, John Derry, from Hatfield on 22 August 1950. Adopted by the Air Ministry for

RAF use, G-5-3 was allocated the serial number WP227 and delivered to Boscombe Down on 3 April 1951 for handling trials. Although pleasant to fly, but lacking the performance required of a night-fighter, WP227 was returned to de Havillands for modification. On its return to Boscombe Down during August with the acorn fairings removed from the base of each fin, WP227 received a more enthusiastic appraisal from A&AEE's pilots, but was considered unacceptable as increases in speed brought about changes in trim. This problem was eventually traced to air flow conditions around the canopy that were eventually cured by the incorporation of symetrical sides to the windscreen.

The first production example of the Venom NF.Mk.2, WL804, was delivered from de Havilland's Hatfield factory on 4 March 1952, before production was switched to Hawarden beginning with WL811. Deliveries of some ninety NF.2s began sometime around September 1952, but it was not until May of the following year that the first aircraft were accepted by Fighter Command. Structural problems in the area of the wheel wells restricted the type's operational ceiling to just 10,000 feet (3,050 metres) and it was not until the late summer of 1954 that this problem was resolved by modification, whereupon the fighter showed itself to be superior to the Meteor NF.11 in terms of its rate of climb and service ceiling.

The need for ejection seats and an improved canopy to aid an escape from the aircraft, brought about an approved set of modifications that resulted in the fitting of a clear view canopy, dorsal fins taken from the Vampire T.Mk.11 and the kidney-shaped fins and rudder of the Venom FB.Mk.4. Designated Venom NF.2A when the modifications were incorporated, the fighter was re-delivered to Fighter Command in April 1955 and used to equip just three squadrons. As with de Havilland's previous night-fighter, the Venom NF.2/2A's service was short, with all three squadrons being disbanded on the fighter by the end of August 1957.

TECHNICAL DATA[14]

Description:	Single-engined, jet-powered night-fighter with a crew of two, of mixed wooden monocoque and stressed metal skin construction.
Manufacturers:	De Havilland Aircraft Company Ltd, Hatfield and Hawarden
Power plant:	One 4,850-lb st (2,200 kg st) de Havilland Ghost 103 turbojet.
Dimensions:	Span, 42 ft 11 ins (13.1 metres). Length, 33 ft 1 ins (10.1 metres). Height, 7 ft 7 ins (2.3 metres). Wing area, 279.75 sq. ft (25 sq. metres).
Weights:	All-up, 13,838 lb (6,275 kg).
Fuel capacity:	Not known.
Performance:	Maximum speed, 630 mph (1,015 km/hr) at altitude. Initial rate-of-climb, 8,762 ft/min (2,670 metres/min). Service ceiling, 44,000 ft (13,410 metres). Maximum range with two 100-gallon external tanks, approximately 1,000 miles (1,610 km).
Armament:	Four 20-mm Hispano cannons in forward lower fuselage with 200 rpg.
Radar:	AI Mk.X (SCR-720B).

UNIT ALLOCATION (see also Appendix 6)

UK Night-fighter:	(NF.2) No. 23 Squadron, (NF.2A) Nos 33, 219 and 253 Squadrons.

de Havilland Venom NF.3

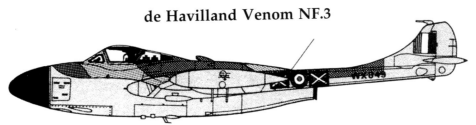

Venom NF.3, WX849 of No. 151 Squadron, Leuchars June 1957.

Following their involvement with the NF.2A upgrade to the Venom, and aware that more performance could be extracted from the airframe, de Havilland's engineers set about the task of improving the night-fighter variant. The NF.2A's modifications were incorporated into a new variant, alongside the more powerful Ghost 104 of 4,950 lb st and the American AN/APS-57 (AI.21) radar enclosed in an aerodynamically improved radome. Although there was a clear need for ejection seats to enhance the crew's safety, none were fitted. The only concession in this regard was a clear view, powered canopy, that could be jettisoned in an emergency. Powered ailerons were also incorporated, but the outer sections of the horizontal tailplanes taken from the NF.2A were removed.

Designated Venom NF.Mk.3 in RAF service, the prototype, WV928, flew from de Havilland's Christchurch airfield for the first time on 22 February 1953. Although un-representative of the production model, it had the framed canopy of the Mk.2 and the pointed fin and rudders of the Vampire NF.10. WV928 was flown to TRE's Defford airfield in June for their evaluation of the APS-57 radar. Handling trials at Boscombe Down with WV928 prior to its transfer to Defford, showed it to be superior to all previous marks of the Venom. Whilst the controls were rated as 'light and effective', the type's handling in poor visibility and turbulance was regarded by the pilots as being in need of improvement. They were also concerned that the improved performance of the American radar would be off-set by the type's poor rate of climb – 16 minutes to 40,000 feet (12,190 metres).[15]

Despite A&AEE's opinion, the NF.3 had already been committed to production in July 1951, with an order for 123 machines, later amended to 129, from de Havilland's Hatfield, Chester and Christchurch lines. The first production aircraft, WX785, and a further thirteen examples were allocated to tests and trials, beginning in September 1953. CS(A) release was granted in February 1955, with the first aircraft being allocated to Fighter Command in the following June. The Venom NF.3 equipped just five squadrons between June 1955 and once again its tenure was short, barely two years, before the last unit returned its aircraft to the MUs in November 1957.

TECHNICAL DATA[16]

Description: Single-engined, jet-powered night-fighter with a crew of two, of mixed wooden monocoque and stressed metal skin construction.

Manufacturers: De Havilland Aircraft Company Ltd, Hatfield, Chester and Christchurch.

Power plant: One 4,950-lb st (2,245 kg st) de Havilland Ghost 104 turbojet.

Dimensions: Span, 42 ft 11 ins (13.1 metres). Length, 36 ft 7 ins (11.15 metres). Height, 6 ft 6 ins (1.98 metres). Wing area, 279.75 sq. ft (25 sq. metres).

Weights: All-up, 14,544 lb (6,600 kg).

Fuel capacity: Not known.

Performance: Maximum speed, 630 mph (1,015 km/hr) at altitude. Initial rate-of-climb, 8,762 ft/min (2,670 metres/min). Service ceiling, 49,200 ft (15,000 metres). Maximum range with two 100-gallon external tanks, approximately 1,000 miles (1,610 km).

Armament: Four 20-mm Hispano cannons in forward lower fuselage with 200 rpg.

Radar: AI Mk.21 (AN/APS-57).

UNIT ALLOCATION (see also Appendix 6)

UK Night-fighter: Nos 23, 89, 125, 141 and 151 Squadrons.

Armstrong Whitworth (Gloster) Meteor NF.11

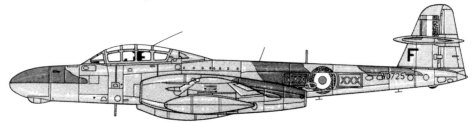

Meteor NF.11 WD725/'F' of No. 29 Squadron, Tangmere 1951.

Designed to fill the gap between the obsolete Mosquito and the F.4/48 specification that would ultimately lead to the Javelin, the Gloster night-fighter was intended to provide the RAF with a jet-powered night-fighter, offering improved performance at limited risk and cost. Gloster Aircraft proposed their two-seat T.7 Meteor as the basis of the new design, to which was added a lengthened nose to accommodate AI Mk.10 radar (taken from Fighter Command's Mosquito force), 3,500 lb st Derwent 5 turbo-jets and four 20 mm Hispano cannon mounted in the outer wings. Gloster's proposal was accepted by the Air Ministry who initiated Specification F.24/48 on 12 February 1949 to cover the manufacture of the type under the designation Meteor NF.Mk.11. Because of their commitment to the manufacture of the single-seat series of Meteors and the up and coming Javelin programme, the design and assembly of the new mark was passed to AWA.

The final design by AWA saw the nose lengthened by 4 feet (1.2 metres) to accommodate the scanner and equipment boxes of the AI.10 radar, the rear fuselage and E.1/44-type tail taken from the Meteor F.8, long-span wings of the Mk.III that contained the cannon and ammunition tanks in the outboard sections and more powerful 3,700 lb (1,680 kg) Derwent 8 engines. The air for cabin pressurisation was taken from the Derwent's compressor and gave an equivalent height of 24,000 feet (7,315 metres) when the aircraft was flying at 40,000 feet (12,190 metres). Ejection seats were not planned, nor fitted, with the crew being required to jettison the heavily framed hood and escape in the traditional fashion. To accommodate the AI.10's scanner mechanism, the radome was bulged below the natural contour of the fuselage. The T.7's 180 gallon (820 litres) ventral fuel tank was also retained.

Meteor T.7, VW413, served as the aerodynamic prototype and following conversion to the night-fighter standard, but with the standard T.7 fin and rudder, was flown for the first time on 28 January 1949 by AWA's test pilot Bill Else. The F.8 rear fuselage and fin was added in March and the prototype was back in the air from 8 April to continue the test programme, which showed no appreciable loss of handling with the longer nose. Later that year the Air Ministry placed an order with AWA for three full prototypes, the first of which, WA546, flew on 31 May 1950 in the hands of Eric Franklin, who alongside J.O.Lancaster, completed much of the type's performance testing and the initial assessment of the radar and IFF capability at TRE, Defford, in October 1950. The second prototype, WA547, which joined the test programme in August 1950, and differed only in it being fitted with AI Mk.9C and instrumented for armament trials. The third and final prototype, WB543, was fitted with production standard strengthened wings and flown for the first time on 23 September 1950.

The first production aircraft, WD585, flew on 19 October 1950 and incorporated a number of improvements as a consequence of the test programme. To enable the pilot to see better in heavy rain, a direct vision (DV) panel was built into the windscreen's port quarter panel, alongside an electrically heated de-misting system. The cabin heating was improved and the *Gee* navigation system's whip aerial was moved to the cabin roof to improve its performance. These modifications were introduced from the sixty-first aircraft and were retrospectively applied to the earlier machines.

Deliveries of the Meteor NF.11 to Fighter Command began in January 1951, where it would eventually equip six squadrons and an OCU until withdrawn from UK service in November 1957. Following their retirement from first line duties, a number of surplus NF.11 airframes were converted to target tugs under the designation TT.Mk.20 and employed by RN and civilian manned target facilities units.

TECHNICAL DATA[17]

Description:	Two-seat, twin engined, turbo-jet powered night-fighter of all metal construction.
Manufacturers:	Armstrong Whitworth Aircraft, Bitteswell and Baginton.
Power plant:	Two 3,700-lb st (1,680 kg st) R-R Derwent 8 turbojets.
Dimensions:	Span, 43 ft (13.1 metres). Length, 48 ft 6 ins (14.8 metres). Height, 13 ft 11 ins (4.24 metres). Wing area, 347 sq. ft (32.24 sq. metres).
Weights:	Empty, 12,019 lb (5,450 kg). Fully loaded with wing and ventral tanks, 20,035 lb (9,090 kg).
Fuel capacity:	Maximum, 750 gallons (3,410 litres) in main, ventral and wing tanks.
Performance:	Maximum speed, 554 mph (885 km/hr) at 10,000 ft (3,050 metres) and 547 mph (875 km/hr) at 30,000 ft (9,145 metres). Initial rate-of-climb, 5,800 ft/min (1,770 metres/min) and 1,500 ft/min (460 metres/min) at 30,000 ft. Service ceiling, 43,000 ft (13,105 metres). Maximum range, 950 miles (1,520 km) at 30,000 ft.
Armament:	Four 20-mm Hispano cannons in outer wing panels with 160 rpg.
Radar:	AI Mk.10.

UNIT ALLOCATION (see also Appendix 6)

UK Night-fighter:	Nos 29, 85, 125, 141, 151 and 264 Squadrons.
Night-fighter training:	Nos 228 and 238 OCUs.

Armstrong Whitworth (Gloster) Meteor NF.12

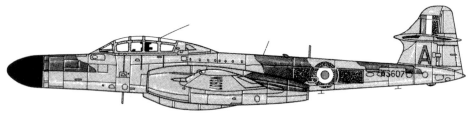

Meteor NF.12 WS607/'A' of No.72 Squadron, Church Fenton, 1958.

The Meteor NF.11's principal failing was the relatively poor performance of its AI.10 radar. To this end, AWA modified two NF.11s, WD670 and WD 687, to trial the Westinghouse AN/APS-57 radar, with a view to its adoption as an upgrade to the Mk.11. The elongated nose of the Mk.11 was lengthened by a further 17 inches (0.43 metres) and the Derwent 8 engines were replaced by the slightly more powerful Mk.9s of 3,800 lb st (1,725 kg st). These did little for the aircraft's overall performance, as the increased thrust was countered by a rise in weight brought about by the equipment units of the APS-57 radar. Nevertheless, the new engines had improved re-lighting capabilities, particularly at high altitudes, and were adopted for what became the Meteor NF.Mk.12. The APS-57 radar also removed the ugly fairing under the aircraft's nose, as the scanner mechanism was contained within the dimensions of the radome. In service the AN/APS-57 was designated AI Mk.21.

Testing with WD670 at Boscombe Down in June 1952 highlighted a tendency towards fin stalling, which the A&AEE pilots regarded as dangerous and in need of rectification. The aircraft was returned to AWA who inserted small fillets in the fin above and below the tailplane's bullet fairing, increasing the side area by approximately one square foot (0.093 sq. metres) and restoring the handling to an acceptable level. Testing also showed the need to further strengthen the type's wings to handle the additional thrust of the Derwent 9s. The first production NF.12, WS590, flew on 21 April 1953, with deliveries to Fighter Command beginning during the early months of the following year. In all, ten UK squadrons were equipped with the Mk.12, which enjoyed a relatively long career until June 1959, when the type was withdrawn from front line use.

TECHNICAL DATA[18]

Description:	Two-seat, twin engined, turbo-jet powered night-fighter of all metal construction.
Manufacturers:	Armstrong Whitworth Aircraft, Bitteswell and Baginton.
Power plant:	Two 3,800-lb st (1,725 kg st) R-R Derwent 9 turbojets.
Dimensions:	As per NF.11 except length, 49 ft 11 ins (15.21 metres).
Weights:	Empty, 12,292 lb (5,575 kg). Fully loaded with wing and ventral tanks, 20,380 lb (9,245 kg).
Fuel capacity:	As per NF.11.
Performance:	As per NF.11.
Armament:	As per NF.11.
Radar:	AI Mk.21 (AN/APS-57).

UNIT ALLOCATION (see also Appendix 6)

UK Night-fighter:	Nos 25, 29, 46, 64, 72, 85, 152, 153 and 264 Squadrons.
Night-fighter training:	Nos 228 and 238 OCUs.

Armstrong Whitworth (Gloster) Meteor NF.14

Meteor NF.14 WS794/'K' of No. 60 Squadron, Tengah 1960.

The final version of the Meteor night-fighter line, and indeed, of the Meteor itself, was a logical development of the Mk.12, whose obvious feature was a fully transparent cockpit canopy. Built from two separate pieces and joined at the centre by a metal joint, the canopy slide rearwards under the power of an electric motor. The opportunity was also taken to install a revised windscreen with a steeper angle to improve the pilot's forward view and fit spring-tab ailerons and an auto stabiliser to eliminate directional instability (snaking) at high altitude. Early aircraft had a cooling vent fitted on the lower part of the radome, but this was transferred to the upper half on later production models. The type retained the Mk.12's AI.21 radar and its Derwent 9 engines.

The new canopy was tested on an NF.11, WM261, before the aircraft was committed to production as the Meteor NF.Mk.14. WS722, the first of 100 production aircraft, flew on 23 October 1953, piloted by Bill Else from Baginton airfield, with deliveries to Fighter Command commencing in March 1954. Regarded by many as the best of the Meteor night-fighter line, the Mk.14 served alongside the Mk.12 as the standard RAF night interceptor until June 1959, when they were withdrawn from front line service.

TECHNICAL DATA[19]

Description:	Two-seat, twin engined, turbo-jet powered night-fighter of all metal construction.
Manufacturers:	Armstrong Whitworth Aircraft, Bitteswell and Baginton.
Power plant:	As per NF.12.
Dimensions:	As per NF.12.
Weights:	Empty, 12,620 lb (5,725 kg). Fully loaded with wing and ventral tanks, 21,200 lb (9,615 kg).
Fuel capacity:	As per NF.12.
Performance:	As per NF.12 except maximum speed, 578 mph (925 km/hr) at 10,000 ft (3,050 metres).
Armament:	As per NF.11.
Radar:	As per NF.12.

UNIT ALLOCATION (see also Appendix 6)

UK Night-fighter:	Nos 25, 29, 33, 46, 64, 72, 85, 152, 153 and 264 Squadrons.
Night-fighter training:	Nos 228 and 238 OCUs.

Notes

1. Owen Thetford, *Aircraft of the Royal Air Force since 1918* (London: Putnam Books, 1962), p. 107; Chaz Bowyer [2], *Bristol Blenheim* (London: Ian Allen, 1984), pp. 124 and 125.

2. In standard day-fighter configuration, with no radar equipment fitted.
3. Owen Thetford, *op cit*, pp.118 and 119; C.H. Barnes, *Bristol Aircraft since 1910* (London: Putnam), p.307; Staff Writer [1], *Beaufighter, Part 1* in Air Enthusiast, January 1974, p.28.
4. C.H. Barnes, *Bristol Aircraft since 1910* (London: Putnam), p.307; Tony Buttler, *Database: Bristol Beaufighter*, in Aeroplane Monthly, August 2003.
5. C.H. Barnes, *Bristol Aircraft since 1910* (London: Putnam), p.307; Tony Buttler, *Database: Bristol Beaufighter*, in Aeroplane Monthly, August 2003.
6. René Francillon, *op cit*, pp.294 and 295.
7. The data is taken from A.J. Jackson, *De Havilland Aircraft Since 1915* (London: Putnam, 1962), pp.383–384; Sharp and Bowyer, *op cit*, pp.393 and 401.
8. A.J. Jackson, *De Havilland Aircraft Since 1915* (London: Putnam, 1962), pp.383–384; Sharp and Bowyer, *op cit*, p.396.
9. The data is taken from A.J. Jackson, *De Havilland Aircraft Since 1915* (London: Putnam, 1962), pp.383–384; Sharp and Bowyer, *op cit*, p.397.
10. Sharp and Bowyer, *op cit*, p.397.
11. Sharp and Bowyer, *op cit*, p.398; A.J. Jackson, *op cit*, p.384.
12. A.J. Jackson, *op cit*, p.457; W.A. Harrison [1], *op cit*, p.80.
13. Taken from Vampire T.Mk.11.
14. Taken from W.A. Harrison, *op cit*, rear cover, Owen Thetford, *op cit*, p.189; David Watkins, *op cit*, p.261.
15. The A&AEE report was comparing the Venom's rate-of-climb by comparison to 'contemporary' bombers, i.e. the B-47, Canberra, and Tu-16.
16. Taken from W.A. Harrison, *op cit*, rear cover, A.J. Jackson, *op cit*, p.443; David Watkins, *op cit*, p.261.
17. Taken from Edward Shacklady, *op cit*, p.167 and 168; Tony Buttler [2], *op cit*, p.16; Derek James, *op cit*, p.305 and 306.
18. Taken from Edward Shacklady, *op cit*, p.167 and 168; Tony Buttler [2], *op cit*, p.16; Derek James, *op cit*, p.305 and 306.
19. Taken from Edward Shacklady, *op cit*, p.167 and 168; Tony Buttler [2], *op cit*, p.16; Derek James, *op cit*, p.305 and 306.

Luftwaffe Aircraft Claimed 'Destroyed' at Night over Great Britain, 1939–1945

This listing is taken from a variety of sources, which the Author does not claim to be definitive. The listing under 'crew' shows the pilot first, followed by his R/O, unless otherwise specified.

Key to abbreviations: AG, air gunner; N/K, not known, Obs, observer.

Date	Claim	Squadron	Aircraft	Crew	Location
1940					
18/19 June	He 111H of 4./KG 4	No. 19	Spitfire I	F/Lt A.G. Malan	In the sea off Foulness Island, Essex.
	He 111H of Stab./KG 4	No. 19	Spitfire I	F/Lt A.G. Malan	Crashed at Writtle, Essex.
	He 111H of Stab./KG 4	No. 29	Blenheim If	P/O J.S. Barnwell	Ditched in the Thames Estuary. P/O Barnwell was killed when his aircraft was hit by return fire.
	He 111H of II./KG 4	No. 23	Blenheim If	Sgt A. Close and LAC L.R. Karasek (AG)	Crashed at Terrington St Clement, near Kings Lynn, Norfolk. Sgt Close's aircraft was shot down by return fire, but both crewmembers survived.
			Blenheim If	F/Lt R. Duke-Wooley and LAC D. Bell (AG)	Crashed at Gt Wilbraham, Cambridgeshire. F/O Petre's aircraft was shot down by return fire, but the crew escaped. S/Ldr O'Brien lost control of his aircraft and it crashed killing all on board.
	He 111H of II./KG 4	No. 19	Spitfire I	F/O G.W. Petre	
		No. 23	Blenheim If	S/Ldr J.S. O'Brien, P/O C. King-Clark (Obs) and Cpl D. Little (AG)	
22/23 June	He 111H of II./KG 4	No. 19	Spitfire I	F/O G. Ball	Crashed into the sea off Margate, Kent.
	Do 17Z possibly of 2./KG 3	FIU	Blenheim If	F/O G. Ashfield, P/O G.C. Morris (Obs) and Sgt R.H. Leyland (R/O)	Crashed into the sea 5 miles south of Bognor Regis. This was the first confirmed AI kill of the war.
25/26 July	He 111H-4 of 1./KG 4	No. 87	Hurricane I	P/O J.R. Cock	Crashed at Smeatharpe, near Honiton, Devon.
17/18 Aug	Unidentified He 111	No. 29	Blenheim If	P/O R.A. Rhodes and Sgt W.J. Gregory (AG)	Crashed into the North Sea 10 miles off Cromer Knoll, Suffolk.

Date	Claim	Squadron	Aircraft	Crew	Location
24/25 Aug	Unidentified Do17Z	No. 29	Blenheim If	P/O J.R.D. Braham, P/O A.A. Wilson (Obs) and AC2 N. Jacobson (R/O)	Crashed into the North Sea off the Humber Estuary.
4/5 Sept	He111H-3 of Stab I./KG1	No. 25	Blenheim If	P/O M.J. Herrick and Sgt J.S. Pugh	Crashed at Rendlesham, Suffolk.
14/15 Sept	He111H-4 of 3./KG4	No. 25	Blenheim If	P/O M.J. Herrick and P/O A.W. Brown (Obs)	Crashed near Sheering, Essex.
15/16 Sept	Unidentified Ju88	No. 600	Blenheim If	F/Lt C.A. Pritchard and P/O H. Jacobs	Crashed into the sea off Bexhill, Sussex.
17/18 Sept	Ju88A-1 of 3./KG54	No. 141	Defiant I	Sgt G. Laurence and Sgt W.T. Chard (AG)	Crashed at Maidstone, Kent.
15/16 Oct	He111H-4 of 2./KGr126	No. 264	Defiant I	P/O F.D. Hughes and Sgt F. Gash (AG)	Crashed at Brentwood, Essex.
19/20 Nov	Ju88A-5 of 3./KG54	No. 604	Beaufighter If	F/Lt J. Cunningham and Sgt J. Philipson	Crashed near East Wittering, Sussex. This was the first Beaufighter kill.
22/23 Dec	He111P-4 of 3./KG55	No. 141	Defiant I	P/O J. Benson and Sgt F. Blain (AG)	Crashed at Etchingham, East Sussex.
1941					
15/16 Jan	Do17Z-2 of 4./KG3	No. 151	Hurricane I	P/O R.P. Stevens	Crashed into the sea off Canvey Island.
4/5 Feb	Do17Z-3 of 7./KG2	No. 151	Defiant I	Sgt H.E. Bodien and Sgt D.E.C. Jonas (AG)	Crashed near Weldon, Northants.
15/16 Feb	He111P-2 of III./KG27	No. 604	Beaufighter If	F/Lt J. Cunningham and Sgt C.F. Rawnsley	Crashed near Harberton, Devon.
25/26 Feb	Do17Z of KG2	No. 85	Hurricane I	S/Ldr P.W. Townsend	Crashed near Sudbury, Suffolk.
4/5 March	He111H-5 of I./KG28	No. 604	Beaufighter If	S/Ldr J. Anderson	Crashed near Beer Head, Dorset.
12/13 March	He111H-4 of 4./KG27	No. 264	Defiant I	F/O T. Welsh and Sgt H. Hayden (AG)	Crashed into the sea off Hastings, Sussex.
	He111P-4 of 5./KG55	No. 264	Defiant I	F/O F.D. Hughes and Sgt F. Gash (AG)	Crashed at Ockley, Surrey.
	He111P-4 of 6./KG55	No. 96	Hurricane I	Sgt McNair	Crashed at Widnes, Lancashire.
	Ju88A-5 of 6.KG76	No. 604	Beaufighter If	F/O K.I. Geddes and Sgt A.C. Cannon	Crashed at Kingston Deverill, Wilts.
	Ju88A-5 of 6./KG76	No. 307	Defiant I	Sgt Jankowiak and unknown AG	Crashed at Wynchbold, Worcs. This was a joint claim with an AA unit.
13/14 March	Do17Z-2 of Stab/KG2	No. 29	Beaufighter If	F/O J.R.D. Braham, Sgt Ross	Crashed into the sea off Skegness, Lincs.
	He111H-5 of 7./KG26	No. 219	Beaufighter If	Sgt J.A. Clandillon and Sgt Dodge	Crashed at Shipley, Sussex.
	He111P-2 of 7./KG55	No. 219	Beaufighter If	P/O A.J. Hodgkinson and Sgt B.E. Dye	Crashed near Bramdean, Hants.
	He111H-3 of KGr100	No. 600	Blenheim If	P/O G.A. Denby and P/O Guest	Crashed at Dunure, Ayrshire.

Date	Claim	Squadron	Aircraft	Crew	Location
	Ju88A-5 of 3./KfGr 106	No.72	Spitfire I	F/Lt D. Sheen	Crashed into the sea off Amble, Northumberland.
	Ju88C-4 of 4./NJG 2	No.29	Beaufighter If	W/Cdr S.C. Widdows and Sgt B. Ryall	Crashed at Dovendale, Lincs.
14/15 March	He111H-3 of 2./KG55	No.604	Beaufighter If	F/Lt G.P. Gibson and Sgt R.H. James	Crashed into the sea off Ingoldmells Point, near Skegness, Lincs.
	He111P-4 of 6./KG55	No.604	Beaufighter If	P/O K.I. Geddes and Sgt A.C. Cannon	Crashed at Falfield, Glos.
	Ju88A-5 of 1./KGr 806	No.219	Beaufighter If	W/Cdr T.G. Pike and unknown R/O	Crashed into the sea off Beachy head, Sussex.
3/4 April	Ju88A-5 of 7./KG1	No.604	Beaufighter If	F/Lt J. Cunningham and Sgt C.F. Rawnsley	Crashed into the sea off the Needles, Isle of Wight.
4/5 April	He111H-5 of III./KG26	No.604	Beaufighter If	F/O E.D. Crew and Sgt N.H. Guthrie	Crashed at West Hewish, Weston-super-Mare.
7/8 April	He111H-5 of 9./KG26	No.87 / No.604	Hurricane I / Beaufighter If	F/Lt D.H. Ward / S/Ldr J. Cunningham and Sgt C.F. Rawnsley	Crashed into the sea off Branscombe, Devon. Shared claim.
	Ju88A-5 of 5./KG54	No.256	Defiant I	F/Lt D.R. West and Sgt Adams (AG)	Crashed near Southport, Lancashire.
	He111P-4 of 1./KG55	No.219	Beaufighter If	P/O A.J. Hodgkinson and Sgt B.E. Dye	Crashed into the sea off Worthing, Sussex.
	Unidentified He111	No.85	Havoc I	F/O G. Howitt and Sgt Reed	Crashed into the North Sea.
8/9 April	He111H-5 of 9./KG26	No.264	Defiant I	S/Ldr A.T.D. Saunders and P/O Sutton (AG)	Crashed near Hitchin, Herts.
	He111H-5 of KG27	No.151	Hurricane I	P/O R.P. Stevens	Crashed near Wellesbourne, Warks.
	He111P-4 of 8./KG55	No.151	Defiant I	F/Lt D.F.W. Darling and P/O J.S. Davidson (AG)	Crashed near Great Windsor Park, Surrey. Shared claim with AA crew.
9/10 April	He111H-2 of II./KG1	N/K	N/K	N/K	Crash details not known.
	He111P of 4./KG27	–	Defiant I	N/K	Crashed near Smethick, Staffs.
	He111P-2 of 5./KG55	No.264	Defiant I	F/Sgt E.R. Thorn and Sgt F.J. Barker (AG)	Crashed near Bushbridge, Surrey.
	Ju88A-5 of 3./KG77	No.151	Defiant I	F/Lt D.A.P. McMullen and Sgt S.J. Fairweather (AG)	Crashed near RAF Bramcote, Warks.
	He111H-3 of 2./KGr100	No.604	Beaufighter If	F/O R. Chisholm and Sgt W.G. Ripley	Crashed near Cranborne, Dorset.
	Ju88C-4 of 4./NJG 2	No.25	Beaufighter If	Sgt S. Bennett and Sgt Curtis	Crashed near Oakham, Rutland.
10/11 April	Ju88A-5 of II.KG1	No.151	Hurricane I	P/O R.P. Stevens	Crashed near Murcott, Oxfordshire.
	He111H-5 of Stab II./KG26	No.264	Defiant I	F/O E.G. Barwell and Sgt A. Martin (AG)	Crashed near Seaford, Sussex.
	He111H-5 of 8./KG26	No.256	Defiant I	F/Lt E.C. Deansley and Sgt W.J. Scott (AG)	Crashed near Radway, Warks.

Date	Claim	Squadron	Aircraft	Crew	Location
	Ju88A-5 of 1./KG54	No. 604	Beaufighter If	F/Lt Watson and unknown R/O	Crashed into the sea off St Aldhelms Head, Dorset.
	He111H-3 of 3./KGr100	No. 604	Beaufighter If	F/Lt G.O. Budd and Sgt G. Evans	Crashed at Chale Green, Isle of Wight.
11/12 April	He111P-2 of 9./KG27	No. 307	Defiant I	Sgt F.O. Jankowiak and Sgt J. Lipinski (AG)	Crashed near Shaftsbury, Dorset. Shared claim.
	Ju88A-5 of Stab II./KG55	No. 604	Beaufighter If	S/Ldr J. Cunningham and Sgt C.F. Rawnsley	Crashed at Holcombe, Devon.
15/16 April	He111P-2 of 8./KG55	No. 604	Beaufighter If	F/Lt Gomm and P/O Curnow	Crashed in Southampton.
	Ju88A-5 of 4./KG1	No. 604	Beaufighter If	S/Ldr J. Cunningham and Sgt C.F. Rawnsley	Crashed at Cranleigh, Surrey.
16/17 April	Ju88A-5 of 2./KG28	No. 219	Beaufighter If	W/Cdr T.G. Pike and Sgt W.T. Clark	Crashed at Wormley, Surrey.
	Ju88A-5 of 3./KG76	No. 219	Beaufighter If	W/Cdr T.G. Pike and Sgt W.T. Clark	Crashed near Wimbledon, Surrey.
19/20 April	He111H-5 of 7./KG4	No. 219	Beaufighter If	F/Lt Dotteridge and Sgt G.T. Williams	Crashed at Stockbury, Kent.
27/28 April	Ju88A-1 of 4(F)./122	No. 151	Hurricane I	P/O R.P. Stevens	Crashed into the Solent, Hampshire.
1/2 May	He111P of 4./KG27	No. 219	Beaufighter If	P/O D.O. Hobbis and unknown R/O	Crashed into the sea off Shoreham, Sussex.
2/3 May	Ju88A-5 of 1./KG30	No. 219	Beaufighter If	P/O A.J. Hodgkinson and Sgt B.E. Dye	Crashed into the sea off Weybourne, Norfolk.
	Ju88A-6 of 8./KG77	No. 151	Defiant I	P/O O.G.A. Edmison and Sgt A.G. Beale (AG)	Crashed near Lyndhurst, Hampshire.
3/4 May	He111H-5 of III./KG26	No. 604	Beaufighter If	F/Lt G.O. Budd and Sgt G. Evans	Crashed near Crowcombe, Taunton, Somerset.
		No. 604	Beaufighter If	F/Lt H. Speke and unknown R/O	
	He111H-5 of 8./KG26	No. 219	Beaufighter If	F/Lt Dotteridge and Sgt Williams	Crashed at Sidlesham, near Chichester, Sussex.
	He111H-5 of 9./KG26	No. 219	Beaufighter If	W/Cdr T.G. Pike and Sgt W.T. Clark	Crashed at Arundel, Sussex.
	He111P-2 of 9./KG27	No. 604	Beaufighter If	S/Ldr J. Cunningham and Sgt C.F. Rawnsley	Crashed near Corton Denham, Somerset.
	He111H-4 of 3./KG53	No. 151	Defiant I	P/O Bodien and Sgt Wrampling (AG)	Crashed landed at Sharrington, Norfolk.
	Ju88A-6 of Stab II./KG54	No. 256	Defiant I	F/Lt E.C. Deansley and Sgt W.J. Scott (AG)	Crashed near Lostock Gralam, Cheshire.
	Unidentified Do215	No. 256	Defiant I	F/Lt E.C. Deansley and Sgt W.J. Scott (AG)	Crashed in North Wales.
	Ju88A-5 of 1./KGr806	No. 600	Beaufighter If	F/O Woodward and Sgt Liscombe	Crashed near Stoke St Michael, Somerset.

Date	Claim	Squadron	Aircraft	Crew	Location
4/5 May	Ju88A-5 8./KG1	No. 25	Beaufighter If	W/Cdr D.F.W. Archerley and F/Lt J. Hunter-Tod	Crashed at Eastgate, near Bourne, Lincs.
	Ju88A-5 of 6./KG54	No. 604	Beaufighter If	P/O P.F. Jackson and Sgt S.N. Hawke	Crashed at East Burton, Dorset.
	Ju88A-5 of 2./KfGr106	N/K	N/K	N/K	Shot down by unidentified night-fighter. Crashed at Idle, near Bradford, Yorks.
5/6 May	He111H-5 of 1./KG4	N/K	N/K	N/K	Crashed at Whorlton Park, near Newcastle.
	Ju88A-5 of 4./KG54	No. 604	Beaufighter If	F/O I.K.S. Joll and Sgt R.W. Dalton	Crashed near Chawleigh, Devon.
6/7 May	He111H-5 of 7./KG27	No. 600	Beaufighter If	S/Ldr C.A. Pritchard and Sgt Gledhill	Crashed at Oborne, near Sherborne, Dorset.
	Ju88A-5 of 2./KG53	No. 141	Defiant I	F/O R.L.F. Day and P/O F.C.A. Lanning (AG)	Force landed on Holy Island, Northumberland.
	Ju88A-5 of 2./KfGr106	No. 141	Defiant I	S/Ldr E.C. Wolfe and Sgt A.E. Ashcroft (AG)	Crashed at Newlands, Lennoxtown, Stirlingshire.
7/8 May	He111H-5 of Stab I./KG4	No. 151	Hurricane I	P/O R.P. Stevens	Force landed at Withernsea, Yorks.
	He111P of Stab/KG27	No. 600	Beaufighter If	F/O Howden and unknown R/O	Crashed at Langford, Weston-super-Mare.
	He111P of Stab II./KG27	No. 600	Beaufighter If	F/O R.S. Woodward and unknown R/O	Crashed at Freshwater, Isle-of-Wight.
	He111H-5 of 7./KG27	No. 604	Beaufighter If	S/Ldr J. Cunningham and Sgt C.F. Rawnsley	Crashed near West Zoyland, Somerset.
	He111H-3 of Stab/KG53	No. 255	Defiant I	Sgt Johnson and Sgt Aitchison (AG)	Crashed near Bawtry, Yorks.
	He111P-4 of 1./KG55	No. 256	Defiant I	F/Lt E.C. Deansley and Sgt W.J. Scott (AG)	Crashed at Hazel Grove, Stockport, Cheshire.
	He111P-4 of 3./KG55	No. 256	Defiant I	P/O D. Tone and F/O R.L. Lamb (AG)	Crashed near the River Dee Marshes, Flintshire.
	He111P-3 of 3./KG55	No. 604	Beaufighter If	Sgt R. Wright and Sgt Vaughan	Crashed into the sea off Portland, Dorset.
	He111P-4 of 6./KG55	No. 256	Defiant I	F/Lt D.R. West and Sgt Adams (AG)	Crashed near Wrexham.
	Ju88A-5 of Stab II./KG76	N/K	N/K	N/K	Crashed near Gradbach, Staffs.
	He111H-2 of 3.KGr100	N/K	Defiant I	N/K	Crashed near Malpas, Cheshire.
	Do17Z-10 of 2./NJG2	No. 25	Beaufighter If	P/O D.W. Thompson and P/O L.D. Britain	Crashed near Boston, Lincs.
8/9 May	He111H-5 of 2./KG27	No. 29	Beaufighter If	F/O J.R.D. Braham and Sgt Ross	Crashed on Wimbledon Common, Greater London.
	He111H-5 of 4./KG53	No. 255	Defiant I	N/K	Crashed near Patrington Hall, Hull.
	He111H-5 of 6./KG53	No. 255	Defiant I	N/K	Crashed near Patrington Hall, Hull.
	He111P-4 of 6./KG55	No. 255	Defiant I	N/K	Force landed at Long Riston, Yorks.

Date	Claim	Squadron	Aircraft	Crew	Location
9/10 May	Ju88A-5 of 9./KG1	No.219	Beaufighter If	P/O A.J. Hodgkinson and Sgt B.E. Dye	Crashed into the sea off Selsey, Sussex.
10/11 May	He111P-2 of 7./KG27	No.29	Beaufighter If	P/O Grout and Sgt Stanton	Crashed into the sea off Seaford, Sussex.
	He111P-2 of 8./KG27	No.219	Beaufighter If	W/Cdr T.G. Pike and Sgt S. Austin	Crashed near Cranleigh, Surrey.
	He111H-5 of 1./KG28	No.85	Havoc I	F/Lt G. Raphael and AC1 W.N. Addison	Crashed at Galleywood, Chemsford, Essex.
	He111H-5 of 3./KG53	N/K	Defiant I	N/K	Crashed near Gillingham, Kent.
	He111H-5 of 5./KG53	No.74	Spitfire I	N/K	Force landed near Ashford, Kent.
	He111P of 9./KG55	N/K	N/K	N/K	Crashed at Withyham, Sussex.
	Unidentified He111	No.85	Havoc I	F/O Evans and Sgt Carter	Crashed somewhere in London.
16/17 May	Ju88A-5 of 7./KG1	No.600	Beaufighter If	F/Lt A.D. McNeil Boyd and F/O A.J. Clegg	Crashed at Plymtree, Devon.
	He111P-2 of 7./KG55	No.219	Beaufighter If	P/O A.J. Hodgkinson and Sgt B.E. Dye	Crashed near Stompting, Sussex.
28/29 May	He111P-2 of 7./KG27	No.604	Beaufighter If	W/Cdr C.H. Appleton and P/O D.A. Jackson	Crashed near Buckley, Flintshire.
31 May/ 1 June	He111H-5 of 1./KG27	No.604	Beaufighter If	F/Lt Gomm and P/O Curnow	Crashed near Tarrant Gunville, Dorset.
	He111P-2 of 8./KG27	No.604	Beaufighter If	S/Ldr J. Cunningham and Sgt C.F. Rawnsley	Crashed near Cranborne, Dorset.
1/2 June	Ju88A-5 of 1./KGr806	N/K	Beaufighter If	N/K	Crashed into the sea off Llandudno.
	Ju88A-5 of 3./KG54	No.219	Beaufighter If	W/Cdr Allen and unkown R/O	Crashed near Saddlescombe, Sussex.
2/3 June	Ju88A-5 of 1./KfGr106	No.317	Hurricane I	N/K	Crashed into the sea off Newcastle.
4/5 June	He111H-5 of 8./KG4	No.25	Beaufighter If	Sgt H. Gigney and Sgt G. Charnock	Force landed near Alford, Lincs.
	He111-5 of 3./KG27	No.604	Beaufighter If	F/O K.I. Geddes and Sgt A. Cannon	Crashed into the sea off Ventnor, Isle of Wight.
13/14 June	He111H-4 of 3./KG28	No.85	Havoc I	F/Lt G. Raphael and AC1 W.D. Addison	Exploded over the Isle of Grain.
	He111H-5 of III./KG4	No.85	Havoc I	F/Lt G. Raphael and AC1 W.D. Addison	Crashed into the North Sea.
	He111H-3 of 3./KGr100	No.604	Beaufighter If	S/Ldr G. Budd and Sgt G. Evans	Crashed into the sea to the south-south-east of Selsey Bill, Sussex.
	Ju88C-4 of 4./NJG 2	No.25	Beaufighter If	P/O D.W. Thompson and P/O L.D. Britain	Crashed near Kings Lynn, Norfolk.
14/15 June	He111H-1 of 2./KGr100	No.604	Beaufighter If	F/O K.I. Gedes and Sgt A.C. Cannon	Crashed near Sturminster Newton, Dorset.
16/17 June	He111H-2 of 1./KGr100	No.604	Beaufighter If	P/O W.M. Gossland and Sgt Philips	Crashed near Maiden Bradley, Wilts.

Date	Claim	Squadron	Aircraft	Crew	Location
	He 111H-3 of 3./KGr 100	No. 68	Beaufighter If	F/Lt D.S. Pain and F/O Davies	Crashed near Bratton, Wilts.
17/18 June	Ju 88A-5 of 1./KG 30	No. 25	Beaufighter If	W/Cdr D.F.W. Archerley and F/Lt J.Hunter-Tod	Crashed into the sea of Sherringham, Norfolk.
21/22 June	Ju 88C-4 of 1./NJG 2	No. 25	Beaufighter If	F/O M.J. Herrick and P/O Yeomans	Crashed near Deeping St James, Peterborough, Northants.
23 June	Ju 88A of 2./KfGr 106	No. 85	Havoc I	F/Lt G. Raphael and AC1 W.N. Addison	Crashed into the sea 10 miles east off Harwich, Essex.
24/25 June	He 111H of Stab III./ KG 4	No. 25	Beaufighter If	P/O D.W. Thompson and P/O L.D. Britain	Crashed near Lullington, Staffordshire.
25/26 June	Ju 88A-6 of 1./KG 30	No. 219	Beaufighter If	P/O A.J. Hodgkinson and Sgt B.E. Dye	Crashed into the sea off Selsey Bill, Sussex.
26/27 June	Unidentified He 111	FIU	Beaufighter If	F/Lt G.Ashfield, P/O Morris (Obs) and F/O Randle	Crashed into the sea off Beachy Head, Sussex.
4/5 July	He 111H-4 of 8./KG 4	No. 604	Beaufighter If	F/O I.S.K. Joll and Sgt R.W. Dalton	Crashed at Oakford, Devon.
	He 111H-5 of 7./KG 26	No. 604	Beaufighter If	F/Lt Pattern and F/Sgt Moody	Crashed near Frome, Somerset.
6/7 July	He 111H-5 of 8./KG 4	No. 29	Beaufighter If	S/Ldr G.P. Gibson and Sgt R.H. James	Crashed into the sea off Sheerness, Kent.
7/8 July	He 111H-5 of 9./KG 4	No. 118	Spitfire II	S/Ldr F.J. Howell	Crashed into the sea to the south-west of Brighton, Sussex.
	He 111H-5 of 3./KG 28	No. 604	Beaufighter If	F/O E.D. Crew and unknown R/O	Crashed near Lymington, Hants.
	He 111H-2 of Stab IV./KG 27	No. 604	Beaufighter If	F/O E.D. Crew and unknown R/O	Crashed into the sea off Selsey Bill, Sussex.
	He 111P-2 of 1./KGr 100	No. 604	Beaufighter If	F/Lt H. Speke and Sgt Dawson	Crashed into the sea off Boscombe, Hants.
	He 111P-3 of 2./KGr 100	No. 604	Beaufighter If	F/Lt H. Speke and P/O D.A. Jackson	Crashed near Lymington, Hants.
8/9 July	He 111H-5 of 7./KG 4	No. 604	Beaufighter If	F/Lt R.A. Chisholm and Sgt W.G. Ripley	Crashed near Kenton, Devon.
	Ju 88A-6 of 3./KG 30	No. 604	Beaufighter If	F/O K.I. Geddes and Sgt A.C. Cannon	Crashed near Alderholt, Dorset.
	Unidentified He 111	FIU	Beaufighter	W/Cdr Evans and Sgt Mitchell	Crashed into the English Channel.
13/14 July	Ju 88A of 2./KfGr 106	No. 85	Havoc I	F/Lt G. Raphael and Sgt W.N. Addison	Crashed into the sea off the Suffolk coast.
27/28 July	Ju 88A-5 of 2./KG 30	No. 219	Beaufighter If	P/O A.J. Hodgkinson and Sgt B.E. Dye	Crashed near Horsham, Sussex.
	Ju 88A-5 of 3.KGr 606	No. 219	Beaufighter If	F/O R.C. Miles and Sgt Hall	Crashed into the sea off Sheerness Boom.

Date	Claim	Squadron	Aircraft	Crew	Location
22/23 Aug	He 111H-5 of 8./KG 40	No. 604	Beaufighter If	W/Cdr J. Cunningham and P/O C.F. Rawnsley	Crashed into the sea off Wells-next-the-Sea, Norfolk.
1/2 Sept	Ju 88A-4 of Stab III./KG 30	No. 406	Beaufighter IIf	F/O R.C. Fumerton and Sgt L.P.S. Bing	Crashed at Bedlington, Northumberland.
16/17 Sept	Ju 88C-4 of 1./NJG 2	No. 85	Havoc II	S/Ldr G. Raphael and Sgt W.N. Addison	Crashed into the sea off Clacton, Essex.
20/21 Sept	He 111H-5 of 7./KG 40	No. 219	Beaufighter If	N/K	Crashed into the sea off Selsey, Sussex.
2/3 Oct	Do 217E-2 of 5./KG 2	No. 406	Beaufighter IIf	W/Cdr D.G. Morris and Sgt A.V. Rix	Crashed into the sea off Blyth, Northumberland.
9/10 Oct	He 111H-6 of 9./KG 40	No. 600	Beaufighter IIf	S/Ldr A.D. McNeil-Boyd and F/O A.J. Clegg	Crashed into the sea off St Ives, Cornwall.
12/13 Oct	He 111H-6 of 8./KG 40	No. 68	Beaufighter If	P/O M.J. Masefield and unknown R/O	Crashed into the sea off Holyhead, Anglesey.
22/23 Oct	He 111H-5 of 9./KG 40	No. 68	Beaufighter If	F/O W.D. Winward and Sgt K.C. Wood	Crashed into the sea off Nefyn, Caernarvonshire.
	Ju 88A-4 of 1./KGr 606	No. 256	Defiant I	F/Lt Coleman and F/Sgt Smith	Joint claim. Crashed at Woore, Shropshire.
1/2 Nov	Do 217E-2 of 4./KG 2	No. 151	Hurricane IIc	P/O R.P. Stevens	Crashed into the sea off Sidmouth, Devon.
	Ju 88A-5 of Erpro/KG 30	No. 307	Beaufighter IIf	Sgt Turzanski and Sgt Ostrowski	Crash site not known.
	He 111H-6 of 7./KG 40	No. 604	Beaufighter If	F/O E.D. Crew and Sgt Facey	Crashed near Gwalchmai, Anglesey.
15 Nov	Ju 88D-1 of 1./KfGr 106	No. 68	Beaufighter If	P/O M.C. Shipard and Sgt D.A. Oxby	Crashed into the sea off Yarmouth, Hants.
23/24 Nov	Ju 88A-4 of 8./KG 30	No. 151	Defiant I	P/O McRitchie and Sgt A.G. Beale	Crashed near Tamerton Folit, near Plymouth, Devon.
	Do 217E-2 of 5./KG 40	No. 307	Beaufighter IIf	P/O Dziegielewski and P/O Swierz	Crashed near Arbury, Nuneaton.
1942					
10/11 Jan	Ju 88A-4 of 3./KfGr 106	No. 456	Beaufighter IIf	S/Ldr Hamilton and P/O Norris-Smith	Crashed into the sea off Land's End, Cornwall.
25/26 Jan	Do 217E-4 of Stab III./KG 2	No. 600	Beaufighter IIf	S/Ldr A.D. McNeil-Boyd and F/Lt A.J. Clegg	Crashed into the sea off Blyth, Northumberland.
15 Feb	He 111H-6 of 9./KG 40	No. 141	Beaufighter If	F/O J.G. Benson and Sgt L. Brandon	Crashed into the Irish Sea to the south-west of Bull Rock Light.
21 March	He 111H-6 of 9./KG 40	No. 256	Beaufighter If	F/Sgt Schand and unkown R/O	Crashed into the sea off Portland, Dorset.
2/3 April	Unidentified Do 217	No. 604	Beaufighter If	F/Lt E.D. Crew and P/O Facey	Crashed into the Thames Estuary.
5/6 April		FIU	Beaufighter If	P/O R. Ryalls and F/Sgt Owen	
23/24 April	Do 217E-2 of 5./KG 2	No. 604	Beaufighter If	P/O Tharp and Sgt King	Crashed near Axminster, Devon.

Date	Claim	Squadron	Aircraft	Crew	Location
25/26 April	Do17Z-2 of 12./KG2	No.219	Beaufighter If	S/Ldr J.G. Topham and F/Lt Strange	Crashed into the sea off Worthing, Sussex.
	Ju88A-6 of 12./KG3	No.255	Beaufighter IIf	F/O Wyrill and Sgt Willins	Crashed near Builth Wells, Breacon.
29/30 April	Ju88D-1 of 1./KfGr106	No.253	Hurricane IIc	W/O Y. Mahe	Crashed near Elvington, Yorks.
3/4 May	Do17Z-2 of 12./KG2	No.604	Beaufighter If	F/Lt E.D. Crew and P/O B. Duckett	Crashed into the sea off Portland, Dorset.
	Ju88A-5 of 10./KG30	N/K	N/K	N/K	Crashed into the sea off Portland, Dorset.
	Ju88A-5 of 10./KG30	No.307	Beaufighter IIf	Sgt Illaszenic and P/O Lissowski	Crashed at Topham Barracks, Exeter.
	Ju88A-5 of 10./KG30	No.307	Beaufighter IIf	Sgt Illaszenic and P/O Lissowski	Force landed near Colyton, Devon.
7/8 May	He111H-6 of 7./KG100	No.29	Beaufighter If	P/O Carr and F/Sgt Whitby	Crashed near Ashford, Kent.
	He111H-6 of 7./KG100	No.219	Beaufighter If	S/Ldr J.G. Topham and F/O H.W. Berridge	Crashed at Patcham, near Brighton, Sussex.
18 May	Ju88D-1 of 3.(F)/123	No.456	Beaufighter IIf	P/O Willis and Sgt Lowther	Crashed into Cardigan Bay.
23 May	He111H-6 of 7./KG100	No.604	Beaufighter If	W/Cdr J. Cunningham and F/Lt C.F. Rawnsley	Crashed at Alvediston, Shaftsbury, Dorset.
23/24 May	Ju88A-4 of 4./KG77	No.604	Beaufighter If	S/Ldr S.H. Skinner and F/Sgt Lacey	Crashed into Studland Bay, Swanage, Dorset.
31 May/ 1 June	Ju88A-4 of 2./KfGr106	No.219	Beaufighter If	F/O J.C. Hooper and F/Sgt S.C. Hubbard	Crashed into the sea off Winchelsea, East Sussex.
2/3 June	Ju88A-4 of 4./KG77	No.219	Beaufighter If	F/Lt Robinson and F/Lt Crowther	Crashed into the sea off Foreness, Kent.
7/8 June	Ju88D-5 of 3.(F)/123	No.600	Beaufighter VIf	P/O A.B. Harvey and F/O B.B. Wicksteed	Crashed into the sea off Land's End, Cornwall. Beaufighter lost to return fire, but crew survived.
15/16 June	Do217E-4 of 1./KG2	FIU	Beaufighter	W/Cdr D.R. Evans and P/O Mitchell	Crashed into the sea off Littlehampton, Sussex.
21/22 June	Unidentified He111	No.604	Beaufighter If	P/O Foster and F/Sgt Newton	Crashed into the sea off Ventnor, Isle of Wight.
24/25 June	Do217E-4 of II./KG40	No.151	Mosquito NF.II	W/Cdr I.S. Smith and F/Lt Sheppard	Crashed into the North Sea.
	Do217E-4 of II./KG40	No.151	Mosquito NF.II	W/Cdr I.S. Smith and F/Lt Sheppard	Crashed into the Wash.
25/26 June	Unidentified Do217	No.151	Mosquito NF.II	F/Lt Moody and P/O March	Crashed in flames into the North Sea.
	Unidentified enemy aircraft	No.151	Mosquito NF.II	P/O Wain and F/Sgt Grieve	Crashed into the North Sea.

Date	Claim	Squadron	Aircraft	Crew	Location
27 June	Ju88D-5 of Wekusta 51	No. 125	Beaufighter IIf	S/Ldr F.D. Hughes and unknown R/O	Crashed into the sea off Hook Head, County Wexford, Eire.
7/8 July	Do217E-4 of 9./KG2	No. 406	Beaufighter VIf	N/K	Crashed into the sea of Middlesborough.
21 July	Do217E-4 of 3./KG2	No. 151	Mosquito NF.II	S/Ldr D.A. Pennington and F/Sgt D.A. Donnett	Crashed into the sea off Cromer, Norfolk.
	Do217E-4 of 1./KG2	No. 151	Mosquito NF.II	N/K	Crashed into the North Sea.
	He111H of IV./KG4	No. 157	Mosquito NF.II	S/Ldr G. Ashfield and unknown R/O	Crashed into the North Sea.
27/28 July	Do217E-4 of 3./KG2	No. 151	Mosquito NF.II	N/K	N/K
29/30 July	He111H-5 of IV./KG4	No. 125	Beaufighter IIf	N/K	N/K
	Do217E-4 of 6./KG2	No. 68	Beaufighter If	F/O Raybould and F/Sgt Mullaley	Crashed on Salthouse Marshes, Sheringham, Norfolk.
	Do217E-4 of 11./KG2	No. 151	Mosquito NF.II	N/K	Crashed into the North Sea off Lowestoft, Suffolk.
	Unidentified Do217E	No. 151	Mosquito NF.II	N/K	Crashed into the North Sea.
	He111H-5 of 12./KG40	No. 456	Beaufighter IIf	W/Cdr Wolfe and P/O Ashcroft	Crashed onto Pwllehi Beach, south-west Wales.
30/31 July	He111H-6 of 7./KG53	No. 604	Beaufighter VIf	N/K	N/K
	Do217E-4 of 7./KG2	N/K	N/K	N/K	Crashed into the sea off Southwold, Suffolk.
	He111H-6 of IV./KG55	No. 604	Beaufighter VIf	P/O B.R. Keele and P/O G.H. Cowles	Crashed near Preston, Dorset.
	Ju88A-4 of 2./KfGr106	No. 264	Mosquito NF.II	S/Ldr C.A. Cook and P/O R.E. McPherson	Crashed at Hornyold Field, near Malvern, Worcs.
	Ju88A-4 of 2./KfGR 106	No. 604	Beaufighter VIf	P/O B.R. Keele and P/O G.H. Cowles	Crashed into the sea off St Alban's Head, Dorset.
4/5 Aug	He111H-6 of 8./KG53	No. 125	Beaufighter If	N/K	Crashed into the sea off Ilfracombe, Devon.
	Ju88A-4 of 3./KfGr106	No. 307	Beaufighter VIf	F/O R.G. Ranoszek and Sgt Trzaskowski	Crashed into the sea off Start Point, Devon.
7/8 Aug	Do217E-4 of 7./KG2	No. 68	Beaufighter If	P/O P.H. Cleaver and F/Sgt Nairn	Crashed at Shire Wood, Revesby, Lincs.
8/9 Aug	Ju88A-4 of 6./KG54	No. 29	Beaufighter If	P/O G. Pepper and Sgt J.H. Troone	Crashed into the sea 10 miles south of Dungeness, Kent.
12/13 Aug	Ju88A-4 of 6./KG54	No. 85	Havoc II	Sgt Sullivan and Sgt Keel	Crashed at Redward Farm, near Burnham-on-Crouch, Essex.
19 Aug	Ju88A-3 of 3./KfGr106	No. 141	Beaufighter If	Sgt Clee and Sgt Grant	Crashed into the sea to the south-east off Selsey Bill, Sussex.
22/23 Aug	Do217E-4 of 6./KG2	No. 157	Mosquito NF.II	W/Cdr R.G. Slade and P/O Truscott	Crashed at Lodge Farm, near Worlingworth, Suffolk.
23/24 Aug	Do217E-4 of 1./KG2	No. 25	Beaufighter If	N/K	Crashed into the sea to the east of Mablethorpe, Lincs.

Date	Claim	Squadron	Aircraft	Crew	Location
	Do217E-4 of 2./KG2	No.25	Beaufighter If	S/Ldr W.J. Alington and F/O D.B. Keith	Crashed at Walton Wood, near East Walton, Norfolk.
28/29 Aug	Ju88A-4 of Stab I./KG77	No.406	Beaufighter VIf	F/Lt J.R.B. Frith and P/O F.G.Harding	Crashed while attempting to land at Scorton, Yorks.
7/8 Sept	He111H-6 of 15./KG6	No.85	Havoc II	Sgt K.R. McCormick and Sgt W. Nixon	Crashed at Buxey Sands, Essex.
8/9 Sept	Do217E-4 of 6./KG40	No.151	Mosquito NF.II	F/O I.A. Ritchie and F/Lt E.S. James	Crashed at Tectory Farm, Orwell, Cambs.
16/17 Sept	Ju88C-4 of 1./NJG 2	No.85	Havoc II	F/Lt G. Raphael and Sgt W.N. Addison	Crashed into the sea off Clacton, Essex.
17/18 Sept	Ju88D-1 of 1(F)/33	No.141	Beaufighter If	W/O R.C. Hammer and unknown R/O	W/O Hammer was killed by return fire from the Ju88, but his R/O survived.
	Do217E-4 of 7./KG2	No.151	Mosquito NF.II	F/Lt H.E. Bodien and Sgt G.B. Brooker	Crashed at Fring Hall, Docking, Norfolk.
19/20 Sept	Do217E-4 of 7./KG2	No.219	Beaufighter VIf	S/Ldr J.G. Topham and F/O H.W. Berridge	Crashed into the sea off Tynemouth.
25/26 Sept	Do217E-4 of 1./KG2	No.406	Beaufighter IIf	S/Ldr D.C. Furse and P/O J.H. Downes	Crashed in the village of St Just, Cornwall.
30 Sept	Unidentified Ju88A-4	No.157	Mosquito NF.II	N/K	Crashed into the sea off the Dutch coast.
19 Oct	Ju88D-1 of 3(F)/33	No.68	Beaufighter If	F/Lt W.D. Winward and P/O C.K. Wood	Crashed into the sea off Cromer, Norfolk.
	Unidentified Ju88A-4	No.157	Mosquito NF.II	N/K	Crashed into the sea off Southwold, Suffolk.
26 Oct	Unidentified Ju88D-1	No.157	Mosquito NF.II	N/K	Crashed into the sea off Beachy Head, Sussex – daylight action.
31 Oct/ 1 Nov	Do217E-4 of 2./KG2	No.29	Beaufighter If	F/O G. Pepper and P/O Toone	Crashed at Great Pett Farm, Bridge, Kent.
	Do217E-4 of 3./KG2	No.29	Beaufighter If	F/O G. Pepper and P/O Toone	Crashed into the sea off Folkstone.
16 Dec	Do217E-4 of 4./KG40	No.141	Beaufighter If	F/O Cook and F/Sgt Warner	Crashed into a gasometer at Bognor Regis after being chased.
1943 15/16 Jan	Do217E-4 of 7./KG2	No.151	Mosquito NF.II	Sgt E.A. Knight and Sgt W.I.L. Roberts	Crashed at Boothby Graffoe, Lincs.
17/18 Jan	Ju88A-14 of 1./KG6	No.29	Beaufighter If	W/Cdr C.M. Wright-Boycott and F/O E.A. Sanders	Crashed at Brenzett, Kent.
	Ju88A-14 of Stab I./KG6	No.29	Beaufighter If	W/Cdr C.M. Wright-Boycott and F/O E.A. Sanders	Crashed at Caterham, Surrey.
	Ju88A-4 of 2./KG6	No.29	Beaufighter If	S/Ldr I.G. Esplin and F/O A.H.J. Palmer	Crashed at Lovelace Place Farm, near Bethersden, Kent.

Date	Claim	Squadron	Aircraft	Crew	Location
20/21 Jan	Do 217E-4 of 4./KG 2	No. 141	Beaufighter If	W/Cdr J.R.D. Braham and F/O W.J. Gregory	Crashed into the sea to the south-west of Dungeness, Kent.
3/4 Feb	Unidentified Do 217	FIU	Beaufighter	F/Lt Davison and F/Lt Clarke	Crashed into the English Channel.
	Do 217E-4 of 3./KG 2	No. 219	Beaufighter VIf	F/Lt J.E. Willson and F/O D.C. Bunch	Crashed near Muston, Yorks.
16/17 Feb	Do 217E-4 of 4./KG 2	No. 125	Beaufighter VIf	P/O Newton and Sgt Rose	Crashed into the sea of the Mumbles, near Gower Peninsular.
	Do 217E-4 of 5./KG 2	No. 125	Beaufighter VIf	W/Cdr R.F.H. Clerke and P/O D.H. Spurgen	Crashed attempting emergency landing near Beaminster, Dorset.
7/8 March	Do 217E-4 of 1./KG 2	No. 29	Beaufighter If	S/Ldr G.H. Goodman and F/O W.F.E. Thomas	Crashed at Van Common, Fernhurst, Sussex.
11/12 March	Do 217E-4 of 3./KG 2	No. 219	Beaufighter VIf	F/Lt J.E. Willson and F/O D.C. Bunch	Crashed near Great Stainton, near Darlington, County Durham.
15/16 March	Do 217E-4 of 6./KG 2	No. 219	Beaufighter VIf	F/Lt J.E. Willson and P/O Holloway	Crashed between Grainsgy Halt and North Thornsby, Lincs.
18/19 March	Do 217E-4 of 1./KG 2	No. 410	Mosquito NF.II	F/O D. Williams and P/O P. Dalton	Force landed on salt marshes near Terrington St Clements, Norfolk.
27 March	Ju 88A-14 of 4./KG 6	No. 604	Beaufighter If	F/Lt T. Wood and F/O Ellis	Crashed into the sea off Worthing Pier, Sussex.
14/15 April	Do 217E-4 of 6./KG 2	No. 157	Mosquito NF.II	F/Lt J.B. Benson and F/O L. Brandon	Crashed near Layer Breton Heath, Essex.
	Do 217E-4 of 4./KG 40	No. 85	Mosquito NF.XII	S/Ldr W.P. Green and F/Sgt A.R. Grimstone	Crashed in the Barrow Deep, Thames Estuary.
	Do 217E-4 of 5./KG 40	No. 85	Mosquito NF.XII	F/Lt G.L. Howitt and F/O G.L. Irving	Force landed near Bockings Elm, Clacton-on-Sea, Essex.
24/25 April	Ju 88A-14 of 8./KG 6	No. 85	Mosquito NF.XII	F/O J.P.M. Lintott and Sgt G.C. Gilling-Lax	Crashed at Bromley, Kent.
16/17 May	Fw 190A-4 of SKG 10	No. 85	Mosquito NF.XII	S/Ldr W.P. Green and F/Sgt A.R. Grimstone	Crashed into the English Channel.
	Fw 190A-4 of 3./SKG 10	No. 85	Mosquito NF.XII	F/O Thwaites and unknown R/O	Crashed into the English Channel.
	Fw 190A-4 of SKG 10	No. 85	Mosquito NF.XII	F/O Thwaites and unknown R/O	Crashed near Ashford, Kent.
	Do 217K-1 of 6./KG 2	No. 604	Beaufighter VIf	F/O Keele and F/O Cowles	Crashed into the sea off Sunderland.
	Fw 190A-4 of 1./SKG 10	No. 85	Mosquito NF.XII	F/O J.D.R. Shaw and F/O A.L. Howton	Crashed at Higham, near Gravesend, Kent.
	Fw 190A-4 of l.SKG 10	No. 85	Mosquito NF.XII	F/O G.L. Howitt and unknown R/O	Crashed into the sea off Hastings, Sussex.
17/18 May	Ju 88A-14 of l./KG 6	No. 151	Mosquito NF.XII	Sgt H.K. Kemp and Sgt R.J. Maidment	Crashed at Timbercombe, Somerset.
21/22 May	Fw 190A-4 of SKG 10	No. 85	Mosquito NF.XII	S/Ldr E.D. Crew and unknown R/O	Crashed into the sea to the north-west of Hardelot, Holland.

Date	Claim	Squadron	Aircraft	Crew	Location
29/30 May	Ju 88S-1 of I./KG 66	No. 85	Mosquito NF.XII	F/Lt J.P.M. Lintott and Sgt G.G. Gilling-Lax	Crashed at Isfield, Sussex.
12/13 June	Do 217E-4 of 1./KG 2	No. 125	Beaufighter VIf	N/K	Crashed into the sea off Plymouth.
	Ju 88A-14 of 9./KG 6	No. 125	Beaufighter VIf	N/K	Crashed into the sea off Plymouth.
	Ju 88A-14 of 2./KG 6	No. 125	Beaufighter VIf	P/O McLachlin and P/O W.E. Pettifer	Crashed at Stoke, Devon.
13/14 June	Ju 88A-14 of 7./KG 6	N/K	N/K	N/K	Crashed at Frittiscombe, Devon.
	Fw 190A-5 of 3./SKG 10	No. 85	Mosquito NF.XII	W/Cdr J. Cunningham and F/Lt C.F. Rawnsley	Crashed near Borough Green, Kent.
21/22 June	Fw 190A-5 of 2./SKG 10	No. 85	Mosquito NF.XII	F/Lt W.H. Maguire and F/O W.D. Jones	Crashed into the River Medway at Strood.
3/4 July	Unidentified Ju 88D-1	FIU	Mosquito	W/Cdr R. Chisholme and F/Sgt Bamford	Crashed into the English Channel.
9 July	Do 217K-1 of 6./KG 2	No. 85	Mosquito NF.XII	F/Lt J.M.P. Linott and Sgt G.C. Gilling-Lax	Crashed near Detling, Kent.
13/14 July	Me 410A-1 of 16./KG 2	No. 85	Mosquito NF.XII	F/Lt E.N. Bunting and P/O F. French	Crashed into the sea 5 miles off Felixstowe, Suffolk.
15/16 July	Me 410A-1 of V./KG 2	No. 85	Mosquito NF.XII	F/Lt Thwaites and unknown R/O	Crashed into the sea off Dunkirk.
25/26 July	Do 217K-1 of Stab/KG 2	No. 604	Beaufighter VIf	F/O B.R. Keele and F/O G.H. Cowles	Crashed in flames into the sea to the east of Spurn Head, Lincs.
	Do 217M-1 of 2./KG 2	No. 604	Beaufighter VIf	F/O B.R. Keele and F/O G.H. Cowles	Crashed into the sea off Spurn Head, Lincs.
29 July	Me 410A-1 of 16./KG 2	No. 256	Mosquito NF.XII	W/Cdr G.R. Parks and unknown R/O	Crashed into the sea to the south of Beachy Head, Sussex.
11/12 Aug	Me 410A-1 of V./KG 2	No. 256	Mosquito NF.XII	N/K	Crashed into the sea to the south of Beachy Head, Sussex.
	Ju 88D-1 of 1(F)/121	No. 125	Beaufighter VIf	Sgt W.F. Millar and Sgt F.C. Bone	Crashed into the sea to the south-west of Plymouth, Devon.
15 August	Do 217M-1 of KG 2	No. 256	Mosquito NF.XII	W/Cdr G.R. Parks and unknown R/O	Crashed into the sea off Ford, Sussex.
	Do 217 of KG 2	No. 256	Mosquito NF.XII	W/Cdr G.R. Parks and unknown R/O	Crashed into the sea off Ford, Sussex.
15/16 Aug	Do 217 of KG 2	No. 125	Beaufighter VIf	N/K	No further details available.
	Do 217 of KG 2	No. 125	Beaufighter VIf	N/K	No further details available.
	Do 217M-1 of 1./KG 2	No. 410	Mosquito NF.II	P/O R.D. Schultz and P/O V.A. Williams	Crashed into the sea off Beachy Head, Sussex.
17/18 Aug	Do 217E-4 of 5./KG 2	N/K	N/K	N/K	Crashed into the sea off the Lincolnshire coast.
22/23 Aug	Me 410A-1 of 15./KG 2	No. 29	Mosquito NF.II	F/Lt C. Kirland and P/O R.C. Raspin	Crashed into the sea 25 miles off Foreness, Kent.
	Me 410A-1 of 15./KG 2	No. 85	Mosquito NF.XII	S/Ldr G.L. Howitt and P/O J.C.O. Medworth	Crashed near Chelmondiston, Suffolk.

Date	Claim	Squadron	Aircraft	Crew	Location
24 Aug	Me 410A-1 of 16./KG 2	N/K	Mosquito	N/K	Crashed into the North Sea off East Anglia.
6/7 Sept	Fw 190A-5 of I./SKG10	No.85	Mosquito NF.XII	F/Lt C.G. Houghton and F/O A.G. Peterson	Crash landed at Hawstead, Bury St Edmunds, Suffolk.
	Fw 190A-5 of 1./SKG10	No.85	Mosquito NF.XII	S/Ldr G.L. Howitt and F/O G.N. Irving	Crashed into the sea off Clacton, Essex.
8/9 Sept	Fw 190A-5 of I./SKG10	No.85	Mosquito NF.XII	W/Cdr J. Cunningham and F/Lt C.F. Rawnsley	Crashed into the sea off Aldeburgh, Suffolk.
	Fw 190A-5 of I./SKG10	Mo.85	Mosquito NF.XII	F/Lt Thwaites and unknown R/O	Crashed into the sea off the Suffolk coast.
	Fw 190A-5 of I./SKG10	No.85	Mosquito NF.XII	N/K	Crashed into the sea off the Suffolk coast.
15/16 Sept	Me 410A-1 of 15./KG 2	No.29	Mosquito NF.XII	N/K	Crashed into the sea to the south-east of Beachy Head, Sussex.
	Ju 88A-14 of 4./KG6	No.488	Mosquito NF.XII	N/K	Crashed into the sea off the Kent coast.
	Do 217M-1 of 9./KG 2	No.488	Mosquito NF.XII	N/K	Crashed into the sea off the Kent coast.
	Ju 88A-14 of 6./KG6	No.85	Mosquito NF.XII	F/O E.R. Hedgecoe and P/O J.R. Whitham	Crashed into the sea off Dungeness. Mosquito hit by return fire and abandoned near Ashford.
7/8 Oct	Me 410A-1 of 16./KG 2	No.85	Mosquito NF.XII	Lt T. Weisteen and F/O F.G. French	Crashed into the sea off Dungeness, Kent.
	Me 410A-1 of 16./KG 2	No.85	Mosquito NF.XII	S/Ldr W.H. Maguire and unknown R/O	Crashed into the sea off Hastings, Sussex.
8/9 Oct	Ju 88S-1 of 7./KG6	No.85	Mosquito NF.XII	F/Lt E.N. Bunting and F/O C.P. Reed	Crashed into the sea to the south of Dover, Kent.
	Ju 88S-1 of 8./KG6	No.85	Mosquito NF.XII	F/O S.V. Holloway and W/O Stanton	Crashed into the Thames Estuary off Foulness, Essex.
	Ju 188E-1 of Erpo Stab./KG6	No.85	Mosquito NF.XII	N/K	Crashed into the sea to the south-east of Dover, Kent.
15/16 Oct	Ju 188E-1 of 1./KG6	No.85	Mosquito NF.XII	F/O H.B. Thomas and W/O C.B. Hamilton	Crashed near Nicholas at Wade, Kent.
	Ju 188E-1 of 1./KG6	No.85	Mosquito NF.XII	S/Ldr W.H. Maguire and F/O D.W. Jones	Crashed into the sea off Clacton, Essex.
	Ju 188E-1 of 3./KG6	No.85	Mosquito NF.XII	S/Ldr W.H. Maguire and F/O D.W. Jones	Crashed at Kirton Creek, Hemley, Suffolk.
17/18 Oct	Me 410A-1 of 15./KG 2	No.85	Mosquito NF.XII	F/Lt E.N. Bunting and F/O C.P. Reed	Crashed near West Hordon, Brentwood, Essex.
30/31 Oct	Ju 88S-1 of 8./KG6	No.85	Mosquito NF.XII	F/O R. Robb and F/O R.C.J. Rye	Crashed into the sea to the south-east of Rye, Sussex.
1/2 Nov	Ju 188E-1 of Erpro./KG6	No.29	Mosquito NF.XII	F/Lt S.F. Hodsman and W/O A.F. Monger	Crashed at Combe, Berkshire.
5/6 Nov	Ju 88S-1 of 8./KG6	No.410	Mosquito NF.II	F/O Green and P/O E.G. White	Crashed into the sea off Dungeness, Kent.

Date	Claim	Squadron	Aircraft	Crew	Location
8/9 Nov	Me 410A-1 of 14./KG 2	No. 488	Mosquito NF.XIII	F/O Reed and P/O Bricker	Crashed into the sea to the east of Clacton, Essex.
	Me 410A-1 of 15./KG 2	No. 85	Mosquito NF.XII	S/Ldr W.H. Maguire and F/O W.D. Jones	Crashed at Eastborne, Sussex.
20/21 Nov	Fw 190A-5 of 1./SKG 10	No. 29	Mosquito NF.XIII	F/Lt R.C. Pargeter and F/Lt R.L. Fell	Crashed at Broadbridge Heath, Sussex.
22 Nov	Unidentified Fw 200C-4	No. 307	Mosquito NF.II	F/Lt Sgt Jaworski and unknown R/O	Crashed into the sea 120 miles north-east of the Shetland Islands.
26 Nov	Unidentified Ju 88	No. 307	Mosquito NF.II	N/K	Crashed into the sea off the Shetland Islands.
10/11 Dec	Do 217M-1 of 2./KG 2	No. 410	Mosquito NF.II	F/O R.D. Schultz and F/O V.A. Williams	Crashed into the sea off Clacton, Essex.
	Do 217M-1 of I./KG 2	No. 410	Mosquito NF.II	F/O R.D. Schultz and F/O V.A. Williams	Crashed into the sea off Clacton, Essex.
	Do 217M-1 of I./KG 2	No. 410	Mosquito NF.II	F/O R.D. Schultz and F/O V.A. Williams	Crashed into the sea off Clacton, Essex.
19/20 Dec	Me 410A-1 of 14./KG 2	No. 488	Mosquito NF.XII	P/O D.N. Robinson and F/O W.T.Clark	Crashed near Iden, Sussex.
1944					
2/3 Jan	Me 410A-1 of 16./KG 2	No. 488	Mosquito NF.XIII	N/K	Crashed in the vicinity of Dover, Kent.
	Fw 190A-5 of 1./SKG 10	No. 96	Mosquito NF.XIII	N/K	Crashed at Camber Sands, Sussex.
4/5 Jan	Ju 88S-1 of 1./KG 66	No. 96	Mosquito NF.XII	W/Cdr E.D. Crew and W/O W.R. Croysdill	Crashed into the sea 3 miles to the south of Hastings, Sussex.
21/22 Jan	Do 217M-1 of I./KG 2	No. 488	Mosquito NF.XIII	F/Lt J.A.S. Hall and F/O J. Cairns	Crashed into the sea 13 miles to the south of Dungeness, Kent.
	He 177A-3 of 1./KG 40	No. 151	Mosquito NF.XIII	W/O H.K. Kemp and F/Sgt J.R. Maidment	Crashed in forced landing at Whitmore Vale, Hindhead, Surrey.
	Ju 88A-14 of 6./KG 54	No. 488	Mosquito NF.XIII	F/Lt J.A.S. Hall and F/O J. Cairns	Crashed near Sellindge, Kent.
	Me 410A-1 of Stab V./KG 2	N/K	N/K	Sub-Lt J.A. Lawley-Wakelin and Sub-Lt H. Williams	Crashed over the Lydd Ranges, Kent.
	Ju 88A-4 to 6./KG 30	No. 96	Mosquito NF.XVII	N/K	Crashed at Paddock Wood, Kent.
	He 177A-3 of 2./KG 40	No. 85	Mosquito NF.XVII	F/O C.K. Nowell and F/Sgt F. Randall	Crashed into the sea to the south-east of Hastings, Sussex.
28/29 Jan	Unidentified Ju 88	No. 410	Mosquito NF.XIII	N/K	Shared kill. Crashed near Biddenden, Kent.
	Ju 88A-4 of 5./KG 6	No. 96	Mosquito NF.XIII	N/K	Damaged and eventually crashed at St Omer, France.
29/30 Jan	Ju 88A-4 of 3./KG 54	No. 96	Mosquito NF.XIII	F/O S.A. Hibbert and F/O G.D. Moody	Crashed near Coddenham, Suffolk.
		No. 68	Beaufighter IVf	F/Sgt L.W. Neal and F/Sgt E. Eastwood	
3/4 Feb	Unidentified Do 217	No. 488	Mosquito NF.XIII	F/Sgt Vlotman AND unknown R/O	Crashed into the sea to the east of Foreness, Kent.

Date	Claim	Squadron	Aircraft	Crew	Location
13/14 Feb	Unidentified Do 217	No. 410	Mosquito NF.XIII	F/O E.S.P. Fox and unknown R/O	Crashed into the sea off Orfordness, Suffolk.
	Ju 88A-4 of 5./KG 6	No. 96	Mosquito NF.XIII	W/Cdr E.D. Crew and W/O W.R. Croysdill	Crashed near Whitstable, Kent.
	Ju 88S-1 of 1./KG 66	No. 406	Mosquito NF.XII	S/Ldr J.D. Sommerville and F/O G.D. Robinson	Crashed at Havering-atte-Bower, Romford, Essex. Also hit by AA fire.
	Unidentified Ju 188E-1	No. 410	Mosquito NF.XIII	N/K	Crashed into the Thames Estuary.
	Unidentified Ju 188E-1	No. 488	Mosquito NF.XIII	F/Lt E.N. Bunting and unknown R/O	Crashed into the sea.
19/20 Feb	Unidentified Me 410A-1	No. 96	Mosquito NF.XIII	N/K	Crashed into the sea to the south of Dungeness, Kent.
20/21 Feb	Ju 188E-1 of 5./KG 2	No. 25	Mosquito NF.XVII	P/O J.R. Brockbank and P/O D. McCausland	Crashed near Wickham St Paul, Essex.
	Unidentified Ju 188E-1	No. 25	Mosquito NF.XVII	F/Lt Singleton and unknown R/O	Crashed into the sea off the East Anglian coast.
22/23 Feb	Unidentified Do 217	No. 25	Mosquito NF.XVII	N/K	Crashed over Norfolk.
	Unidentified Me 410A-1	No. 85	Mosquito NF.XII	N/K	Crashed into the English Channel.
	Unidentified Me 410A-1	No. 85	Mosquito NF.XII	N/K	Crashed into the English Channel.
	Ju 88A-4 of 8./KG 66	No. 410	Mosquito NF.XIII	S/Ldr C.A.S. Anderson and unknown R/O	Crashed near Earls Colne, Essex.
	Ju 188E-1 of 2./KG 6	No. 410	Mosquito NF.XIII	S/Ldr C.A.S. Anderson and unknown R/O	Crashed at Bullers Farm, Shopland, near Romford, Essex.
24/25 Feb	Do 217M-1 of 3./KG 2	No. 29	Mosquito NF.XII	S/Ldr C. Kirkland and F/O Raspin	Crashed at Westcott, near Dorking, Surrey.
	Do 217M-1 of 3./KG 2	No. 29	Mosquito NF.XIII	F/Lt E. Barry and F/O G. Hopkins	Crashed near Ashford, Kent.
	Ju 188E-1 of 6./KG 2	No. 29	Mosquito NF.XIII	F/Lt R.C. Pargeter and F/Lt R.L. Fell	Crashed at Queens Farm, near Shorne, Kent.
	Ju 188E-1 of 2./KG 66	No. 29	Mosquito NF.XIII	F/O Provan and W/O Nichol	Crashed at Great Steel Farm, near Framfield, Sussex.
	He 177A-3 of 2./KG 100	No. 488	Mosquito NF.XIII	F/Lt P.F.L. Hall and F/O R.D. Marriott	Crashed at Chequers Farm, near Lamberhurst, Kent.
25/26 Feb	Unidentified Ju 188E-1	No. 25	Mosquito NF.XVII	N/K	Crashed into the sea off Great Yarmouth, Norfolk.
27/28 Feb	Unidentified Ju 88	No. 456	Mosquito NF.XVII	N/K	Crashed at Beer, near Portland, Dorset.
	Unidentified Ju 88	No. 456	Mosquito NF.XVII	N/K	Crashed into the sea off the south-west coast.
29 Feb/ 1 March	Unidentified He 177A-3	No. 85	Mosquito NF.XII	N/K	Crashed into the English Channel.
	Fw 190A-5 of SKG 10	No. 96	Mosquito NF.XIII	N/K	Crashed into the sea off Dieppe.
1/2 March	Unidentified He 177A-3	No. 151	Mosquito NF.XII	W/Cdr G.H. Goodman and F/O W.F.E. Thomas	Crashed at Hammer Wood, near East Grinstead, Sussex.
	Unidentified Me 410A-1	No. 96	Mosquito NF.XIII	F/O Gough and unknown R/O	Crashed on the French mainland.

Date	Claim	Squadron	Aircraft	Crew	Location
	Unidentified Ju 188E-1	No.151	Mosquito NF.XII	W/Cdr G.H. Goodman and F/O W.F.E. Thomas	Crashed into the English Channel.
	Unidentified Ju 188E-1	No.151	Mosquito NF.XII	S/Ldr Harrison and unknown R/O	Crashed into the English Channel.
	Unidentified Ju 88	No.151	Mosquito NF.XII	F/Lt Stevens and unknown R/O	Crashed into the English Channel.
14/15 March	Ju 188E-1 of 4./KG2	No.488	Mosquito NF.XIII	S/Ldr Bunting and F/Lt C.P. Reed	Crashed at White House Farm, Great Leighs, Chelmsford, Essex.
	Ju 88A-14 of 6./KG6	No.68	Beaufighter VIf	W/Cdr D. Hayley-Bell and F/O H.W. Uezzell	Crashed at Gants Hill, near Ilford, Essex.
	Ju 88A-4 of 6./KG30	No.96	Mosquito NF.XIII	F/Lt N.S. Head and F/O A.C. Andrews	Crashed near Holmwood Common, Dorking, Surrey.
	Unidentified Ju 188	No.96	Moquito NF.XIII	F/O Gough and unknown R/O	Crashed into the English Channel off Sussex.
	Ju 88A-14 of 2./KG54	No.410	Mosquito NF.XIII	Lt Harrington (USAAF) and Sgt D.G. Tongue	Crashed at Merchants Farm near Hildenborough, Kent.
	Unidentified Ju 88	No.410	Mosquito NF.XIII	S/Ldr Green and unknown R/O	Crashed into the sea off the East Anglian coast.
19/20 March	Do 217M-1 of 2./KG2	No.264	Mosquito NF.XII	F/O R.L.J. Barbour and F/O Paine	Crashed near Legbourne, Lincs.
	He 177A-3 of 2./KG100	No.307	Mosquito NF.XII	P/O J. Bruchoci and F/Lt Ziolkowski	Crashed into the sea off Skegness, Lincs.
	Unidentified Ju 188	No.25	Mosquito NF.XVII	F/Lt Singleton and unknown R/O	Crashed off the East Anglian coast.
	Unidentified Ju 188	No.25	Mosquito NF.XVII	F/Lt Singleton and unknown R/O	Crashed off the East Anglian coast. Following this action Singleton made a forced landing near Coltishall. He and his R/O escaped uninjured.
21/22 March	Ju 188E-1 of 2./KG6	No.488	Mosquito NF.XIII	S/Ldr E.N. Bunting and F/Lt C.P. Reed	Crashed at Butlers Farm, Shopland, Essex.
	Ju 88A-14 of 7./KG6	No.410	Mosquito NF.XIII	N/K	Crashed into the sea off the East Anglian coast.
	Ju 88A-14 of 8./KG6	No.488	Mosquito NF.XIII	F/Lt J.A.S. Hall and F/O Cairns	Crashed on Earls Colne Airfield, Essex, destroying three B-26s.
	Ju 88A-4 of 4./KG30	No.410	Mosquito NF.XIII	F/O S.B. Huppert and P/O J. Christie	Crashed near Latchington, Essex.
	Ju 88A-4 of 9./KG30	No.488	Mosquito NF.XIII	S/Ldr E.N. Bunting and F/Lt C.P. Reed	Crashed near Cavendish, Suffolk.
	Ju 88A-4 of 4./KG54	No.604	Mosquito NF.XIII	F/Lt J.C. Surman and F/Sgt C.E. Weston	Crashed into the sea off the Isle of Sheppey, Kent.
	Unidentified Ju 188	No.25	Mosquito NF.XVII	F/Lt L.R. Davies and unknown R/O	Crashed into the sea off the East Anglian coast.

Date	Claim	Squadron	Aircraft	Crew	Location
22/23 March	Unidentified Ju 188	No. 25	Mosquito NF.XVII	F/Lt L.R. Davies and unknown R/O	Crashed into the sea off the East Anglian coast.
	Fw 190G-3 of Stab I./SKG 10	No. 85	Mosquito NF.XVII	F/Lt N.S. Head and unknown R/O	Crashed into the sea to the south-west of Pevensey, Sussex.
	Fw 190G-3 of I./SKG 10	No. 85	Mosquito NF.XVII	S/Ldr B.J. Thwaites and F/O W.P. Clemo	Crashed into the sea off Hastings, Sussex.
24/25 March	Ju 88A-4 of 6./KG 6	No. 456	Mosquito NF.XVII	W/Cdr K.M. Hampshire and F/O T. Condon	Crashed near Arundel, Sussex.
	Unidentified Ju 188	No. 25	Mosquito NF.XVII	N/K	Crashed into the sea 45 miles east of Great Yarmouth, Norfolk.
27/28 March	Unidentified Ju 188	No. 85	Mosquito NF.XVII	N/K	Crashed into the English Channel.
	Unidentified Ju 188	No. 85	Mosquito NF.XVII	N/K	Crashed into the English Channel.
	Ju 188E-1 of 5./KG 2	No. 68	Beaufighter VIf	F/O Russell and F/Lt Weir	Crashed near Coxely, Somerset.
	Ju 88A-4 of 9./KG 6	No. 456	Mosquito NF.XVII	W/Cdr K.M. Hampshire and F/O T. Condon	Crashed at Beer, Devon.
	Ju 88A-4 of 2./KG 30	No. 406	Beaufighter VIf	F/Lt H.D. McNabb and P/O J.N.L. Hall	Crashed near Berkley, Gloucester.
	Ju 88A-4 of 2./KG 54	No. 219	Mosquito NF.XVII	S/Ldr Ellis and F/Lt Craig	Crashed at Hestar Combe, Somerset.
	Ju 88A-4 of 9./KG 6	No. 456	Mosquito NF.XVII	W/Cdr K. Hampshire and F/O T.Condon	Crashed at Isle Brewers, near Ilminster, Somerset.
13/14 April	Unidentified Fw 190	No. 96	Mosquito NF.XIII	N/K	Crashed into the English Channel.
	Unidentified Ju 88	No. 96	Mosquito NF.XIII	N/K	Crashed near Le Touquet.
	Unidentified Me 410A	No. 96	Mosquito NF.XIII	N/K	Crashed into the English Channel.
	Unidentified Me 410A	No. 96	Mosquito NF.XIII	N/K	Crashed into the English Channel.
18/19 April	Ju 188E-1 of 4./KG 2	No. 25	Mosquito NF.XVII	F/Lt Carr and F/Lt Saunderson	Crashed into the sea 3 miles to the south of Southwold, Suffolk.
	Ju 188E-1 of 5./KG 2	No. 85	Mosquito NF.XVII	W/Cdr C.M. Miller and Capt L. Lovestad, RNorAF	Crashed near Ivychurch, Kent.
	Ju 88A-4 of 6./KG 6	No. 96	Mosquito NF.XIII	P/O Allen and F/Sgt Patterson	Crashed near Cranbrook, Kent.
	Me 410A-1 of 1./KG 51	No. 96	Mosquito NF.XIII	W/Cdr E.D. Crew and W/O W.R. Croysdill	Crashed at St Nicholas Churchyard, Brighton, Sussex.
	He 177A-3 of 2./KG 100	No. 410	Mosquito NF.XIII	F/O S.B. Huppert and P/O J.S. Christie	Crashed near Saffron Walden, Essex.
19/20 April	Me 410A-1 of 1./KG 51	No. 456	Mosquito NF.XVII	F/Lt C.L. Brooks and W/O R.J. Forbes	Crashed near Nuthurst, Sussex.
	He 177A-3 of 2./KG 100	No. 264	Mosquito NF.XIII	N/K	Crashed into the sea 40 miles to the east of Spurn Head, Lincs.
23/24 April	Unidentified Ju 88	No. 25	Mosquito NF.XVII	N/K	Crashed into the North Sea.
	Ju 88A-14 of 4./KG 30	No. 125	Mosquito NF.XVII	S/Ldr E.G. Barwell and F/Lt D. Haige	Broke up and crashed near Hill Deverill, Wilts.
	Unidentified Ju 88	No. 456	Mosquito NF.XVII	N/K	Crashed into the sea off Swanage.

Date	Claim	Squadron	Aircraft	Crew	Location
25/26 April	Me 410B-4 of 1.(F)/122	No. 85	Mosquito NF.XVII	F/Lt B.A. Burbridge and F/Lt S. Skelton	Crashed into the sea off Portsmouth, Hants.
29/30 April	Do 217K-3 of StabIII./KG 100	No. 406	Mosquito NF.XII	S/Ldr D.J. Williams and F/O C.J. Kirkpatrick	Crashed into the sea off Plymouth Harbour.
	Do 217K-3 of 9./KG 100	No. 406	Mosquito NF.XII	S/Ldr D.J. Williams and F/O C.J. Kirkpatrick	Crashed near Blackawton, Devon.
	Do 217K-3 of KG 100	No. 456	Mosquito NF.XVII	N/K	Crashed into the sea off the south coast.
14/15 May	Ju 188A-2 of 1./KG 2	No. 456	Mosquito NF.XVII	F/O A.S. McEvoy and F/O M.N. Austin	Crashed on Greenlands Artillery range, Larkhill, Wilts.
	Do 217K-1 of 7.?KG 2	No. 488	Mosquito NF.XIII	F/O R.G. Joffs and F/O E. Spedding	Crashed at West Camel, near Yeovilton, Somerset.
	Ju 188A-2 of 1./KG 6	No. 264	Mosquito NF.XIII	F/Lt C.M. Ramsay and F/O Edgar	Broke up over Manor Farm, West Worldland, near Selborne, Hants. The Mosquito was hit by return fire and crashed, killing F/O Edgar.
	Ju 188A-2 of 3./KG 6	No. 488	Mosquito NF.XIII	F/O R.G. Joffs and F/O E. Spedding	Crashed at Inwood House, near Henstridge, Somerset.
	Unidentified Ju 88	No. 125	Mosquito NF.XVII	N/K	Crashed into the sea off Cherbourg.
	Unidentified Me 410	No. 125	Mosquito NF.XVII	N/K	Crashed into the sea off Portland Bill.
	Unidentified Ju 88	No. 406	Mosquito NF.XII	N/K	Crashed into the sea off the south coast.
	Unidentified Do 217	No. 604	Mosquito NF.XIII	F/Lt J. Surman and unknown R/O	Crashed into the sea between the Isle of Wight and Portland Bill.
	Unidentified Ju 88	No. 604	Mosquito NF.XIII	W/Cdr M.H. Maxwell and F/Lt J. Quinton	Crashed to the south off the Isle of Wight.
15/16 May	Ju 88A-4 of 9./KG 54	No. 456	Mosquito NF.XVII	F/O D.W. Arnold and F/O J.B. Stickley	Crashed at Medstead, Hants.
22 May	Unidentified Ju 88	No. 125	Mosquito NF.XVII	N/K	Crashed into the sea off Southampton.
	Unidentified Ju 88	No. 125	Mosquito NF.XVII	N/K	Crashed into the sea off Southampton.
	Unidentified Ju 88	No. 456	Mosquito NF.XVII	N/K	Crashed into the sea off Southampton.
	Unidentified Ju 88	No. 456	Mosquito NF.XVII	N/K	Crashed into the sea off Southampton.
27 May	Unidentified Me 410	No. 456	Mosquito NF.XVII	N/K	Crashed into the sea off Cherbourg.
28 May	Unidentified Ju 88	No. 410	Mosquito NF.XIII	N/K	Chased into France and shot down near Lille.
29 May	Unidentified Ju 88	No. 25	Mosquito NF.XVII	N/K	Crashed into the sea 50 miles to the east of Cromer, Norfolk.
8 June	Unidentified Me 410	No. 25	Mosquito NF.XVII	N/K	Crashed into the sea off Southwold, Suffolk.
23/24 June	Ju 188F-1 of 3.(F)122	N/K	N/K	N/K	Crashed at padley Water, Chillesford, Suffolk.
25 Sept	He 111H-22 of III./KG 3	No. 409	Mosquito NF.XIII	N/K	V-1 carrier, over the North Sea.
29 Sept	He 111H-22 of KG 3	No. 25	Mosquito NF.XVII	N/K	V-1 carrier, over the North Sea.

Date	Claim	Squadron	Aircraft	Crew	Location
5 Oct	He 111H-22 of KG 3	No. 25	Mosquito NF.XVII	N/K	V-1 carrier, over the North Sea.
25 Oct	He 111H-22 of KG 3	No. 125	Mosquito NF.XVII	N/K	V-1 carrier, over the North Sea.
30 Oct	He 111H-22 of KG 3	No. 125	Mosquito NF.XVII	N/K	V-1 carrier, over the North Sea.
5 Nov	He 111H-22 of KG 3	No. 68	Mosquito NF.XIX	N/K	V-1 carrier, over the North Sea.
10 Nov	He 111H-22 of KG 3	No. 125	Mosquito NF.XVII	N/K	V-1 carrier, over the North Sea.
19 Nov	He 111H-22 of KG 3	No. 456	Mosquito NF.XVII	N/K	V-1 carrier, 75 miles to the east of Lowestoft.
25 Nov	He 111H-22 of KG 3	No. 456	Mosquito NF.XVII	N/K	V-1 carrier, 10 miles off Texel.
23/24 Nov	He 111H-22 of KG 3	No. 125	Mosquito NF.XVII	N/K	V-1 carrier, over the North Sea.
1945					
6 Jan	He 111H-22 of KG 3	No. 68	Mosquito NF.XIX	N/K	V-1 carrier, over the North Sea.
3/4 March	Unidentified Ju 188	No. 125	Mosquito NF.30	N/K	Shot down over the North Sea.
20/21 March	Unidentified Ju 88	No. 125	Mosquito NF.XVII	F/Lt Kennedy and F/O Morgan	Crashed in flames into the sea 10 miles north-east of Cromer.

APPENDIX 5

Description of AI Mk.VI Installation in Hurricanes Z2509 and BN288

This description is drawn from the letters, reports and memoranda contained in the National Archives File AVIA7/2676, *AI Mk.VI in Hurricane*, dated 1942. The following was recorded at the meeting of a Design Review Board held at Hawker's Kingston factory on 20 June 1942 to inspect the mock-up of an AI Mk.VI installation in Hurricane Z2905, a standard Mk.II fighter. To enable the various units of the Mk.VI system to be accommodated within the aircraft, it proved necessary to locate the transmitter, receiver and modulator on mountings usually reserved for the desert survival equipment. The Control Panel Type 4 was positioned on the starboard side, underneath and slightly forward of the transmitter, receiver and modulator, with the responder unit and its junction boxes being located in a tray behind the pilot and over the top of the longerons. The pilot's indicator unit took the place of the undercarriage indicator, which was in turn transferred to the position usually occupied by the aircraft clock. The Control Unit Type 96 was mounted on the port side of the cockpit, on the inboard face of the electrical panels. No definite position for the Control Unit Type 67 was agreed, but it was stipulated that it had to be within easy reach of the pilot. The aircraft's normal engine driven generator was replaced with a Generator Type RLX, which required the repositioning of a number of cooling pipes, as it was somewhat longer than the standard unit.

Rough positions for the aerial system were agreed, namely:

- The transmitter dipole on the leading edge of the port wing, 5 feet (1.5 metres) outboard of the landing light.
- The receiver azimuth dipoles on the leading edge of the wing between the gun-bays and landing lights.
- The receiver elevation unipoles above and below the starboard wing, 1 foot (30.5 cm) inboard of the landing light and 2 feet 6 inches (76.2 cm) behind the leading edge.

It was estimated the installation would increase the Hurricane's empty weight by some 200 lb (90.72 kg) and shift the CG aft by 1.5 inches (3.81 cm), from 59.3 inches (15.06 cm) aft of the datum, to 60.8 inches (15.44 cm). The 200 lb figure was calculated using data from Hurricane BN288, a standard Brookland's built Mk.IIc, and information contained in the type's maintenance manual.

305

The Board also considered an alternative proposal to fit the transmitter and modulator inside a container similar in outline to a standard long-range fuel tank. This would then be suspended on the wing pylon. In this location the transmitter dipole was to be located on the nose of the container, with the azimuth dipoles located on the wing's leading edge, five feet (1.5 metres) inboard of the landing lights and the elevation unipoles in the same position as on Z2905. It was estimated the 'containerisation' of the AI equipment would drive the CG rearwards by a further half inch (1.27 cm) to 59.8 inches (15.19 cm) aft of the datum. It should be noted that this approach was adopted for the installation of AI Mk.VI in Hawker Typhoon Mk.Ib, R7881 in 1942.

A second Hurricane, the aforementioned BN288, a standard Mk.IIc, was modified at Defford by TFU personnel in line with that of Z2509 and flown to Hawker's Langley airfield for inspection on 18 June 1942.[1] Having been flown briefly the previous May, BN288 was now regarded as the 'standard' for the manufacture of eleven Hurricanes fitted with AI Mk.VI. Apart from a recommendation that the modulator be moved forward and its mounting structure be improved to withstand forces up to 12 g, as did almost every other piece of the radar equipment, and that long range tanks (and desert equipment) should not be carried on AI Hurricanes, BN288 appeared to pass its inspection test.

The aerial system on BN288 comprised a Type 69 transmitting dipole taken from the Mosquito AI installation, which needed moving inboard of the landing light to clear the guns, Defiant Type 29 unipole arrays for the elevation elements and vertically polarised azimuth dipoles taken from the night-fighter version of the Fairey Fulmar. With a requirement that the aforementioned transmitting aerial be moved inboard 2 feet 6 inches (76.2 cm) and the pilot's control unit be moved to a better location, the AI Mk.VI installation was approved for the Hurricane Mk.IIb and Mk.IIc and the Sea Hurricane.

Note

1. The inspection team comprised officers from TRE, TFU, MAP, the Air Ministry, HQ No. 43 Group, HQ Fighter Command and Hawkers.

APPENDIX 6

Night-Fighter Squadrons and Airfields, 1939–1961[1]

The charts below list the UK-based squadrons and their aircraft from the Second World War and the post-war periods that operated in the defensive and offensive (intruder) night-fighter roles, at home and abroad.

Squadron	Aircraft	Period	Remarks
No. 5	Meteor NF.11	Jan 1959– Aug 1960	R/F, Laarbruch, 21 Jan 1959. Converted to Javelin FAW.5.
No. 11	Meteor NF.11	Jan 1959– Mar 1960	R/F, Geilenkirchen, 21 Jan. Converted to Javelin FAW.4 1959.
No. 23	Blenheim If	Dec 1938–April 1941	
	Mosquito NF.30	Aug 1945–Sept 1945	
	Mosquito NF.30	Sept 1946–Nov 1946	
	Vampire NF.10	Sept 1951–Jan 1954	
	Venom NF.2	Nov 1953–Nov 1954	
	Venom NF.2A	Aug 1954–Dec 1955	
	Venom NF.3	Oct 1955–May 1957	Converted to Javelin FAW.4.
No. 25	Blenheim If	Dec 1938–Jan 1941	
	Beaufighter If	Sept 1940–Jan 1943	
	Mosquito NF.II	Oct 1942–Jan 1944	
	Mosquito NF.XVII	Dec 1943–Oct 1944	
	Mosquito NF.30	Sept 1944–Sept 1946	
	Mosquito NF.36	Sept 1946–Oct 1951	
	Vampire NF.10	July 1951–Mar 1954	
	Meteor NF.12	Mar 1954–June 1958	
	Meteor NF.14	April 1954–June 1958	
	Meteor NF.12	July 1958–April 1959	
	Meteor NF.14	July 1958–April 1959	Converted to Javelin FAW.7.
No. 29	Blenheim If	Dec 1938–Feb 1941	
	Beaufighter If	Sept 1940–May 1943	
	Beaufighter VIf	Mar 1943–May 1943	
	Mosquito NF.XII	May 1943–April 1944	
	Mosquito NF.XIII	Oct 1943–Feb 1945	
	Mosquito NF.30	Feb 1945–Aug 1946	
	Mosquito NF.36	July 1946–Oct 1950	
	Mosquito NF.30	Oct 1950–Aug 1951	
	Meteor NF.11	Aug 1951–Nov 1957	
	Meteor NF.12	Feb 1958–July 1958	Converted to Javelin FAW.9.
No. 33	Venom NF.2A	Oct 1955–July 1957	R/F, Leeming, 30 Sept 1955.

Squadron	Aircraft	Period	Remarks
No. 46	Meteor NF.14	Oct 1957–Sept 1958	Converted to Javelin FAW.7.
	Meteor NF.12	Aug 1954–Feb 1956	R/F, Odiham, 15 Aug 1954.
No. 60	Meteor NF.14	Aug 1954–Feb 1956	Converted to Javelin FAW.1.
	Meteor NF.14	Oct 1959–	Det R/F, Leeming, 27 May 1959.
		Aug 1961	Converted to Javelin FAW.9.
No. 64	Meteor NF.12	Sept 1956–Sept 1958	
	Meteor NF.14	Sept 1956–Sept 1958	Converted to Javelin FAW.7.
No. 68	Blenheim If	Jan 1941–May 1941	
	Beaufighter If	May 1941–Feb 1943	
	Beaufighter VIf	Feb 1943–July 1944	
	Mosquito NF.XVII	July 1944–Feb 1945	
	Mosquito NF.XIX	July 1944–Feb 1945	
	Mosquito NF.30	Feb 1945–April 1945	
	Meteor NF.11	Mar 1952–	R/F, Wahn, 1 Jan 1952.
		Jan 1959	D/B, Laarbruch, 21 Jan 1959.
No. 72	Meteor NF.12	Feb 1956–June 1959	Last U/K Meteor N/F squadron.
	Meteor NF.14	Feb 1956–June 1959	Converted to Javelin FAW.4/5.
No. 85	Hurricane I	Sept 1940–April 1941	
	Defiant I	Jan 1941–Feb 1941	
	Havoc I	Feb 1941–Nov 1941	
	Havoc II	July 1941–Sept 1942	
	Mosquito NF.II	Aug 1942–June 1943	
	Mosquito NF.XV	Mar 1943–Aug 1943	
	Mosquito NF.XII	Mar 1943–Feb 1944	
	Mosquito NF.XIII	Oct 1943–May 1944	
	Mosquito NF.XVII	Nov 1943–Nov 1944	
	Mosquito NF.30	Nov 1944–Jan 1946	
	Mosquito NF.36	Jan 1946–Oct 1951	
	Meteor NF.11	Sept 1951–May 1954	
	Meteor NF.12	April 1954–Nov 1958	
	Meteor NF.14	April 1954–Nov 1958	D/B, Church Fenton, 30 Nov 1958.
No. 87	Hurricane I	Sept 1940–Sept 1942	
	Hurricane IIc	June 1941–Nov 1942	
	Meteor NF.11	Mar 1952–	R/F, Wahn, 1 Jan 1952.
		Feb 1958	Converted to Javelin FAW.1.
No. 89	Beaufighter If	Sept 1941–April 1943	R/F, Colerne, 25 Sept 1941.
	Beaufighter VIf	Aug 1942–June 1945	
	Mosquito NF.XIX	Mar 1945–Mar 1946	D/B, Seletar, 1 May 1946.
	Venom NF.3	Jan 1956–	R/F, Stradishall, 15 Dec 1955.
		Nov 1957	Converted to Javelin FAW.6.
No. 96	Hurricane I	Dec 1940–May 1941	
	Defiant I	Mar 1941–Feb 1942	
	Hurricane IIc	Sept 1941–Jan 1942	
	Defiant Ia	Feb 1942–May 1942	
	Defiant II	April 1942–July 1942	
	Beaufighter IIf	May 1942–Feb 1943	
	Beaufighter VIf	Sept 1942–Nov 1943	
	Mosquito NF.XIII	Oct 1943–Dec 1944	
	Meteor NF.11	Oct 1952–	R/F, Ahlhorn, 17 Nov 1952.
		Jan 1959	Converted to Javelin FAW.4.
No. 125	Defiant I	June 1941–May 1942	
	Defiant II	Oct 1941–May 1942	
	Beaufighter IIf	Feb 1942–Sept 1942	

Squadron	Aircraft	Period	Remarks
	Beaufighter VIf	Sept 1942–Mar 1944	
	Mosquito NF.XVII	Feb 1944–Mar 1945	
	Mosquito NF.30	Feb 1945–Nov 1945	D/B, Church Fenton, 20 Nov 1945.
	Meteor NF.11	April 1955–Feb 1956	R/F, Stradishall, 31 Mar 1955.
	Venom NF.3	Nov 1955–May 1957	D/B, Stradishall, 10 May 1957.
No. 141	Defiant I	Sept 1940–Sept 1941	
	Beaufighter If	Aug 1941–June 1943	
	Beaufighter VIf	June 1943–Jan 1944	
	Mosquito NF.II	Nov 1943–Aug 1944	
	Mosquito NF.30	April 1945–Sept 1945	D/B, Little Snoring, 7 Sept 1945.
	Mosquito NF.36	June 1946–Sept 1951	R/F, Wittering, 17 June 1946.
	Meteor NF.11	Sept 1951–Aug 1955	
	Venom NF.3	June 1955–Feb 1957	Converted to Javelin FAW.4.
No. 151	Hurricane I	Nov 1940–June 1941	
	Defiant I	Dec 1940–Oct 1941	
	Hurricane IIc	April 1941–Jan 1942	
	Defiant II	Sept 1941–July 1942	
	Mosquito NF.II	April 1942–July 1943	
	Mosquito NF.XII	June 1943–Mar 1944	
	Mosquito NF.XIII	Dec 1943–Sept 1944	
	Mosquito NF.30	Sept 1944–Oct 1946	D/B, Weston Zoyland, 10 Oct 1946.
	Vampire NF.10	Feb 1952–Aug 1953	R/F, Leuchars, 15 Sept 1952.
	Meteor NF.11	Mar 1953–Oct 1955	
	Venom NF.3	July 1955–June 1957	Converted to Javelin FAW.5.
No. 152	Meteor NF.12	July 1954–July 1958	R/F, Wattisham, 30 June 1954.
	Meteor NF.14	July 1954–July 1958	D/B, Stradishall, 31 July 1954.
No. 153	Defiant I	Oct 1941–April 1942	R/F, Ballyhalbert, 24 Oct 1941.
	Beaufighter If	Jan 1942–Jan 1943	
	Beaufighter VIf	Oct 1942–Sept 1944	Converted to Hurricane IIc.
	Meteor NF.12	Mar 1955–June 1958	R/F, West Malling, 28 Feb 1955.
	Meteor NF.14	Feb 1955–June 1958	D/B, Waterbeach, 2 July 1958.
No. 157	Mosquito NF.II	Jan 1942–May 1944	R/F, Debden, 15 Dec 1941.
	Mosquito NF.XIX	May 1944–May 1945	
	Mosquito NF.30	Mar 1945–Aug 1945	D/B, Swannington, 16 Aug 1954.
No. 219	Blenheim If	Oct 1939–June 1941	
	Beaufighter If	Sept 1940–May 1943	
	Beaufighter VIf	May 1943–Jan 1944	
	Mosquito NF.XVII	Feb 1944–Nov 1944	
	Mosquito NF.30	June 1944–Sept 1946	D/B, Wittering, 1 Sept 1946.
	Mosquito NF.36	Mar 1951–	R/F, Kabrit, 1 Mar 1951.
		Mar 1953	D/B, Kabrit, 1 Mar 1953.
	Venom NF.2A	Sept 1955–	R/F, Driffield, 5 Sept 1955.
		July 1957	D/B, Driffield, 31 July 1957.
No. 239	Beaufighter If	Oct 1943–Jan 1944	
	Mosquito NF.II	Dec 1943–Sept 1944	
	Mosquito NF.30	Jan 1945–July 1945	D/B, West Raynham, 1 July 1945.
No. 253	Hurricane IIa	Oct 1941–Sept 1942	
	Venom NF.2A	April 1955–	R/F, Waterbeach, 18 April 1955.
		Aug 1957	D/B, Waterbeach, 2 Sept 1957.
No. 255	Defiant I	Nov 1940–Sept 1941	R/F, Kirton-in-Lindsey, 23 Nov 1941.
	Hurricane I	Mar 1941–July 1941	
	Beaufighter IIf	July 1941–May 1941	
	Beaufighter VIf	Mar 1942–Feb 1945	

Squadron	Aircraft	Period	Remarks
No. 256	Mosquito NF.XIX	Feb 1945–April 1946	
	Mosquito NF.30	Jan 1946–April 1946	D/B, Gianaclis, 30 April 1946.
	Defiant I	Nov 1941–May 1942	
	Hurricane I	May 1941–July 1941	
	Defiant II	May 1942–June 1942	
	Beaufighter If	May 1942–Nov 1942	
	Beaufighter VIf	Oct 1942–May 1943	
	Mosquito NF.XII	May 1943–July 1944	
	Mosquito NF.XIII	Jan 1944–Sept 1945	Converted to Spitfire Mk.VIII.
	Mosquito NF.XIX	Oct 1945–Sept 1946	D/B, Nicosia, 12 Sept 1946.
	Meteor NF.11	Nov 1952–	R/F, Ahlhorn, 17 Nov 1952.
		Jan 1959	D/B, Geilenkirchen, 21 Jan 1959.
No. 264	Defiant I	Aug 1940–Sept 1941	
	Defiant II	Sept 1941–July 1942	
	Mosquito NF.II	May 1942–Jan 1944	
	Mosquito NF.XIII	Dec 1943–Aug 1945	D/B, Twente, 25 Aug 1945.
	Mosquito NF.30	Nov 1945–May 1946	R/F, Church Fenton, 20 Nov 1945.
	Mosquito NF.36	Mar 1946–Jan 1952	
	Meteor NF.11	Nov 1951–Oct 1954	
	Meteor NF.14	Oct 1954–Sept 1957	
	Meteor NF.12	Jan 1957–Sept 1957	D/B, Leeming, 30 Sept 1957.
No. 307	Defiant I	Sept 1940–Aug 1941	Formed, Kirton-in-Lindsey, 5 Sept 1940.
	Beaufighter IIf	Aug 1941–May 1942	
	Beaufighter VIf	May 1942–Jan 1943	
	Mosquito NF.II	Dec 1942–Mar 1944	
	Mosquito NF.XII	Jan 1944–Jan 1945	
	Mosquito NF.30	Oct 1944–Nov 1946	D/B, Horsham St Faith, 6 Jan 1947.
No. 406	Blenheim If	May 1941–June 1941	Formed, Acklington, 5 May 1941.
	Blenheim IVf	May 1941–June 1941	
	Beaufighter IIf	June 1941–July 1942	
	Mosquito NF.XII	April 1944–July 1944	
	Mosquito NF.30	July 1944–Aug 1945	D/B, Predannack, 1 Sept 1945.
No. 409	Defiant I	July 1941–Sept 1941	Formed, Digby, 16 June 1941.
	Beaufighter IIf	Aug 1941–June 1942	
	Beaufighter VIf	June 1942–April 1944	
	Mosquito NF.XIII	Mar 1944–June 1945	D/B, Twente, 11 June 1945.
No. 410	Defiant I	June 1941–May 1942	Formed, Ayr, 30 June 1941.
	Beaufighter IIf	April 1942–Jan 1943	
	Mosquito NF.II	Oct 1942–Dec 1943	
	Mosquito NF.XIII	Dec 1943–Aug 1944	
	Mosquito NF.30	Aug 1944–June 1945	D/B, Gilze-Rijen, 6 June 1945.
No. 456	Defiant I	June 1941–Nov 1941	Formed, Valley, 30 June 1941.
	Beaufighter IIf	Sept 1941–Mar 1943	
	Beaufighter VIf	July 1942–Jan 1943	
	Mosquito NF.II	Dec 1942–April 1944	
	Mosquito XVII	Jan 1944–Feb 1945	
	Mosquito NF.30	Dec 1944–June 1945	D/B, Bradwell Bay, 15 June 1945.
No. 488	Beaufighter IIf	June 1942–May 1943	R/F, Church Fenton, 25 June 1942.
	Beaufighter VIf	Mar 1943–Sept 1943	
	Mosquito NF.XII	Aug 1943–Mar 1944	
	Mosquito NF.XIII	Oct 1943–Oct 1944	
	Mosquito NF.30	Oct 1944–April 1945	D/B, Gilzen-Rijen, 26 April 1945.
No. 500	Mosquito NF.19	Feb 1947–Aug 1947	R/F, West Malling, 10 May 1946.

Squadron	Aircraft	Period	Remarks
	Mosquito NF.30	April 1947–Oct 1948	Converted to Spitfire F.22.
No. 502	Mosquito NF.30	Dec 1947–	R/F, Aldergrove, 10 May 1946.
		Oct 1948	Converted to Spitfire F.22.
No. 504	Mosquito NF.30	May 1947–	R/F, Syerton, 10 May 1946.
		May 1948	Converted to Spitfire F.22.
No. 515	Defiant II	Oct 1942–Dec 1943	Formed, Northolt, 1 Oct 1942.
	Beaufighter IIf	June 1943–April 1944	
	Mosquito NF.II	Feb 1944–April 1944	Converted to Mosquito FB.VI.
No. 530	Havoc II (Turbin)[2]	Sept 1942–Jan 1943	Formed, Hunsdon, 8 Sept 1942.
	Boston III (Turbin)	Sept 1942–Jan 1943	
	Hurricane IIc	Sept 1942–Jan 1943	D/B, Hunsdon, 25 Jan 1943.
No. 531	Havoc I (Turbin)	Sept 1942–Jan 1943	Formed, West Malling, 8 Sept 1942.
	Havoc I	Sept 1942–Jan 1943	
	Boston III (Turbin)	Sept 1942–Jan 1943	
	Hurricane IIc	Sept 1942–Jan 1943	D/B, West Malling, 25 Jan 1943.
No. 532	Havoc I (Turbin)	Sept 1942–Jan 1943	Formed, Wittering, 2 Sept 1942.
	Boston III (Turbin)	Sept 1942–Jan 1943	
	Hurricane IIb	Sept 1942–Jan 1943	
	Hurricane IIc	Sept 1942–Jan 1943	
	Boston III	Sept 1942–Jan 1943	D/B, Hibaldstow, 25 Jan 1943.
No. 533	Havoc I (Turbin)	Sept 1942–Jan 1943	Formed, Charmy Down, 8 Sept 1942.
	Havoc I	Sept 1942–Jan 1943	
	Havoc II (Turbin)	Sept 1942–Jan 1943	
	Boston III (Turbin)	Sept 1942–Jan 1943	
	Hurricane IIb	Sept 1942–Jan 1943	
	Hurricane IIc	Sept 1942–Jan 1943	
	Boston III	Sept 1942–Jan 1943	D/B, Charmy Down, 25 Jan 1943.
No. 534	Havoc I (Turbin)	Sept 1942–Jan 1943	Formed, Tangmere, 2 Sept 1942.
	Havoc II (Turbin)	Sept 1942–Jan 1943	
	Boston III (Turbin)	Sept 1942–Jan 1943	
	Hurricane IIc	Sept 1942–Jan 1943	
	Boston I	Sept 1942–Jan 1943	D/B, Tangmere, 25 Jan 1943.
No. 535	Havoc I (Turbin)	Sept 1942–Jan 1943	Formed, High Ercall, 2 Sept 1942.
	Havoc II (Turbin)	Sept 1942–Jan 1943	
	Boston III (Turbin)	Sept 1942–Jan 1943	
	Hurricane IIc	Sept 1942–Jan 1943	D/B, High Ercall, 25 Jan 1943.
No. 536	Havoc II (Turbin)	Sept 1942–Jan 1943	Formed, Predannack, 8 Sept 1942.
	Hurricane IIc	Sept 1942–Jan 1943	D/B, Fairwood Common, 25 Jan 1943.
No. 537	Havoc I (Turbin)	Sept 1942–Jan 1943	Formed, Middle Wallop, 8 Sept 1942.
	Havoc I	Sept 1942–Jan 1943	
	Boston III (Turbin)	Sept 1942–Jan 1943	
	Hurricane IIb	Sept 1942–Jan 1943	
	Hurricane IIc	Sept 1942–Jan 1943	D/B, Middle Wallop, 25 Jan 1943.
No. 538	Havoc I (Turbin)	Sept 1942–Jan 1943	Formed, Hibaldstow, 2 Sept 1942.
	Havoc II (Turbin)	Sept 1942–Jan 1943	
	Boston III (Turbin)	Sept 1942–Jan 1943	
	Hurricane I	Sept 1942–Jan 1943	
	Hurricane IIc	Sept 1942–Jan 1943	D/B, Hibaldstow, 25 Jan 1943.
No. 539	Havoc I (Turbin)	Sept 1942–Jan 1943	Formed, Acklington, 2 Sept 1942.
	Havoc II (Turbin)	Sept 1942–Jan 1943	
	Boston III (Turbin)	Sept 1942–Jan 1943	
	Hurricane IIc	Sept 1942–Jan 1943	
	Boston I	Sept 1942–Jan 1943	D/B, Acklington, 25 Jan 1943.

Squadron	Aircraft	Period	Remarks
No. 600	Blenheim If	Jan 1939–Oct 1941	
	Beaufighter If	Sept 1940–June 1941	
	Beaufighter IIf	April 1941–April 1942	
	Beaufighter VIf	Mar 1942–Feb 1945	
	Mosquito NF.XIX	Dec 1944–Aug 1945	D/B, Aviano, 21 Aug 1945.
No. 601	Blenheim If	Jan 1939–Feb 1940	Converted to Hurricane I.
No. 604	Blenheim If	Jan 1939–May 1941	
	Beaufighter If	Sept 1940–April 1943	
	Beaufighter VIf	April 1943–April 1944	
	Mosquito NF.XIII	Feb 1944–April 1945	
	Mosquito NF.XII	Mar 1944–May 1944	D/B, Lille/Vendeville, 18 April 1945.
No. 605	Mosquito NF.30	May 1947–Sept 1948	Converted to Vampire F.1.
No. 608	Mosquito NF.30	July 1947–June 1948	Converted to Spitfire F.22.
No. 609	Mosquito NF.30	April 1947–Sept 1948	Converted to Spitfire LF.16E.
No. 616	Mosquito NF.30	Sept 1947–May 1949	Converted to Meteor F.3.

Notes

1. Compiled from C.G. Jefford, *op cit* and John Rawlings, *op cit.*
2. Turbinlite.

Abbreviations

2TAF	2nd Tactical Air Force
AA	Anti-aircraft
A&AEE	Aeroplane and Armament Experimental Establishment
AAF	Auxiliary Air Force
AAM	Air-to-air missile
AASF	Advanced Air Striking Force
ac	Alternating current
AC1	Aircraftsman 1st Class
AC2	Aircraftsman 2nd Class
ACAS	Assistance Chief of the Air Staff
ACI	Air control of interception
ADD	*Aviatsiya Dal'nevo Deistviya* – the long range aviation division of the Red Air Force
ADC	Air Defence Commander
ADEE	Air Defence Experimental Establishment
ADGB	Air Defence of Great Britain
AEAF	Allied Expeditionary Air Force
AEW	Airborne early warning
AFC	Air Force Cross
AI	Air intercept radar
AIF	Air intercept follow
AIH	High powered air intercept radar and AI homing
AIL	Air intercept radar with locking timebase
AIS	Air intercept 'sentimetric'
ALF	Automatic lock follow
AMES	Air Ministry Experimental Station
AMRE	Air Ministry Research Establishment
ANS	Air Navigation School
AOC-in-C	Air Officer Commander-in-Chief
APC	Armament practice camp
ASH	Air and surface homing (AN/APS-4) radar
ASI	Air speed indicator
ASM	Air-to-surface missile
ASV	Air-to-surface vessel radar
AVMF	*Aviatsiya Voenno-morskoi flot* – the air branch of the Soviet Navy
AWA	Armstrong Whitworth Aircraft
BABS	Bomber Approach Blind-Landing System

BAe	British Aerospace
BBC	British Broadcasting Corporation
BAFO	British Air Force of Occupation in Germany
BS	Bombardment Squadron (US)
CA	Controller of Aircraft in the Ministry of Supply
CAS	Chief of the Air Staff
CATCS	Central Air Traffic Control School
CS(A)	Controller of Supplies (Air) in the Ministry of Supply
CFE	Central Fighter Establishment
CH	Chain home radar
CHEL	Chain home extra low radar
CHL	Chain home low radar
CID	Committee of Imperial Defence
C-in-C	Commander-in-Chief
CCOS	Combined Chiefs of Staff
CO	Commanding Officer
C&R	Control and reporting
CRO	Civilian Repair Organisation
CRT	Cathode ray tube
CTE	Controller of Telecommunications Equipment at the Ministry of Aircraft Production
CVD	Communications Valve Development Committee, otherwise known as the 'Committee on Valve Development'
CW	Continuous wave radar
DBAP	*dahl'nebombardiro-vochnyy aviapolk* – Soviet heavy bomber regiment
DBAD	*dahl'nebombardiro-voch-naya aviadiveeziya* – Soviet heavy bomber division
dc	Direct current
DCAS	Deputy Chief of the Air Staff
DCD	Director of Communications Development at MAP
DDCD	Deputy Director of Communications Development – see above
DFC	Distinguished Flying Cross
D/F	Direction finding
DFM	Distinguished Flying Medal
DOR(Air)	Director of Operational Requirements (Air)
D per T	'D' Performance and Testing Flight of A&AEE
DSigs(Air)	Director of Signals in the Air Ministry
DSO	Distinguished Service Order
DSR	Director of Scientific Research in the Air Ministry and the Admiralty
DV	Direct vision panel
ECM	Electronic counter-measures
EMC	Electro-magnetic compatibility
EMI	Electrical & Musical Industries Ltd
FAW	Fighter all-weather
FEE	Fighter Experimental Establishment – a Fighter Command unit
FDT	Fighter direction tender
FIDS	Fighter Interception Development Squadron – FIU's name from the autumn of 1944
FIU	Fighter Interception Unit

Flak	*Fliegerabwehrkannone* – the German abbreviation for anti-aircraft gun
GCI	Ground control of interception
GE	General Electric (USA)
GEC	General Electric Company (UK)
GGS	Gyro gunsight
GHZ	GigaHertz (1,000,000,000 cycles per second)
GL	Gunlaying radar
GOC	General Officer Commanding
GR	General reconnaissance
HF	High frequency
HQ	Headquarters
HT	High tension (high voltage)
Hz	Hertz (one cycle per second)
H2S	Code name for Bomber Command ground mapping radar
ICBM	Intercontinental ballistic missile
ICI	Imperial Chemical Industries Ltd
IF	Intermediate frequency
IFF	Identification friend or foe
IFR	In-flight refuelling
IR	Infra-red
JG	*Jadgeschwader* – a *Luftwaffe* fighter unit
Jabo	A *Luftwaffe* fighter-bomber unit
JSTU	Joint Services Trials Unit
KG	*Kampfgeschwader* – a *Luftwaffe* bomber unit
KGr	*Kampfgruppe*
KHz	KiloHertz (1,000 cycles per second)
kW	KiloWatt (1,000 Watts)
LAM	Long aerial mine
LST	Landing strip tank
MAP	Ministry of Aircraft Production
MDAP	Mutual Defence Air Programme
Metro-Vick	Metropolitan Vickers Ltd
MHz	MegaHertz (1,000,000 cycles per second)
MIT	Massachusetts Institute of Technology
Mk	Mark number
MoS	Ministry of Supply – the post-war successor to MAP
MRAF	Marshal of the Royal Air Force
MRS	Master radar station
MTB	Motor torpedo boat
MU	Maintenance Unit
mW	MilliWatt (1,000th of a Watt)
NATO	North Atlantic Treaty Organisation
NCO	Non-Commissioned Officer
NDRC	National Defence Research Council – a branch of the United States Government
NFIU	Naval Night Fighter Interception Unit
NPL	National Physical Laboratory

NKAP	*Narodnyy komissariaht aviatsionnoy promyshlennosti* – the Soviet People's Commissariat of the Aircraft Industry
N/R	Navigator (Radio) – the successor title to Observer (Radio) – see below
NRC	National Research Council of Canada
NRL	Naval Research Laboratory – a branch of the USN
OC	Officer Commanding
OCU	Operational Conversion Unit
OKB	*Opytno konstruktorskoe byuro* – a Soviet aircraft design bureau
OR	Operational requirement
OSS	Orientable sector scan – a function of AN/APS-57
OTU	Operational training unit
PDS	Post-Design Services
PoW	Prisoner of War
PPS	Pulses per second
PRF	Pulse repetition frequency – measured in Hertz
PRR	Pulse repetition rate
QRA	Quick Reaction Alert
RAE	Royal Aircraft Establishment
RAF	Royal Air Force
RCA	Radio Corporation of America
RCAF	Royal Canadian Air Force
R&D	Research & development
RDF	Radio direction finding
RDF1R	Radio direction finding type 1R
RDF2	Radio direction finding type 2
RDS-1 to 5	A series of Soviet nuclear weapons of varying yield
RF	Radio frequency
R/O	RDF operator, later Observer (Radio)
ROC	Royal Observer Corps
RM	Royal Marines
R-R	Rolls Royce
RSRE	Royal Signals & Radar Establishement – the succesor to TRE
R/T	Radio telephony
SAC	Strategic Air Command of the USAF
SAM	Surface-to-air missile
SBA	Standard beam approach
SD	Special Duty (Flight)
SHAPE	Supreme Headquarters Allied Powers, Europe
SL/SLC	Searchlight control radars
SKG	*Schnellkampfgeschwader* – a *Luftwaffe* fast bomber unit
SOC	Sector operations centre
SOR	Sector operations room
SSBN	Submersible ship (ballistic) nuclear
SR	Special Reserve
TFU	Telecommunications Flying Unit – a unit of TRE
TRE	Telecommunications Research Establishment
UK	United Kingdom
UN	United Nations

US	United States
USAAC	Unites States Army Air Corps
USAAF	United States Army Air Forces
USAF	United States Air Force
USN	United States Navy
V-1	*Vergeltungswaffe 1* – Revenge Weapon No. 1, also known as 'Doodlebug'
VHF	Very high frequency
V-VS	*Voenno-vozdushnye sily* – the Military Air Forces of the Red Army
W/Op	Wireless operator
W/OpAG	Wireless operator/air gunner

Bibliography

Books

Adkin, Fred, *Through the Hanger Doors, RAF Ground Crew since 1945*, Shrewsbury: Airlife Publishing Ltd, 1983.

Alexander, Jean, *Russian Aircraft Since 1940*, London: Putnam, 1975.

Andrews, C.F. and Morgan, E.B, *Vickers Aircraft since 1908*, London: Putnam, 1988.

Armitage, Michael, *The Royal Air Force, An Illustrated History*, London: Brockhampton Press, 1996.

Baker, David, *The Rocket, The History and Development of Rocket & Missile Technology*, London: New Cavendish Books, 1978.

Barnes, C.H., *Bristol Aircraft since 1910*, London: Putnam, 1964.

Batt, Reg, *The Radar Army, Winning the War of the Airwaves*, London: Robert Hale, 1991.

Bowen, E.G., *Radar Days*, Bristol: Adam Hilger, 1987, an imprint of Routledge/Taylor & Francis Group, LLC, New York.

Bowers, Peter M., *Boeing Aircraft Since 1916*, London: Putnam Books, 1966.

Bowyer, Chaz [1], *Fighter Command 1936–1968*, London: J.M. Dent & Sons Ltd, 1980.

[2], *Bristol Blenheim*, Shepperton: Ian Allen Ltd, 1984.

[3], *Beaufighter*, London: William Kimber, 1987.

[4], *Mosquito Squadrons of the Royal Air Force*, London: Ian Allan Ltd, 1984.

Bowyer, Michael J.F., *The Boulton-Paul Defiant*, Leatherhead: Profile Publications, 1966.

Braham, J.R.D., *Scramble*, London: William Kimber, 1985.

Buderi, Robert, *The Invention That Changed The World*, New York: Touchstone Books, 1997.

Bushby, John R., *Air Defence of Great Britain*, London: Ian Allan, 1973.

Buttler, Tony [1], *Gloster Javelin*, in Warpaint Series No. 17, Milton Keynes: Hall Park Books Ltd, undated.

[2], *Gloster Meteor*, Warpaint Series No. 22, Milton Keynes: Hall Park Books Ltd, undated.

Chisholm, R.A., *Cover of Darkness*, Morley: Elmfield Press, 1953.

Dierich, Wolfgang, *Kampfgeschwader 'Edelweiss', The History of a German Bomber Unit, 1939–1945*, London: Ian Allan Ltd, 1975.

Dobinson, Colin, *AA Command, Britain's Anti-Aircraft Defences of the Second World War*, London: Methuen, 2001.

Francillon, Reñe, J, *McDonnell Douglas Aircraft since 1920, Volume 1*, London: Putnam, 1988.

Gething, Michael, *Sky Guardians, Britain's Air Defences 1918–1993*, London: Arms & Armour Press, 1993.

Gordon, Yefim, *Myasishchev M-4 and 3M, The First Soviet Strategic Bomber*, Hinckley: Midland Publishing, 2003.

Gordon, Yefim, and Rigmant, Vladimir [1], *Tupolev Tu-4, Soviet Superfortress*, Hinckley: Midland Publishing, 2002.

[2], *Tupolev Tu-16 Badger*, Hinckley: Midland Publishing, 2004.

[3], *Tupolev Tu-95/-142 'Bear'*, Hinckley: Midland Publishing, 1997.

Gough, Jack, *Watching the Skies, A History of Ground Radar for the Air Defence of the United Kingdom by the Royal Air Force from 1946 to 1975*, London: HMSO, 1993.

Green, William, *Warplanes of the Third Reich*, Macdonald & Co Ltd, London, 1970.

Green, William and Swanborough, Gordon, *The Observer's Soviet Aircraft Directory*, London: Frederick Warne, 1972.

Greer, Bob, *Bristol Blenheim in action*, Carrollton: Squadron/Signal Publications Inc, 1988.

Gunston, Bill [1], *Night Fighters, A Development and Combat History*, Cambridge: Patrick Stephens Ltd, 1976.

Hanbury Brown, R., *Boffin, A Personal Story of the Early Days of Radar, Radio Astronomy & Quantum Optics*, Bristol: Adam Hilger, 1991, an imprint of Routledge/Taylor & Francis Group, LLC, New York.

Hartley, Harold, *Tizard, Sir Henry Thomas (1885–1959)*, in 'The Dictionary of National Biography, 1951–1960'.

Harrison, W., *Fairey Firefly*, Shrewsbury: Airlife Publishing Ltd, 1992.

Harrison, W.A. [1], *De Havilland Vampire*, Warpaint Series No. 27, Milton Keynes, Hall Park Books Ltd, undated.

[2], *De Havilland Venom*, Warpaint Series No. 44, Milton Keynes, Hall Park Books Ltd, undated.

Hazell, Steve, *De Havilland Sea Vixen*, Warpaint Series No. 11, Milton Keynes: Hallpark Books Ltd, undated.

Hichens, Mark, *The Troubled Century, British & World History, 1914–1993*, Durham: Pentland Press, 1994.

Hodgkin, Alan, *Chance & Design, Reminiscences of Science in Peace and War*, Cambridge: Cambridge University Press, 1992.

Holmes, Richard, *The Little Field Marshal, A Life of Sir John French*, London: Weidenfeld & Nicolson, 2004.

Hooton, E.R., *Phoenix Triumphant, The Rise & Rise of the Luftwaffe*, Brockhampton Press, 1999.

Howard Williams, Jeremy [1], *Night Intruder, A personal account of the radar war between the Luftwaffe and the RAF night fighter forces*, London: David & Charles, 1976.

Howse, Derek, *Radar at Sea, The Royal Navy in World War 2*, Macmillan Press Ltd, 1994.

Jackson, A.J., *De Havilland Aircraft Since 1915*, London: Putnam, 1962.

James, Derek [1], *Gloster Aircraft Since 1917*, London: Putnam, 1971.

Jefford, C.G. [1], *RAF Squadrons, A Comprehensive Record of the Movement and Equipment of all RAF Squadrons and their Antecedents Since 1912*, Shrewsbury: Airlife Publishing Ltd, 2001.

Jones, Barry, *Gloster Meteor*, Crowood Press, 1998.

Jones, R.V., *Most Secret War, British Scientific Intelligence, 1939–1945*, London: Hamish Hamilton, 1978.

Nowarra, Heinz J., *The Focke-Wulf 190, A Famous German Fighter*, Kings Langley: Harleyford Publications, 1973.

Latham, Colin and Stubbs, Anne, *Pioneers of Radar*, Thrupp: Sutton Publishing, 1999.

Lindsay, Roger [1], *De Havilland Vampire NF.Mk.10 in Royal Air Force Service*, Luton: Alan W. Hall (Publications) Ltd, undated.

[2], *de Havilland Venom*, Privately published in 1974.

[3], *Service History of the Gloster Javelin Marks 1 to 6*, Privately published in 1975.

Lovell, Sir Bernard, *Echoes of War, The Story of H2S Radar*, Bristol: Adam Hilger, 1991, an imprint of Routledge/Taylor & Francis Group, LLC, New York.

Martel, Gordon, *The Origins of the First World War*, London: Longmans, 1996.

Mason, Francis K, *The Hawker Typhoon & Tempest*, Bourne End: Aston Publications Ltd, 1988.

Middlebrook, Martin and Everitt, Chris, *The Bomber Command War Diaries, An Operational Reference Book*, London: Viking/Penguin Books, 1985.

Murray, William, *Luftwaffe, Strategy for Defeat, 1993–45*, London: George Allen & Unwin, 1985.

Overy, R.J., *The Origins of the Second World War*, London: Longmans, 1998.

Philpott, Bryan [1], *Meteor*, Wellingborough: Patrick Stephens Ltd, an imprint of Hayes Publishing, Sparkford, Yeovil, Somerset, 1984.

[2] *English Electric/BAC Lightning*, Wellingborough: Patrick Stephens Ltd, 1984.

Price, Alfred [1], *The Luftwaffe Data Book*, London: Greenhill Books, 1997.

[2], *The Blitz on Britain, 1939–1945*, London: Ian Allen, 1977.

[3], *Luftwaffe Handbook, 1939–1945*, London: Ian Allan Ltd, 1986.

[4], *Instruments of Darkness, The History of Electronic Warfare*, London: Macdonald & Janes, 1977.

Radcliffe, J.A., *Robert Alexander Watson-Watt*, The Dictionary of National Biography.

Ramsey, Winston G. (Ed.) [1], *The Blitz Then & Now*, Volume 1, London: Battle of Britain Prints Ltd, 1987.

[2], *The Blitz Then & Now*, Volume 2, London: Battle of Britain Prints Ltd, 1988.

[3], *The Blitz Then & Now*, Volume 3, London: Battle of Britain Prints Ltd, 1990.

Rawlings, John, *Fighter Squadrons of the RAF and their Aircraft*, Crecy Books, 1993.

Richards, Denis, *The Royal Air Force 1939–1945, Volume 1, The Fight at Odds*, London: HMSO, 1993.

Sayer, Brigadier A.P., *Army Radar*, London: The War Office, 1950.

Shacklady, Edward, *The Gloster Meteor*, London: Macdonald & Co Ltd, 1962.

Streetly, Martin [1], *Confound & Destroy, 100 Group and The Bomber Support Campaign*, London: MacDonald & James, 1978.

[2], *Aircraft of No. 100 Group, a Historical Guide for the Modeller*, London: Robert Hale Ltd, 1984.

Sturtivant, Ray, *The Squadrons of the Fleet Air Arm*, Tonbridge: Air Britain (Historians) Ltd, 1984.

Sturtivant, Ray, Hamlin, John and Halley, James J., *Royal Air Force Flying Training & Support Units*, Tunbridge Wells: Tonbridge: Air Britain (Historians) Ltd, 1997.

Swanborough, Gordon and Bowers, Peter M., *United States Military Aircraft Since 1909*, London: Putnam, 1989.

Taylor, Bill, *Royal Air Force Germany Since 1945*, Hinckley: Midlands Publishing, 2003.

Terraine, John, *The Right of the Line, the Royal Air Force in the European War, 1939–1945*, London: Hodder & Stoughton, 1985.

Thetford, Owen, *Aircraft of the Royal Air Force Since 1918*, London: Putnam, 1962.

Wakefield, Ken, Pfadfinder, *Luftwaffe Pathfinder Operations Over Britain, 1940–44*, Stroud: Tempus Publishing, 1999.

Watkins, David [1], *De Havilland Vampire, The Complete History*, Thrupp: Sutton Publishing Ltd, 1996.

[2], *Venom, De Havilland Venom & Sea Venom, The Complete Story*, Thrupp: Sutton Publishing, 2003.

Weinberg, Gerhard L., *A World at Arms, a Global History of World War II*, Cambridge: Cambridge University Press, 1999.

White, Ian [1], *If You Want Peace, Prepare for War, a History of No. 604 (County of Middlesex) Squadron, RAuxAF, in Peace and in War*, London: 604 Squadron Association, 2005.

Wynn, Kenneth G., *Men of the Battle of Britain*, Croydon: CCB Associates, 1999.

Zaloga, Steven J., *The Kremlin's Nuclear Sword, The Rise and Fall of Russia's Strategic Forces, 1945–2000*, Washington: Smithsonian, 2002.

Documents Held in the National Archive, Kew, London

Air Publication, *AP1093D, Notes on the History of AI, IFF & Radar Beacons*.

Air Publication, *AP2892E, Volume 1, Part 1, AI Mk.9B*, dated June 1950.

Air Publication, *AP2913D, Volume 1, 2nd Edition, AI Mk.21*, dated June 1955.

AIR9/27, *The Operational Record Book of the Fighter Interception Unit*.

AIR10/5485, Ministry of Defence (MoD), Air Historic Branch, *Signals History of the Second World War, Volume 5*.

AIR64/30, *CFE Report No. 228, Trials of Pre-Production Mk.IXB in a Mosquito XVII*, dated 27 April 1945.

AIR64/255, *CFE Report No. 247, Trials of AI Mk.21 in the Gloster Meteor NF.12*, dated 1 May 1954.

AIR64/132, *CFE Report No. 129 on The Comparative Performance of AI Mks.9 & 10, in Mosquito Aircraft*, 22 February 1948.

AIR64/207, *CFE Report No. 168, The Tactical Trials of the De Havilland Venom NF.2*, dated February 1951.

AVIA7/99, *AIS in Blenheim Aircraft*, 1940.

AVIA7/100, *AIS in Blenheim Aircraft*, 1940.

AVIA7/344, *The Fitting of AI, ASV & IFF to Aircraft at St Athan*.

AVIA7/1465, *The Installation of AI in Fulmar Aircraft*, June 1941–January 1944.

AVIA7/2627, *AI Mk.VI in Hurricane, 1942*.

AVIA10/66, *AI MLX (SCR-720) Panel, Meetings & Papers, 1943*.

AVIA13/1047, *AI Mk.VI, 1940–41*.

AVIA13/1048, *AI Mk.VI, 1940–41*.

AVIA13/1050, *AI Mk.V, 1940–41*.

AVIA13/1052, *1940–1941, AI Receivers*.

AVIA13/1054, *1940–1942, AI Mk.IIIA*.

AVIA15/136, *1939–1942, The Early Development of AI Equipment*.

AVIA26/5, *3 cm AI-ASV in Firefly Aircraft*, dated 2 February 1942.

AVIA26/6, *AIS Ground Rejector*, dated 1942.

AVIA26/315, *TRE Report on Centimetre Wave Homing Beacon Type TR3145 for use with AI Mk.VII & VIII*, dated 1942.

FIU Report No. 40, *Preliminary Testing of Pilot's Indicator (AI)*, dated 10 October 1940.

FIU Report No. 63, *AI Mk.V Pilot's Indicator in Douglas Boston BJ468*, dated 23 May 1941.

FIU Report No. 65, *Pilot's Indicator AI Mk.V in Beaufighter R2195*, dated 28 May 1941.

FIU Report No. 72, *Use of the Operator's Indicator on AI Mk.V*, dated 25 June 1941.

Other Documents, Journals and Publications

AT&T, *Chapter Two, Radar*, in 'Engineering & Science in the Bell System', New Jersey: AT&T, undated.

Australian Department of Defence, *No. 456 Squadron, RAAF, Preliminary Narrative, 30 June 1941–15 June 1945*, undated.

Buttler, Tony, *Database, the Bristol Beaufighter*, in Aeroplane Monthly, August 2003.

Flypast, *Airfields of World War Two*, in Flypast Magazine, May 2001.

Freeman, Tony, *The Post-War Royal Auxiliary Air Force*, in the 'Royal Air Force Reserve & Auxiliary Forces', a seminar by the Royal Air Force Historical Society, 2003.

Gunston, Bill [2], *The AAM Story*, in Air International, Stamford: Key Publishing, December 1986.

Hayward, Lawrence, *Radar Pioneer, The story of Wellington R1629*, in Aviation World, Spring 2004, pp. 24–26.

Howard-Williams, Jeremy [2], *The Development of Air Intercept Radar 1935–45*, The American Airpower Museum & Midland College, International Symposium, 11–13 November 1993.

[3], *Head-Up, 50 – UP*, in Flypast, Stamford: Key Publishing, December 1993.

James, Derek [2], *Database, the Gloster Javelin*, in Aeroplane Monthly, January 2004.

Jefford, C.G. [2], *The Post-War RAF Reserves to 1960*, in the 'Royal Air Force Reserve & Auxiliary Forces', a seminar by the Royal Air Force Historical Society, 2003.

Jones, Barry, *Trials and Testbed Meteors*, in Aeroplane Monthly, January 1996.

Kuhn, H.G. and Hartley, Sir Christopher, *Derek Ainsley Jackson*, a Biographical Memoir, London: The Royal Society.

King-Brewster, L.M., *Turbinlite Operations*, photocopy from unknown source.

Paterson, C.C., *Report No. 8717, The High Powered Pulsed Magnetron: Notes on the Contribution of GEC Research Laboratories to the Initial Development*, GEC Hirst Laboratories, 30 August 1945.

Pheasant, V.A., *The Sixtieth Anniversary of Window, 1943–2003*, The Chemring Group, 2003.

Price, Alfred [5], *Operation Steinbock, The Baby Blitz of 1944*, in Aeroplane Monthly, September 2002, London: IPC Media, 2002.

Staff writer [1], *Beaufighter, Part 1*, in 'Air Enthusiast International', January 1974.

[2], *Beaufighter, Part 2*, in 'Air Enthusiast International', February 1974.

White, Ian [2], *The Origins & Development of Allied IFF During World War Two*, in The History of Identification Friend or Foe & Radio Direction Finding, Proceedings of the Fourth Annual Colloquium, Centre for the History of Defence Electronics (CHiDE), Bournemouth University, 16 September 1998.

[3], *Nocturnal & Nautical, Fairey Firefly Night-Fighters*, in Air Enthusiast, Issue 107, September/October 2003, Stamford, Key Publishing, 2003.

Internet Sources

www.designationsystems.net, *The Designations of US Military Electronic Equipment.*
www.doramusic.com/soundmirrors.htm, *Sound Mirrors on the South Coast.*
www.vectorsite.net, *The Douglas F3D Skyknight.*

Index

323

OTHER UNITS & ESTABLISHMENTS